Springer Complexity

Springer Complexity is an interdisciplinary program publishing the best research and academic-level teaching on both fundamental and applied aspects of complex systems – cutting across all traditional disciplines of the natural and life sciences, engineering, economics, medicine, neuroscience, social and computer science.

Complex Systems are systems that comprise many interacting parts with the ability to generate a new quality of macroscopic collective behavior the manifestations of which are the spontaneous formation of distinctive temporal, spatial or functional structures. Models of such systems can be successfully mapped onto quite diverse "real-life" situations like the climate, the coherent emission of light from lasers, chemical reaction-diffusion systems, biological cellular networks, the dynamics of stock markets and of the internet, earthquake statistics and prediction, freeway traffic, the human brain, or the formation of opinions in social systems, to name just some of the popular applications.

Although their scope and methodologies overlap somewhat, one can distinguish the following main concepts and tools: self-organization, nonlinear dynamics, synergetics, turbulence, dynamical systems, catastrophes, instabilities, stochastic processes, chaos, graphs and networks, cellular automata, adaptive systems, genetic algorithms and computational intelligence.

The two major book publication platforms of the Springer Complexity program are the monograph series "Understanding Complex Systems" focusing on the various applications of complexity, and the "Springer Series in Synergetics", which is devoted to the quantitative theoretical and methodological foundations. In addition to the books in these two core series, the program also incorporates individual titles ranging from textbooks to major reference works.

Understanding Complex Systems

Founding Editor: J.A. Scott Kelso

Future scientific and technological developments in many fields will necessarily depend upon coming to grips with complex systems. Such systems are complex in both their composition – typically many different kinds of components interacting simultaneously and nonlinearly with each other and their environments on multiple levels – and in the rich diversity of behavior of which they are capable.

The Springer Series in Understanding Complex Systems series (UCS) promotes new strategies and paradigms for understanding and realizing applications of complex systems research in a wide variety of fields and endeavors. UCS is explicitly transdisciplinary. It has three main goals: First, to elaborate the concepts, methods and tools of complex systems at all levels of description and in all scientific fields, especially newly emerging areas within the life, social, behavioral, economic, neuro- and cognitive sciences (and derivatives thereof); second, to encourage novel applications of these ideas in various fields of engineering and computation such as robotics, nano-technology and informatics; third, to provide a single forum within which commonalities and differences in the workings of complex systems may be discerned, hence leading to deeper insight and understanding.

UCS will publish monographs, lecture notes and selected edited contributions aimed at communicating new findings to a large multidisciplinary audience.

Fatihcan M. Atay

Editor

Complex Time-Delay Systems

Theory and Applications

 Springer

Editor
Fatihcan M. Atay
Max Planck Institute for Mathematics
  in the Sciences
Inselstr. 22-26
04103 Leipzig
Germany
atay@member.ams.org

ISBN 978-3-642-02328-6   e-ISBN 978-3-642-02329-3
DOI 10.1007/978-3-642-02329-3
Springer Heidelberg Dordrecht London New York

Library of Congress Control Number: 2010920732

Cover design: WMXDesign GmbH, Heidelberg

Printed on acid-free paper

Springer is part of Springer Science+Business Media (www.springer.com)

Preface

One of the major contemporary challenges in both physical and social sciences is modeling, analyzing, and understanding the self-organization, evolution, behavior, and eventual decay of complex dynamical systems ranging from cell assemblies to the human brain to animal societies. The multi-faceted problems in this domain require a wide range of methods from various scientific disciplines. There is no question that the inclusion of time delays in complex system models considerably enriches the challenges presented by the problems. Although this inclusion often becomes inevitable as real-world applications demand more and more realistic models, the role of time delays in the context of complex systems so far has not attracted the interest it deserves. The present volume is an attempt toward filling this gap.

There exist various useful tools for the study of complex time-delay systems. At the forefront is the mathematical theory of delay equations, a relatively mature field in many aspects, which provides some powerful techniques for analytical inquiries, along with some other tools from statistical physics, graph theory, computer science, dynamical systems theory, probability theory, simulation and optimization software, and so on. Nevertheless, the use of these methods requires a certain synergy to address complex systems problems, especially in the presence of time delays.

The following series of chapters combine expertise from mathematics, physics, engineering, and biology to address several current issues from the forefront of research in the field. To unify the various problems and approaches presented, the language of dynamical systems is heavily used throughout the book. Dynamical systems, be it in isolation or in interaction with other systems, can display a rich spectrum of behavior. At one end of the spectrum is the simplest point attractor, namely a stable equilibrium. Despite its dynamical simplicity, it commands considerable interest from a control perspective, since it represents the desired behavior in many applications. However, stable equilibria can also spontaneously arise as a result of coupling, particularly in delayed networks. The so-called amplitude death refers to the emergence of such stability in an otherwise oscillatory or even chaotic network. In some fields like chaos control, the aim may be to stabilize a periodic solution instead of a fixed point. Further along the ranks is synchronization phenomena in its many forms, where the attractor is typically a subset of the diagonal of the system's state space. Nearby such orderly behavior are regimes of cluster formation and incoherence, as well as the co-existence of several attractors. Order and disorder can even exist in the same system at different spatial locations, as in the case of chimera solutions. Though sequences of bifurcations as the parameter values or inputs are varied, the system can visit the whole plethora of dynamical regimes, only a fraction of which may be known or amenable to existing techniques of analysis. This is certainly a challenging field both from theoretical and applied perspectives.

The present volume starts with a chapter on the collective dynamics of coupled oscillators, authored by Sen, Dodla, Johnston, and Sethia, which already exhibits the range of dynamics from equilibrium to chimera solutions. Chapter 2 by Atay focuses on the suppression of oscillations by time delays in feedback systems and complex networks, in particular making the connection between stability and network topology. Chapter 3 by Niculescu, Michiels, Gu, and Abdallah examines stability of equilibria by delayed output feedback from a control-theoretical point. In Chap. 4, authored by Schöll, Hövel, Flunkert, and Dahlem, the emphasis is shifted to the stabilization of periodic solutions, studying a range of applications from lasers to coupled neurons. The investigation of neural systems is continued in Chap. 5 by Hutt, this time with a different network model, namely a continuum field description of collective neural activity. Chapter 6 by Longtin addresses stochastic dynamics of neurons, after a discussion of stochastic delay differential equations. Chapter 7 by Lu and Chen gives a comprehensive coverage of the stability of neural networks. Chapter 8 by Crauste looks at stability in systems with distributed delays, with an application to oscillations in stem cell populations. Finally, Chap. 9 by Sipahi and Niculescu gives a survey and latest results on a novel application to complex systems, namely time-delayed traffic flow.

This book is aimed at researchers and students from all disciplines who are interested in time-delay systems. The chapters contain the state-of-the-art in their respective fields, in addition to the current research of the contributors. However, the emphasis has been to make the book a self-contained volume by providing sufficient introductory material in every chapter, as well as ample references to the relevant literature. In this way, the reader will be exposed to the recent results and at the same time be provided with directions for further research.

Leipzig,   Fatihcan M. Atay
January 2010

Contents

Contributors

C.T. Abdallah ECE Department, University of New Mexico, Albuquerque, NM 87131-1356, USA, chaouki@eece.unm.edu

F.M. Atay Max Planck Institute for Mathematics in the Sciences, 04103 Leipzig, Germany, atay@member.ams.org

T. Chen Shanghai Key Laboratory for Contemporary Applied Mathematics, Fudan University, Shanghai, China, tchen@fudan.edu.cn

F. Crauste Université de Lyon; Université Lyon 1, CNRS UMR5208 Institut Camille Jordan, 43 blvd du 11 novembre 1918, F-69622 Villeurbanne-Cedex, France, crauste@math.univ-lyon1.fr

M.A. Dahlem Institut für Theoretische Physik, Technische Universität Berlin, Hardenbergstraße 36, 10623 Berlin, Germany, dahlem@physik.tu-berlin.de

R. Dodla Department of Biology, University of Texas at San Antonio, San Antonio, TX 78249, USA, Ramana.Dodla@utsa.edu

V. Flunkert Institut für Theoretische Physik, Technische Universität Berlin, Hardenbergstraße 36, 10623 Berlin, Germany, flunkert@itp.tu-berlin.de

K. Gu Department of Mechanical & Industrial Engineering, Southern Illinois University at Edwardsville, Edwardsville, Illinois 62026, USA, kgu@siue.edu

P. Hövel Institut für Theoretische Physik, Technische Universität Berlin, Hardenbergstraße 36, 10623 Berlin, Germany, phoevel@physik.tu-berlin.de

A. Hutt INRIA CR Nancy - Grand Est Équipe CORTEX, CS20101, 54603 Villers-ls-Nancy Cedex, France, axel.hutt@loria.fr

G.L. Johnston EduTron Corp., 5 Cox Road, Winchester, MA 01890, USA, gljohnston@gmail.com

A. Longtin Department of Physics, University of Ottawa, Canada, alongtin@uottawa.ca

W. Lu Lab. of Nonlinear Sciences, Fudan University, 200433, Shanghai, China, wenlian.lu@gmail.com

W. Michiels Department of Computer Science, K.U. Leuven, Celestijnenlaan 200A, B-3001 Heverlee, Belgium, Wim.Michiels@cs.kuleuven.be

S.-I. Niculescu Laboratoire des Signaux et Systèmes (L2S), CNRS-Supélec, 3, rue Joliot Curie, 91190 Gif-sur-Yvette, France, Silviu.Niculescu@lss.supelec.fr

E. Schöll Institut für Theoretische Physik, Technische Universität Berlin, Hardenbergstraße 36, 10623 Berlin, Germany, schoell@physik.tu-berlin.de

A. Sen Institute for Plasma Research, Bhat, Gandhinagar 382428, India, abhijit@ipr.res.in

G.C. Sethia Institute for Plasma Research, Bhat, Gandhinagar 382428, India, gautam@ipr.res.in

R. Sipahi Department of Mechanical and Industrial Engineering, Northeastern University, Boston, MA 02115, USA, rifat@coe.neu.edu

Chapter 1
Amplitude Death, Synchrony, and Chimera States in Delay Coupled Limit Cycle Oscillators

Abhijit Sen, Ramana Dodla, George L. Johnston, and Gautam C. Sethia

1.1 Introduction

In this chapter we will discuss the effects of time delay on the collective states of a model mathematical system composed of a collection of coupled limit cycle oscillators. Such an assembly of coupled nonlinear oscillators serves as a useful paradigm for the study of collective phenomena in many physical, chemical, and biological systems and has therefore led to a great deal of theoretical and experimental work in the past [1–6]. Examples of practical applications of such models include simulating the interactions of arrays of Josephson junctions [7, 8], semiconductor lasers [9, 10], charge density waves [11], phase-locking of relativistic magnetrons [12], Belousov–Zhabotinskii reactions in coupled Brusselator models [2, 13–15], and neural oscillator networks for circadian pacemakers [16]. One of the most commonly studied phenomena is that of synchronization of the diverse frequencies of an oscillator assembly to a single common frequency. Synchrony was highlighted by Winfree [1] in a simple model of weakly coupled limit cycle oscillators and further developed by Kuramoto and others in the context of phase transition models [17, 18]. Research on synchrony in a variety of coupled and complex systems has seen an explosive growth in the past few years and has also captured the popular imagination [19] due to its application to such natural phenomena as the synchronous flashing of a swarm of fire flies, the chirping of crickets in unison, and the electrical synchrony in cardiac cells. Apart from synchrony, coupled limit cycle oscillator models are capable of exhibiting other interesting behavior. For example, if the strength of the interaction between the oscillators is comparable to the attraction to their own individual limit cycles, then the original phase-only model of Winfree or Kuramoto is no longer valid and the amplitudes of the individual oscillators begin to play a role [20–23]. For sufficiently strong coupling and a broad spread in the natural frequencies of the oscillators, the assembly can suffer an amplitude quenching or death [5, 24, 25] in which all the oscillators cease to oscillate and have zero amplitudes. Such

A. Sen (✉)
Institute for Plasma Research, Bhat, Gandhinagar 382428, India
e-mail: abhijit@ipr.res.in

F.M. Atay (ed.), *Complex Time-Delay Systems,* Understanding Complex Systems,
DOI 10.1007/978-3-642-02329-3_1, © Springer-Verlag Berlin Heidelberg 2010

behavior has been observed in experiments of coupled chemical oscillator systems, e.g., coupled Belousov–Zhabotinskii reactions carried out in coupled tank reactors [26]. Other collective phenomena that these coupled oscillator models display include partial synchronization, phase trapping, large amplitude Hopf oscillations, and even chaotic behavior [25, 27]—all of which have been discussed widely in the literature.

The question we wish to address now is what happens to the collective properties of the coupled oscillator system when one introduces time delay in the coupling. The physical motivation for such a modification of the coupling is to simulate the situation in real-life systems where the interaction between individual oscillators may not be instantaneous but may be delayed due to finite propagation time of signals. Time delays can similarly occur in chemical systems due to finite reaction times, and in biological systems like neuron assemblies, the synaptic integration mechanisms may provide a natural delay. From a mathematical point of view, one can expect time delays to have a profound effect on the dynamical characteristics of a single oscillator. This is well known from the study of single delay differential equations which show fundamental changes in the nature of solutions and novel effects that are absent in a non-delayed system. What happens to the collective modes of a coupled system in the presence of time delay? Surprisingly, there has not been a great deal of work in this area despite the vast literature on single delay differential equations and the considerable recent developments in the field of coupled oscillator research. Some notable exceptions are the works of Schuster and Wagner [28], Niebur et al. [29], Nakamura et al. [30], and Kim et al. [31], who in the past have looked at time delay effects in the context of the simple phase-only coupled oscillator models and found interesting effects like the existence of higher frequency states and changes in the onset conditions and nature of synchronization. More recently we have investigated a variety of model systems starting from a simple case of just two oscillators with a discrete time-delayed coupling to a large number of oscillators with time-delayed global, local, and non-local couplings [32–37]. Time delay is found to introduce significant changes in the character and onset properties of the various collective regimes such as amplitude death and phase-locked states. Some of the results are novel and somewhat surprising—such as time delay-induced death in an assembly of identical oscillators or the existence of clustered chimera states—and may have important applications. With a growing recognition of the significance and prevalence of time delay in various systems, there is now a considerable increase in the number of investigations on this topic. The aim of this chapter is to provide some appreciation of this interesting area of nonlinear dynamical systems through an exposition of the basic concepts of the field followed by a discussion of some research results. It is not meant to be a review of the field and the choices of topics and research results are heavily influenced by our own work.

The chapter is organized as follows. In Sect. 1.2 we develop the basics of the subject by introducing a minimal collective model consisting of two coupled limit cycle oscillators that are close to a supercritical Hopf bifurcation. After identifying the fundamental collective states of this system in the undelayed case we discuss the effects of a finite time delay on the existence and stability of phase-locked states

and amplitude death. In Sect. 1.3 we introduce a more complex model consisting of
N-coupled oscillators ($N > 2$). The oscillators in such a system can be coupled in
various ways: all-to-all (global), nearest neighbor (local), or spatially varying coupling (non-local). We discuss time-delayed collective states in all three systems and
discuss their essential characteristics. Our primary emphasis is on the exploration
of the *amplitude death state* in all three coupling scenarios. We also investigate
the effect of time delay on a novel state of the non-local system—the so-called
chimera state—consisting of co-existing regions of coherent and incoherent states.
Time delay is seen to impose a spatial modulation of such a state leading to a *clustered chimera* state. Section 1.4 provides a summary of the main results and some
perspective on the future directions and potential developments of the field.

1.2 A Minimal Collective Model

We begin our exploration of time delay effects on the collective states of coupled
oscillator systems by investigating the dynamics of a minimal model system consisting of just two coupled limit cycle oscillators that are close to a supercritical Hopf
bifurcation. The individual oscillator of this model is chosen to have the nonlinear
normal form of a van der Pol-type equation. The van der Pol equation has a parameter that can be varied to take the solution state from a fixed point (a steady state) to
an oscillatory state via a supercritical Hopf bifurcation. To illustrate this, consider
the van der Pol equation in the following form:

$$\ddot{x} - (a - x^2)\dot{x} + \omega^2 x = 0. \tag{1.1}$$

In this and other equations below, the variables x, y, z, ξ, ϕ, ρ, and θ are functions of
time, though, for simplicity, such a dependence is not explicitly written down, and
a and ω are real parameters. For $a < 0$, $x = 0$ is a stable steady state of (1.1), and
a periodic solution emerges as a is increased past 0. For large a these oscillations
acquire the character of relaxation oscillations. Thus, $a = 0$ is a bifurcation point.
The eigenvalues of the system for a linear perturbation around the origin ($\lambda_{1,2} = \frac{a}{2} \pm$
$i\sqrt{\omega^2 - a^2/4}$) acquire pure imaginary values ($\pm i\omega$) with $\frac{dRe(\lambda_{1,2})}{da} = \frac{1}{2} > 0$ at $a = 0$.
A normal form of the equation that preserves these properties can be obtained by
doing an appropriate averaging over the fast periodic behavior near the critical point
$a = 0$. The resulting nonlinear equation has a simpler structure (that is easier to work
with analytically and numerically) and yet maintains the bifurcation characteristics
of the original oscillator equation. To carry out such a reduction we rewrite (1.1) as
a set of two first-order differential equations, $\dot{x} = \omega y$ and $\dot{y} = -\omega x + (a - x^2)y$.
Defining a complex variable $z = x + iy$, these equations may be written as a single
equation in terms of z:

$$\dot{z} = -i\omega z + \frac{1}{2}\left[a - \frac{1}{4}(z + \bar{z})^2\right](z - \bar{z}).$$

Let $z = \xi e^{-i\phi}$. (The negative sign introduced in the phase is of no physical significance. The phase flow of the van der Pol is clockwise, but in the biological
phase oscillators where equations of the type $\dot{\theta} = \omega$ are common, the phase flow is
anti-clockwise and is in the direction of increasing angle.) Then the following two
equations emerge for the amplitude and the phase:

$$\dot{\xi} = (a - \xi^2 \cos^2 \phi)\xi \sin^2 \phi,$$

$$\dot{\phi} = \omega + \frac{1}{2}(a - \xi^2 \cos^2 \phi)\sin(2\phi).$$

We now average these two equations over the phase ϕ. The right-hand side of the
$\dot{\xi}$ equation is an even function of ϕ, and the average over ϕ from 0 to π results in
$(a - \xi^2/4)\xi/2$. The second term on the right-hand side of the $\dot{\phi}$ equation is an odd
function of ϕ, and an average over a period of ϕ makes it zero. We call the averaged
amplitude and the phase ρ and θ. Hence,

$$\dot{\rho} = \left(a - \frac{1}{4}\rho^2\right)\frac{\rho}{2}, \tag{1.2}$$

$$\dot{\theta} = \omega. \tag{1.3}$$

The behavior of the two eigenvalues of this reduced set of equations at $a = 0$ is
the same as that of (1.1) mentioned before. We can also see this by noting from
(1.2) and (1.3) that the growth of amplitude for very small perturbations around
$\rho = 0$ is proportional to $a/2$ and the frequency of such a growth is ω, identical to
the eigenvalue behavior of (1.1). For a positive but near 0, the phase plane orbits
of (1.1) are circular just as those for (1.2) and (1.3), but will become distorted for
large a. Using ρ and θ, a second-order approximation for the amplitude and phase
can be derived that reveals the dependence of the amplitude on the phase and the
phase evolution on the amplitude. But we confine our description to the first-order
approximation. By redefining $Z(t) = \frac{\rho}{2}e^{i\theta}$, (1.2) and (1.3) can be written as a single
equation in the complex variable $Z(t)$, namely,

$$\dot{Z}(t) = (a + i\omega - |Z(t)|^2)Z(t) \tag{1.4}$$

and which is the final normal form of the equation that we will work with. Equation (1.4) is also widely known as the Stuart–Landau oscillator. Note that this equation shows stable oscillations ($Z(t) = \sqrt{a}e^{i\omega t}$) for $a > 0$ with amplitude $\sqrt{a}$ and
a stable rest state ($Z(t) = 0$) for $a < 0$. The value $a = 0$ is the supercritical
Hopf bifurcation point. Our minimal model consists of two such oscillators that are
linearly (diffusively) coupled to each other and where the coupling is time delayed.
The model equations are

$$\dot{Z}_1(t) = (1 + i\omega_1 - |Z_1(t)|^2)Z_1(t) + K[Z_2(t - \tau) - Z_1(t)], \tag{1.5}$$
$$\dot{Z}_2(t) = (1 + i\omega_2 - |Z_2(t)|^2)Z_2(t) + K[Z_1(t - \tau) - Z_2(t)], \tag{1.6}$$

where we have chosen $a = 1$ (so that each oscillator in the uncoupled state is in the stable limit cycle state), K is the strength of coupling, and τ is a discrete and constant delay time. As an approximation to a physical system, (1.5) and (1.6) can be viewed as two nonlinear electronic circuits that are resistively coupled to each other. We will study the dynamics of the above system in terms of the frequency difference ($\Delta = | \omega_1 - \omega_2 |$), the average frequency ($\bar{\omega} = (\omega_1 + \omega_2)/2$), and the coupling strength (K) as a function of the time delay parameter τ. In the absence of time delay, the above set of equations (and its generalizations) have been studied in detail by Aronson et al. [24] to delineate the bifurcation structures and the existence of various collective states. In Fig. 1.1 we have redrawn their bifurcation diagram to illustrate the main features of their analyses. Broadly, the bifurcation diagram may be divided into three regimes: (1) frequencies are identical, $\omega_1 = \omega_2 = \omega$, that is $\Delta = 0$, (2) frequencies are weakly dissimilar (i.e., $0 < \Delta < 2$), and (3) frequencies are very dissimilar (i.e., $\Delta > 2$). For identical intrinsic frequencies ($\Delta = 0$), the coupled oscillators are always synchronized (with no phase delay between the oscillations). This is the only stable solution for any positive and finite coupling strength. The level of K determines how fast the synchronized state is attained from any given set of initial conditions. For weakly dissimilar frequencies ($0 < \Delta < 2$), the oscillators can be found in two different states: phase drift or phase-locked states. A critical value of coupling ($K > \Delta/2$) is required to phase-lock the oscillators. In

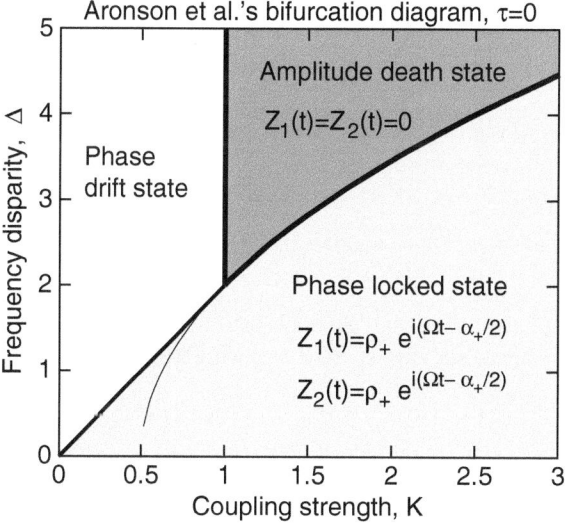

Fig. 1.1 Aronson et al.'s [24] bifurcation diagram of (1.5) and (1.6) for $\tau = 0$. The boundaries of the death state are determined from the eigenvalues of the linearized equations at $Z_{1,2} = 0$. For $\Delta > 2$, $K = 1$ and $\kappa \equiv K = \frac{1}{2}(1 + \Delta^2/4)$ define the boundaries. The stable phase-locked state is the node of symmetric solutions that form a saddle-node pair that emerges on $\Delta = 2K$ and $K < 1$. The saddle merges with the origin on the thin line, κ. The node has the amplitude $\rho_+ = \sqrt{1 - K + \sqrt{K^2 - \Delta^2/4}}$, $\Omega = \bar{\omega}$, and $\alpha_+ = \sin^{-1}(\Delta/2K)$

such phase-locked state, the phase difference between the oscillators is a constant of time and is determined by the coupling strength and the frequency difference. For weak coupling in this regime, a phase drift occurs (i.e., $\theta_1 - \theta_2$ is a function of time and runs from 0 to 2π). In these two regimes, the amplitude of the oscillators does not offer any particularly interesting feature. In the third regime ($\Delta > 2$), however, when the frequency disparity is strong, a third solution, a stable *amplitude death* state, can exist in addition to phase drift and phase-locked states. The level of coupling strength determines the stability of each of these states. For $K < 1$, phase drift occurs. For $K > (1 + \Delta^2/4)/2$, phase-locking occurs. For intermediate K, the amplitude of the oscillators becomes zero. This state is nothing but the stable fixed point state ($Z_{1,2}(t) = 0$) and does not exist for phase-only coupled oscillators (i.e., when the amplitudes of the two oscillators are forced to assume a value of unity while letting the phases to evolve). This state is a reflection of the effect of amplitude on the collective oscillations of the coupled oscillators. The boundaries of various regions can be determined by a stability analysis of these states [24]. We will now study the effect of finite time delay ($\tau \neq 0$) on the characteristics of this phase diagram.

1.2.1 Time delay effects

We rewrite the model Equations (1.5) and (1.6) in polar coordinates by letting $Z_{1,2}(t) = r_{1,2}e^{i\theta_{1,2}}$ to get

$$\dot{r}_1(t) = (1 - K - r_1(t)^2)r_1(t) + Kr_2(t - \tau)\cos\left[\theta_2(t - \tau) - \theta_1(t)\right], \quad (1.7)$$

$$\dot{\theta}_1(t) = \omega_1 + K\frac{r_2(t - \tau)}{r_1(t)}\sin\left[\theta_2(t - \tau) - \theta_1(t)\right], \quad (1.8)$$

$$\dot{r}_2(t) = (1 - K - r_2(t)^2)r_2(t) + Kr_1(t - \tau)\cos\left[\theta_1(t - \tau) - \theta_2(t)\right], \quad (1.9)$$

$$\dot{\theta}_2(t) = \omega_2 + K\frac{r_1(t - \tau)}{r_2(t)}\sin\left[\theta_1(t - \tau) - \theta_2(t)\right]. \quad (1.10)$$

This form is more useful for analysis of periodic states whereas the Cartesian form is convenient for linear stability studies, and we will utilize either form as per our needs.

1.2.1.1 Phase-Locked States

Let us consider identical oscillators, that is, $\Delta = 0$, and hence $\omega_1 = \omega_2 \equiv \omega_0$. This is also trivially the average frequency $\bar{\omega}$. Without time delay ($\tau = 0$), the in-phase-locked state is a stable solution. The interactions are instantaneous and do not depend on time history. Hence, any phase mismatch introduced by way of perturbation is transmitted to both the oscillators instantly. Once the perturbation ceases, the oscillators resume their oscillations with their natural frequencies. The coupling coefficient K determines the recovery time window before their natural oscillations

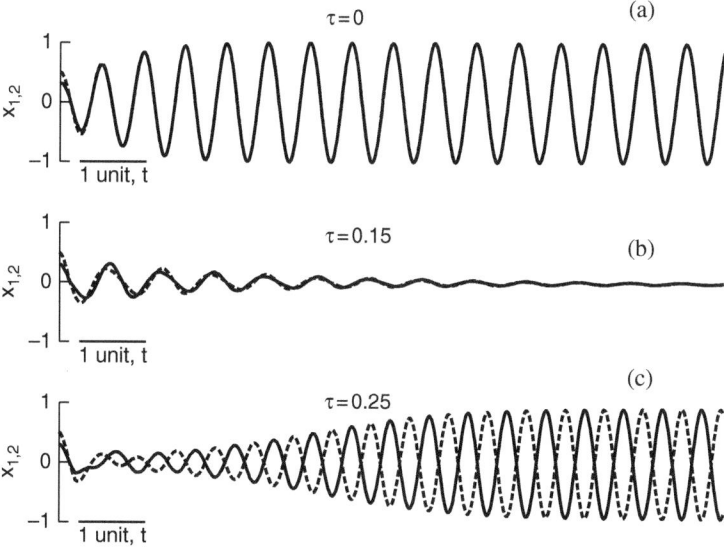

Fig. 1.2 Numerical solutions of coupled identical ($\Delta = 0$) oscillators ((1.5) and (1.6)) at $K = 2$, $\omega_0 = 10$. The initial conditions are $x_1 = 0.3$, $x_2 = 0.5$, $y_1 = 0$, $y_2 = 0$. The delay vectors $Z_{1,2}(t) = 0$, $\tau < t < 0$. Stable in-phase-locked state is reached quickly for very small or no time delay. Amplitude death is encountered as τ is increased, and the system emerges into an anti-phase-locked state for further increments. Our analysis will reveal multiple regions of amplitude death at larger values of τ

are synchronized. This stable state is also seen in the bifurcation diagram (Fig. 1.1). The evolution of the real components of the two oscillators in this in-phase-locked state are illustrated in Fig. 1.2(a).

For finite τ the interactions are non-instantaneous. Earlier studies on phase coupled oscillators that included time delays predicted multiple frequency states where the oscillators could possess any of several stable frequency states allowed for the given parameter set. The possibility of such multiple states arising due to time delay may be seen directly from (1.7), (1.8), (1.9), (1.10), where the dependence of the derivatives on phases involves sinusoidal functions of τ. We will show that this dependence will lead to transcendental equations for oscillation frequency and thus result in multiple frequency states. For identical oscillators, such multiple frequency states are either in-phase or anti-phase. Multi-stability can occur between in-phase states, anti-phase states, or between in-phase and anti-phase states. For any given state, the frequency of oscillation decreases with increasing τ (Fig. 1.3(b)) as also predicted earlier by other studies [29]. In our example simulating the parameters allow both in-phase and anti-phase states and they exist in different parameter regions for $\tau < 0.3$ (Fig. 1.2(c) and Fig. 1.3(b) and (c)).

The fact that we have amplitude evolutions along with the phase evolutions (see (1.7), (1.8), (1.9), (1.10)) has significance for the existence of these states. The in-phase state at $\tau = 0$ continues to exist for slightly higher levels of τ, but the

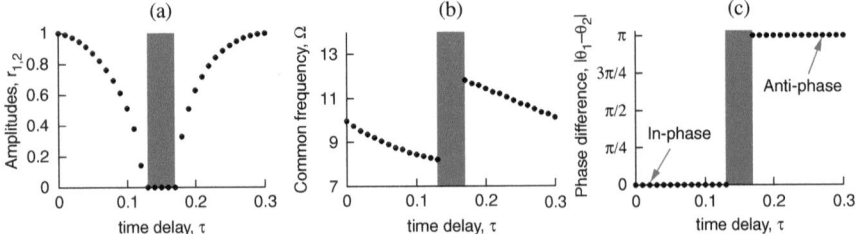

Fig. 1.3 Numerical simulation results of coupled identical ($\Delta = 0$) oscillators ((1.5) and (1.6)) at $K = 2$, $\bar{\omega} = 10$. After initial transients, the amplitudes and frequencies of both the oscillators become constants in time. The solutions are sinusoidal, but the plotted frequencies are the peaks of the Fourier spectrum of the time course of x_1. The phases of the oscillators are computed using $\theta_1 = \tan^{-1}(y_1/x_1)$ and $\theta_2 = \tan^{-1}(y_2/x_2)$. The shaded region is the parameter region of amplitude death along τ

amplitude of the oscillations decreases until it completely becomes zero (Figs. 1.2(b) and 1.3(a)). In this state any damped oscillations might still show an in-phase relationship, but they are transient in nature, and the long-time steady state is the zero amplitude steady state. In fact as τ is increased, these damped oscillations become anti-phase, and above a critical τ, the amplitude of these oscillations becomes non-zero, and a stable anti-phase state emerges (Fig. 1.3). The zero-amplitude state is the region of stability of $Z_{1,2} = 0$ and is the death state. The boundaries of the death state seen in Fig. 1.3 (shaded regions) are the boundaries of the 'death island' which will be discussed later.

We will later on derive the boundaries of the death state by using the eigenvalue analysis. But the emergence of the death and anti-phase solutions and their stability may be derived simply from an empirical observation of the numerical results of Fig. 1.3. Numerical simulations reveal the symmetry of the system. In the in-phase and anti-phase states the amplitudes of both the oscillators are identical and independent of time ($r_1(t) = r_2(t) = r^*$). Their phase evolutions (not shown) are linear growths ($\theta_1(t) = \Omega t + c_1$, $\theta_2(t) = \Omega t + c_2$) of time with a frequency (Ω) that may differ from their intrinsic frequency (ω_0). The quantities c_1 and c_2 are constants in time and depend on the initial conditions. The phase difference of the oscillators is measured by $|c_1 - c_2|$. In the in-phase state the phase difference is zero and in the anti-phase state it is π. For ease of analysis, we will assume that $c_1 = -\alpha/2$ and $c_2 = \alpha/2$, so that in the in-phase state $\alpha = 0$ and in the anti-phase state $\alpha = \pi$. Let us first substitute in (1.8) and (1.10) the above observations on the amplitudes:

$$\dot{\theta}_1(t) = \omega_0 + K \sin[\theta_2(t - \tau) - \theta_1(t)], \tag{1.11}$$

$$\dot{\theta}_2(t) = \omega_0 + K \sin[\theta_1(t - \tau) - \theta_2(t)]. \tag{1.12}$$

Define $\phi(t) = \theta_2(t) - \theta_1(t)$ and take the difference of the above two equations to get

$$\dot{\phi}(t) = -2K \cos(\Omega \tau) \sin \phi, \tag{1.13}$$

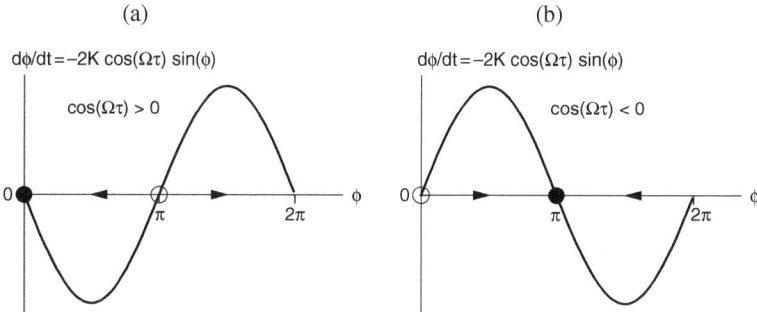

Fig. 1.4 Stability of in-phase and anti-phase solutions is decided by the sign of $\cos(\Omega\tau)$. The filled phase point is the stable solution in each case. For $\cos(\Omega\tau) > 0$ the in-phase solution is stable, and for $\cos(\Omega\tau) < 0$ the anti-phase solution is stable

where we have used the relations $\theta_1(t-\tau) - \theta_1(t) = \theta_2(t-\tau) - \theta_2(t) = -\Omega\tau$ and $\theta_2(t-\tau) - \theta_1(t-\tau) = \theta_2 - \theta_1$. These relations are again derived from the observations that we made above on the phase evolutions in the phase-locked states. We have not yet specified the phase difference α between the oscillators. So this phase difference evolution equation applies for both in-phase and anti-phase states alike. This equation also helps us in understanding the stability of both in-phase and anti-phase states. For $\cos(\Omega\tau) > 0$ (Fig. 1.4(a)), the slope of $\dot{\phi}$ (i.e., the rate of change of the phase difference) is negative, signifying that any brief perturbation from this state will decay to that state in time, as also indicated by the directional flow. Hence, the in-phase state is stable, and the anti-phase state is unstable. But for $\cos(\Omega\tau) < 0$ (Fig. 1.4(b)), the anti-phase state acquires stability and the in-phase state loses its stability.

Does the in-phase state have to become unstable for the anti-phase state to become stable, and vice versa? No. In fact both states can co-exist. This is possible because the frequencies of these two states can be different while still obeying the stability relations shown in Fig. 1.4. The frequencies are in fact determined by solving different transcendental equations involving Ω and τ for the two states. To see this, substitute the in-phase states $\theta_{1,2}(t) = \Omega t$ (i.e., $\alpha = 0$) in (1.11) to obtain a transcendental equation for the in-phase frequencies:

$$f_{\text{in}}(\Omega) - \Omega - \omega_0 + K\sin(\Omega\tau) = 0. \tag{1.14}$$

Similarly, for the anti-phase states, substituting $\theta_{1,2}(t) = \Omega t \mp \pi/2$ in (1.11), another transcendental equation for Ω is obtained for anti-phase frequencies:

$$f_{\text{anti}}(\Omega) = \Omega - \omega_0 - K\sin(\Omega\tau) = 0. \tag{1.15}$$

f_{in} and f_{anti} are plotted as a function of Ω in Fig. 1.5 for sample parameter values. The zeros of these curves are the allowed in-phase or anti-phase frequencies. For small τ only one solution for each state (marked by dots) could be found, but at

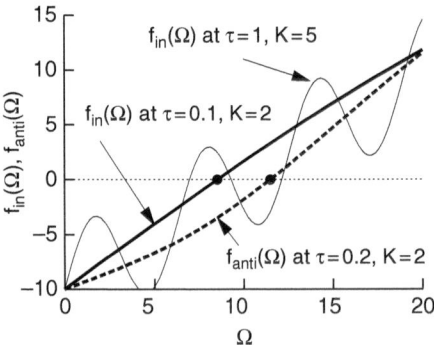

Fig. 1.5 The allowed values of frequencies for in-phase and anti-phase states are determined by solving for the zeros of the functions $f_{in}(\Omega)$ and $f_{anti}(\Omega)$. These two curves plotted as a function of Ω at $\tau = 0.1$, $K = 2$, show single solutions for in-phase and anti-phase states (marked by filled bullets). At large K, the curves show bigger amplitude oscillations and at longer τ they show more wiggles in any given Ω-window. Both these features help establish multiple frequency states for both in- and anti-phase states as seen, for example, by the multiple zeros of f_{in} at $\tau = 1$ and $K = 5$:

larger τ and/or K, multiple zeros could result, correspondingly yielding multiple frequencies.

The stability of any one state with frequency Ω, as stated above, depends on the sign of $\cos(\Omega\tau)$. This stability analysis is illustrated pictorially for an in-phase and an anti-phase branch in Fig. 1.6. The frequencies of the in-phase and anti-phase states are obtained by solving $f_{in} = 0$ and $f_{anti} = 0$ and are plotted as a function of τ. The stable in-phase branch corresponding to $\cos(\Omega\tau) > 0$ (Fig. 1.6(a)) and the stable anti-phase branch corresponding to $\cos(\Omega\tau) < 0$ (Fig. 1.6(b)) are marked with filled dots. The unfilled dots indicate unstable portions of the frequency branches. We might already have here a bistable region between in-phase and anti-phase states. However, the amplitudes of each of these states must also be considered and verified whether these states assume physically acceptable (i.e., real and positive) values.

The amplitude of the in-phase state is obtained by substituting $r_1(t) = r_2(t) = r_{in}$ and $\theta_{1,2}(t) = \Omega t$ in (1.7):

$$r_{in}^2 = 1 - K + K\cos(\Omega\tau). \tag{1.16}$$

The right-hand side of this equation is real and positive only when $\cos(\Omega\tau) > 1 - 1/K$. Since $\cos(\Omega\tau)$ is a smooth function of its argument, in fact, as τ is increased the amplitude gradually decreases to 0 on the curve

$$\cos(\Omega\tau) = 1 - \frac{1}{K} \tag{1.17}$$

and r_{in}^2 remains unphysically negative for $0 < \cos(\Omega\tau) < 1/K - 1$ (shown by the guiding lines in Fig. 1.6(a)). This boundary where the amplitude becomes 0, however, marks the transition of the in-phase state with finite amplitude to a

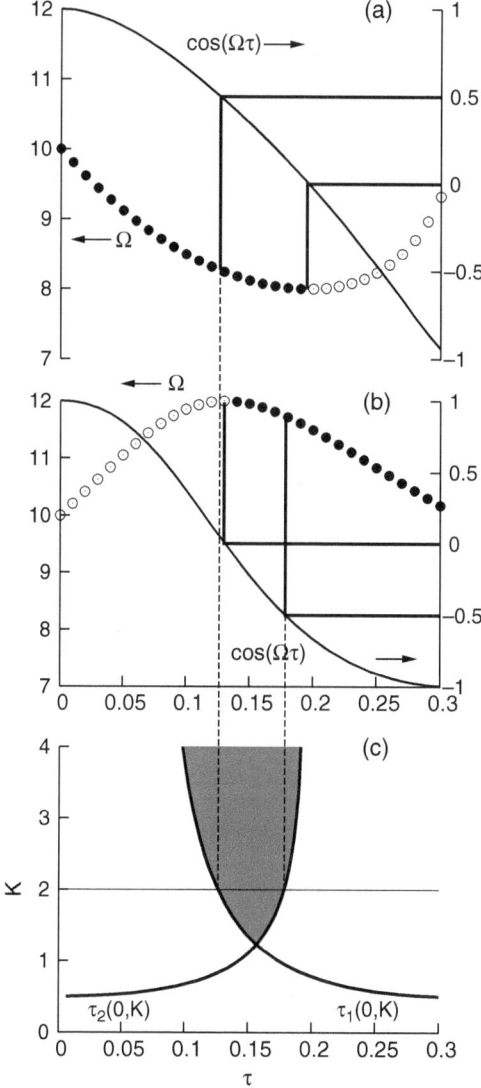

Fig. 1.6 Frequency branches of in-phase-locked (**a**) and anti-phase (**b**)-locked states obtained by solving $f_{in} = 0$ and $f_{anti} = 0$ at $K = 2$ and $\omega_0 = 10$. The *filled circles* indicate the stable states in each case corresponding to the stability conditions shown in Fig. 1.4. The *guiding lines* enclose parameter regions where the amplitudes of the in-phase and anti-phase states have unphysical values. The left boundary of this region in the in-phase state and the right boundary in the anti-phase state correspond to the parameter values where the amplitudes become 0. These boundaries match exactly with those (**c**) ($\tau_1(0, K)$ and $\tau_2(0, K)$) obtained from eigenvalue analysis. Here the match must be at $K = 2$. The shaded region is the death island region and extends beyond the ordinate boundary

zero-amplitude state or the death state. We will in fact see from eigenvalue analysis that this is one of the death region (death island) boundaries. We plot the death boundary $\tau_1(0, K)$ corresponding to $\omega_0 = 10$ (to be derived later) and find an exact match between a death island boundary and that predicted by the above equation (Fig. 1.6(c)). Similarly, we examine the amplitude of the anti-phase state by substituting $r_1(t) = r_2(t) = r_{anti}$ and $\theta_{1,2}(t) = \Omega t \mp \pi/2$ in one of the phase evolution equations, say (1.8), and obtain

$$r_{anti}^2 = 1 - K - K \cos(\Omega\tau). \tag{1.18}$$

The right-hand side of this equation is real and positive only when $\cos(\Omega\tau) < 1 - 1/K$. Again, this function varies smoothly with its argument and becomes zero on the curve

$$\cos(\Omega\tau) = \frac{1}{K} - 1 \tag{1.19}$$

and remains unphysically negative for $1/K - 1 < \cos(\Omega\tau) < 0$ (shown by the guiding lines in Fig. 1.6(b)). This boundary where the amplitude of the anti-phase oscillations becomes 0 matches exactly with the boundary $\tau_2(0, K)$ derived from the eigenvalue analysis (Fig. 1.6(c)).

1.2.1.2 Amplitude Death

We will now study the amplitude death region and show how to derive these boundaries systematically from the characteristic equation. The amplitude death region is the region of stability of the trivial solution: $Z_{1,2} = 0$, and the eigenvalues of this state determine the boundaries of the amplitude death both for identical and for non-identical oscillators. The characteristic equation we obtain will be transcendental in nature and can possess an infinite number of eigenvalues. The amplitude death region is determined from the parameters in the characteristic equation by insisting that all the eigenvalues have negative real parts. For example, if $\lambda = \alpha + i\beta$ (where α and β are real) represents all the eigenvalues of the system, the stable death region is determined by the condition $\alpha < 0$ and the boundary of the death region is determined by $\alpha = 0$. Owing to the fact that we will have a transcendental characteristic equation (and hence multiple solutions; see, for example, Fig. 1.5), this death boundary condition results in multiple curves in the parameter space, leading to the possibility of multiple regions of death state.

The characteristic equation of (1.5) and (1.6) is obtained by linearizing these equations around $Z_{1,2} = 0$ and substituting $Z_{1,2}(t) = Z_{1,2}(0)e^{\lambda t}$. The resultant matrix on the right-hand side of the equations is

$$A = \begin{bmatrix} 1 - K + i\omega_1 & Ke^{-\lambda\tau} \\ Ke^{-\lambda\tau} & 1 - K + i\omega_2 \end{bmatrix}. \tag{1.20}$$

The characteristic equation is nothing but $\det(A - \lambda I) = 0$, where I is the 2×2 unit matrix. By expanding this equation, we can write it as the following transcendental equation:

$$(a - \lambda + i\omega_1)(a - \lambda + i\omega_2) - K^2 e^{-2\lambda\tau} = 0, \tag{1.21}$$

where $a = 1 - K$. Complete analytical solutions of such transcendental equations are not generally available, but we can use the equations to obtain critical curves bounding the stable (death) region. We first show the stable regions between non-identical oscillators and then derive boundaries for identical oscillators. As outlined above, we obtain critical curves by seeking that the real parts of the eigenvalues are zero. On these curves, a pair of eigenvalues cross into right half of the eigenvalue plane (i.e., a stability switch could take place). Since we already know the region of stability in the absence of time delay, we increase τ slightly and look for the critical curves that are nearest to this region. Across these curves, stability of the rest state is lost, and thus they provide the boundaries of the death region. On the critical curves, let $\lambda = i\beta$. Substituting this in the above equation and separating the real and imaginary components, we obtain $(\beta - \bar{\omega})^2 - \Delta^2/4 - a^2 + K^2 \cos(2\beta\tau) = 0$ and $2a(\beta - \bar{\omega}) - K^2 \sin(2\beta\tau) = 0$. These two equations may be used to compute critical curves in (K, Δ) plane by eliminating β. For convenience we write them as follows:

$$F = (\beta - \bar{\omega})/\sin(2\beta\tau), \tag{1.22}$$

$$K \equiv K_{\pm} = -F \pm \sqrt{F^2 + 2F}, \tag{1.23}$$

$$\Delta^2 = -4a^2 + 4(\beta - \bar{\omega})^2 + 4K^2 \cos(2\beta\tau). \tag{1.24}$$

By choosing β from intervals $I_n = (n\pi/2\tau, (n+1)\pi/2\tau)$, portions of curves are obtained in (K, Δ) plane. We term these curves as $S_{\pm}$ depending on the sign being used to compute K in (1.23). For $\bar{\omega} = 10$ we show in Fig. 1.7 these critical curves in the (K, Δ) plane using the interval I_0. S_+ curves are drawn in dashes and S_- in continuous lines. The shaded region is the amplitude death region. For τ very small, the region is closer to that of Aronson et al.'s (Fig. 1.1). But as τ is increased, the region expands toward smaller values of Δ, and for a range of τ values, it displays amplitude death state along $\Delta = 0$ axis. That is, identical oscillators can exhibit amplitude death state if appropriate time delay is introduced in their interactions.

We will now be interested in this phenomenon of amplitude death for $\Delta = 0$ and wish to quantify the region of death for various τ by finding the critical curves that define the boundaries of the death region in (K, τ) plane. To do this, it is best to consider (1.21) and substitute $\omega_1 = \omega_2 = \omega_0$ there. This gives a set of two simpler characteristic equations

$$\lambda = 1 - K + i\omega_0 \pm K e^{-\lambda\tau}, \tag{1.25}$$

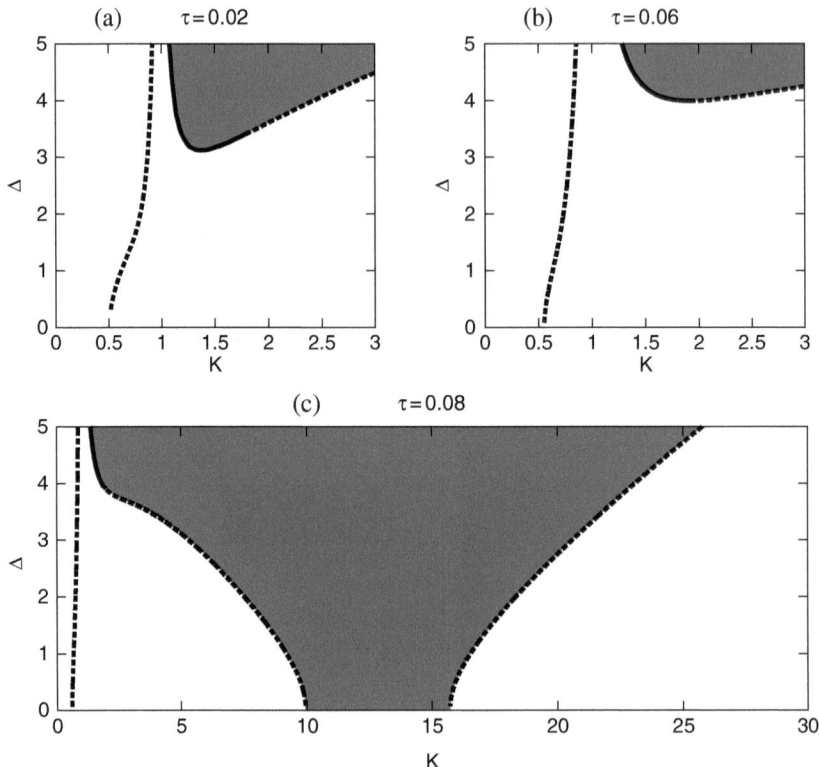

Fig. 1.7 Amplitude death region (*shaded*) shown for three values of τ at $\bar{\omega} = 10$. The boundaries are the curves S_- (*solid lines*) and S_+ (*dashed lines*) drawn by eliminating β from (1.22), (1.23) and (1.24)

which can now be analyzed for critical curves. Again for criticality, substitute $\lambda = i\beta$. We obtain, by separating the real and imaginary parts, the following two equations which can be used to eliminate β and obtain curves, or death island boundaries, in (τ, K) plane. Using + sign,

$$\cos(\beta\tau) = 1 - \frac{1}{K}, \tag{1.26}$$

$$\beta - \omega_0 + K\sin(\beta\tau) = 0. \tag{1.27}$$

These are exactly the same equations we obtained earlier for in-phase-locked states (1.14), and the condition for amplitude of that state to be zero (1.17), confirming that the in-phase state indeed emerges from the death state on this critical curve. Using − sign, we obtain

$$\cos(\beta\tau) = \frac{1}{K} - 1, \tag{1.28}$$

$$\beta - \omega_0 - K\sin(\beta\tau) = 0. \tag{1.29}$$

These are exactly the same equations we obtained earlier for anti-phase-locked states (1.15), and the condition for amplitude of that state to be zero (1.19), confirming that an anti-phase state emerges from the death state on this critical curve. The value of β can be expressed independently of the sinusoidal function in both the above cases. Inverting the cosine functions results in multiple values for τ as a function of K. Each of these curves is a critical curve across which pairs of eigenvalues cross into the right half eigenvalue plane. A numerical ordering of these curves reveals [32, 33] that the death island boundaries are given by

$$\tau_1 \equiv \tau_1(n, K) = \frac{n\pi + \cos^{-1}(1 - 1/K)}{\omega_0 - \sqrt{2K - 1}}, \qquad (1.30)$$

$$\tau_2 \equiv \tau_2(n, K) = \frac{(n + 1)\pi - \cos^{-1}(1 - 1/K)}{\omega_0 + \sqrt{2K - 1}}. \qquad (1.31)$$

We plot these curves in Fig. 1.8 for $\omega_0 = 30$ and shade the regions of amplitude death. The death regions are multiple in number at this value of ω_0. A more detailed analysis [32, 33] also reveals that the eigenvalues indeed cross these boundaries from inside to outside as we expect.

1.3 *N*-Oscillator Models

With the insights gained from the analysis of the minimal model of just two coupled oscillators, we will now try to study the collective states of a more complex system where we have a large number N of coupled oscillators. When the number of oscillators is large ($N > 2$), the mutual coupling can occur in a variety of ways.

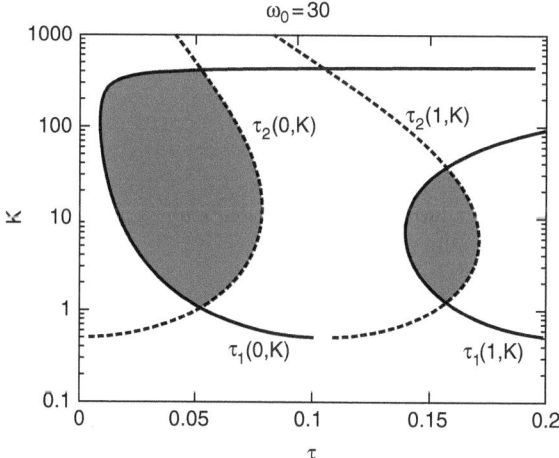

Fig. 1.8 Two of the three death islands are shown for $\omega_0 = 30$

One simple way is to connect each oscillator to every other oscillator with a constant coupling strength. Such a system is called a globally coupled system or more commonly as a mean field model. Another widely used coupling scheme is that of nearest neighbor coupling which is often referred to as diffusive coupling. A more generalized form of coupling that reduces to the above two in limiting cases is that of non-local coupling in which the coupling extends over a wider spatial extent but with a varying strength. In this section we will introduce time delay coupled versions of all these three models and discuss the effect of time delay on their collective states. We will restrict ourselves to one-dimensional configurations (for the locally and non-locally coupled systems) which are simpler to analyze. Our present discussion will be primarily based on the work carried out in [33, 35–37].

1.3.1 Global Coupling

Making a straightforward generalization of (1.5) and (1.6), one can describe a system of N globally coupled Stuart–Landau oscillators (with a linear time-delayed coupling) by the following set of model equations:

$$\dot{Z}_j(t) = (1 + i\omega_j - | Z_j(t) |^2)Z_j(t) + \frac{K'}{N} \sum_{k=1}^{N} [Z_k(t - \tau) - Z_j(t)]$$

$$- \frac{K'}{N}[Z_j(t - \tau) - Z_j(t)], \qquad (1.32)$$

where $j = 1, \ldots, N$, $K' = 2K$, is the coupling strength and τ is the delay time. The coupling term on the right-hand side now has a summation up to N, ensuring coupling of an individual oscillator to every other one in the system, whereas the last term has been introduced to subtract the self-coupling term from the summation. In the absence of time delay, such a mean field model has been studied extensively in the past in [5, 20, 25, 27] in order to delineate the various stationary and non-stationary states of the system including phase-locked states, amplitude death, phase drift states, Hopf oscillations, and even chaotic states. We will examine the effect of time delay on some of these states. For this it is convenient to define an order parameter, defined as,

$$\bar{Z} = Re^{i\phi} = \frac{1}{N} \sum_{j=1}^{N} Z_j(t), \qquad (1.33)$$

where R and ϕ denote the amplitude and phase of the centroid. In a large N model the order parameter provides a time-asymptotic measure of the coherence (collective aspects) of the system in both a qualitative and a quantitative fashion. When $R = 0$ (in the large time limit) the system can be considered to be in an incoherent state whereas $R = 1$ marks a totally synchronized or 'phase-locked' state. Any value in between indicates that the system is in a partially synchronized state. The order parameter is also capable of displaying signatures of non-stationary states like chaos and large amplitude Hopf oscillations through its temporal behavior [25, 27]. Using the order parameter, the model equations can be rewritten more compactly as

$$\dot{Z}_j(t) = (1 - K'd + i\omega_j - |Z_j(t)|^2)Z_j(t) + K'\bar{Z}(t - \tau) - \frac{K'}{N}Z_j(t - \tau), \quad (1.34)$$

where $d = 1 - 1/N$. We will now study the stability of the origin of these equations (for examining the amplitude death state) and also discuss the phase-locked states of the system.

1.3.1.1 Amplitude Death

The stability of the origin can be determined as before by doing a linear perturbation analysis around $Z_j = 0$ in (1.32). Assuming the perturbations to vary in time as $e^{\lambda t}$, the characteristic matrix of (1.32) is given by

$$B = \begin{bmatrix} l_1 & f & \cdots & f \\ f & l_2 & \cdots & f \\ \vdots & \vdots & \ddots & \vdots \\ f & f & \cdots & l_N \end{bmatrix}, \quad (1.35)$$

where $l_n = 1 - K'd + i\omega_n$ and $f = \frac{K'}{N}e^{-\lambda\tau}$. The eigenvalue problem can also be cast in terms of another matrix $C = B + (K'd - 1)I$ (with I being the identity matrix), such that if μ is the eigenvalue of C then it is related to λ by the relation $\mu = \lambda + (K'd - 1)$. The eigenvalue equation $\det(C - \mu I) = 0$ can be compactly expressed as a product of two factors:

$$\left[\prod_{k=1}^{N}(i\omega_k - \mu - f)\right]\left[1 + f\sum_{j=1}^{N}\frac{1}{i\omega_j - \mu - f}\right] = 0. \quad (1.36)$$

As discussed by Matthews and Strogatz [38, 27], for the no-delay case, solutions of the first factor represent the continuous spectrum of the system whereas the second factor provides the discrete spectrum. The characteristic Equation (1.36) is difficult to solve analytically or even numerically when N is large. We will confine ourselves to the case of N identical oscillators where a simple analysis is possible and which will also allow us to seek a generalization of the $N = 2$ result discussed in the previous section. For N identical oscillators the frequency distribution of the system can be expressed as a delta function

$$g(\omega) = \delta(\omega - \omega_0), \quad (1.37)$$

where ω_0 is the natural frequency of each oscillator. With such a distribution the eigenvalue equation can be simplified to the form

$$\lambda = \left\{ 1 - K'd + i\omega_0 + K'de^{-\lambda\tau}, \quad 1 - K'd + i\omega_0 - \frac{K'}{N}e^{-\lambda\tau} \right\}, \qquad (1.38)$$

in which the second eigenvalue has a degeneracy of $N - 1$. Using both the eigenvalue equations and the procedure described for the $N = 2$ case, one can obtain the following set of critical curves:

$$\tau_a(n, K) = \frac{2n\pi + \cos^{-1}\left[1 - \frac{1}{K'd} \right]}{\omega_0 - \sqrt{2K'd - 1}}, \qquad (1.39)$$

$$\tau_b(n, K) = \frac{2(n + 1)\pi - \cos^{-1}\left[1 - \frac{1}{K'd} \right]}{\omega_0 + \sqrt{2K'd - 1}}, \qquad (1.40)$$

$$\tau_c(n, K) = \frac{2(n + 1)\pi - \cos^{-1}\left[\frac{1 - K'd}{K'(1 - d)} \right]}{\omega_0 - \sqrt{[K'(1 - d)]^2 - (K'd - 1)^2}}, \qquad (1.41)$$

$$\tau_d(n, K) = \frac{2n\pi + \cos^{-1}\left[\frac{1 - K'd}{K'(1 - d)} \right]}{\omega_0 + \sqrt{[K'(1 - d)]^2 - (K'd - 1)^2}}. \qquad (1.42)$$

In contrast to the $N = 2$ case, the family of curves is now four instead of two, and they are functions of the parameter N. It is easy to check that for $N = 2$, the curves $\tau_a(n, K)$ and $\tau_c(n, K)$ combine to give $\tau_1(n, K)$ and $\tau_b(n, K)$ and $\tau_d(n, K)$ combine to give $\tau_2(n, K)$ which were obtained in the earlier section. We show some typical death island regions in Fig. 1.9(a) for various values of N as obtained from the critical curves (1.39), (1.40), (1.41) and (1.42) with $n = 0$. The sizes of the islands are seen to vary as a function of N and to approach a saturated size as $N \to \infty$. The existence of these regions has also been independently confirmed by direct numerical solution of the coupled oscillator equations [33]. Thus, the phenomenon of time delay-induced death for identical oscillators happens even for an arbitrarily large number of oscillators and is a generic property of our coupled oscillator system. As in the $N = 2$ case these death islands can also show multiple connectedness for higher values of ω_0 which is a characteristic feature of delay equations.

1.3.2 Nearest Neighbor Coupling

We now look at a local coupling model in which each oscillator is coupled only to its next nearest neighbor. The summation term on the right-hand side of (1.32) then collapses to just two terms and the model equations have the form

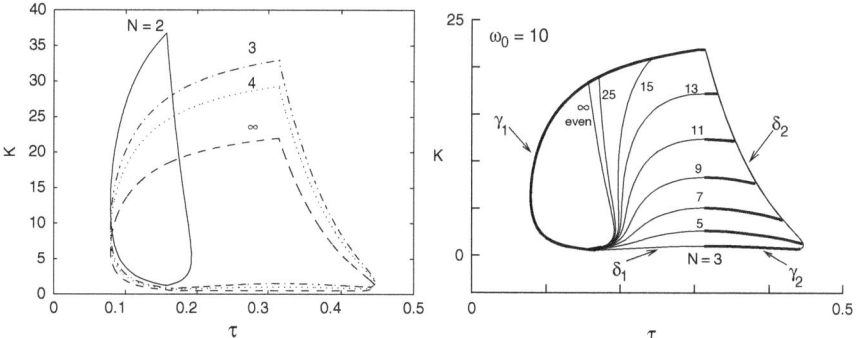

Fig. 1.9 Death islands for global and nearest neighbor couplings. ($\omega_0 = 10$) (a) Death island regions for oscillators with global coupling [32, 33]. (b) Death island regions for oscillators with nearest neighbor coupling. All even number of oscillators have a single death island region that is independent of the number of oscillators. The odd number of oscillators are bounded by four curves when $N \leq 13$ and two curves otherwise. These two curves merge in the infinite limit with the curves that represent the even number of oscillators [35]. For $N \leq 13$, $\gamma_1 = \tau_b(0, K)$ at $R_j = 1$, $\gamma_2 = \tau_b(0, K)$ at $R_{j=(N+1)/2}$, $\delta_1 = \tau_a(0, K)$ at $R_{j=(N+1)/2}$, $\delta_2 = \tau_a(1, K)$ at $R_{j=1}$, and for $N \geq 15$, $\gamma_1 = \tau_b(0, K)$ at $R_{j=1}$, and $\delta_1 = \tau_a(0, K)$ at $R_{j=(N+1)/2}$

$$\frac{\partial Z_j}{\partial t} = (1 + i\omega_j - |Z_j|^2)Z_j + K[Z_{j+1}(t - \tau) - Z_j(t)]$$
$$+ K[Z_{j-1}(t - \tau) - Z_j(t)], \qquad (1.43)$$

where the notation is as before. We will simplify this model further by considering only identical oscillators (setting all $\omega_j = \omega_0$) and assuming the oscillators to be arranged in a closed ring. We choose identical oscillators in order to continue our exploration of the death state for the case where there is no frequency dispersion and the ring configuration allows application of periodic boundary conditions when considering phase-locked states. The model equation then has the form

$$\frac{\partial Z_j}{\partial t} = (1 + i\omega_0 - |Z_j|^2)Z_j + K[Z_{j+1}(t - \tau) - Z_j(t)]$$
$$+ K[Z_{j-1}(t - \tau) - Z_j(t)]. \qquad (1.44)$$

Note that unlike the global coupling case there is now a spatial dependence in the coupling and therefore the geometrical arrangement of the oscillators matters. This additional dimension introduces new equilibrium states in the system such as traveling waves and in higher dimensions spiral patterns or scroll waves. For $\tau = 0$ one can also make an interesting connection to a well-known nonlinear dynamical equation, namely the complex Ginzburg–Landau equation (CGLE). To see this one can take the limit where the spacing between two oscillators, a, goes to zero, so that $Z_j = Z(ja) \to Z(x)$ where x is a continuum variable. Then (1.44) reduces to

$$\frac{\partial Z(x,t)}{\partial t} = (1 + i\omega_0 - |Z(x,t)|^2)Z(x,t) + K\frac{\partial^2 Z(x,t)}{\partial x^2}, \qquad (1.45)$$

where K has been rescaled as K/a^2. This equation has been widely studied and has a rich variety of nonlinear solutions that have found a number of interesting applications.[1] We will return to a more generalized form of this equation later in the chapter. We continue now with the exploration of the amplitude death and phase-locked solutions of (1.44).

1.3.2.1 Amplitude Death

To obtain the conditions for the existence of the death region and to determine its location in parameter space, we resort as usual to a linear perturbation analysis about the origin. Applying periodic boundary conditions (because of the assumed closed chain configuration) we can get the following eigenvalue equation:

$$\prod_{j=1}^{N} \left(\lambda + 2K - 1 - i\omega_0 - Ke^{-\lambda\tau} U_j - Ke^{-\lambda\tau} U_j^{N-1} \right) = 0,$$

where $U_j = e^{i2\pi(j-1)/N}$ are the Nth roots of unity. Since $U_j + U_j^{N-1} = U_j + U_j^{-1} = 2\cos\left[(j-1)2\pi/N\right]$, the above equation can be further simplified to

$$\prod_{j=1}^{N} \left(\lambda + 2K - 1 - i\omega_0 - 2K\cos\left[(j-1)2\pi/N\right]e^{-\lambda\tau} \right) = 0. \qquad (1.46)$$

The above equation has to be complemented by its conjugate equation in order to obtain a complete set of eigenvalue equations. We notice that for $\tau = 0$, (1.46) always admits at least one unstable eigenvalue, namely $\lambda = 1 + i\omega_0$. This means that identical oscillators that are locally coupled cannot have an amplitude death state in the absence of time delay, an echo of our earlier results for $N = 2$ and N globally coupled oscillators. In the presence of finite delay, we adopt the same standard procedure that we used earlier for the global coupling case, namely that of determining the marginal stability condition to identify the critical curves. Before we do that, it is worthwhile pointing out another essential and interesting difference from the global coupling case, namely, if N is a multiple of 4 then some of the factors of (1.46) have no explicit τ dependence since for them $R_j = 2\cos\left[(j-1)2\pi/N\right] = 0$. Consider the case of $N = 4$ and $j = 2, 4$, for which the eigenvalue equation becomes $\lambda = 1 - 2K \pm i\omega_0$, and the only criticality condition is then given by $K = 1/2$. Thus, the stable region lies on the side of the parameter space that obeys $K > 1/2$. For other values of R_j, the death island boundaries can be determined by a marginal

---

[1] In its most general form, the CGL equation is of the form $\frac{\partial Z(x,t)}{\partial t} = (1 + i\omega_0 - (1 + ib)|Z(x,t)|^2)Z(x,t) + K(1 + ia)\frac{\partial^2 Z(x,t)}{\partial x^2}$, where a and b are real quantities.

stability analysis, and the expressions for the critical curves in the (τ, K) plane are given by

$$
\tau_a(n, K) = \begin{cases} \dfrac{2n\pi - \cos^{-1}\left[(2K-1)/KR_j\right]}{\omega_0 + \sqrt{K^2 R_j^2 - (2K-1)^2}}, & R_j > 0, \\[2em] \dfrac{(2n+1)\pi - \cos^{-1}\left[(2K-1)/K\left|R_j\right|\right]}{\omega_0 + \sqrt{K^2 R_j^2 - (2K-1)^2}}, & R_j < 0, \end{cases} \tag{1.47}
$$

$$
\tau_b(m, K) = \begin{cases} \dfrac{2m\pi + \cos^{-1}\left[(2K-1)/KR_j\right]}{\omega_0 - \sqrt{K^2 R_j^2 - (2K-1)^2}}, & R_j > 0, \\[2em] \dfrac{(2m+1)\pi + \cos^{-1}\left[(2K-1)/K\left|R_j\right|\right]}{\omega_0 - \sqrt{K^2 R_j^2 - (2K-1)^2}}, & R_j < 0, \end{cases} \tag{1.48}
$$

where n and m are integers. For a detailed analysis of these curves, which includes determination of useful bounds on K for ordering and finding the degeneracies of the critical curves, we refer the reader to [35]. The essential features of the death islands in this system can be gathered from Fig. 1.9(b), where they have been plotted for different values of N. One striking difference is that the size and shape of the death island is now determined by the odd or even property of the number of oscillators N. For an even number of oscillators there is a single death region, whereas for an odd number of oscillators the boundary of the death region depends on the value of N. We illustrate the death states for a sample number of even ($N = 4$) and odd ($N = 5$) nearest neighbor coupled oscillators in Fig. 1.10. The death state for $N = 4$ is surrounded by an in-phase state on the left and an anti-phase state that has neighboring oscillators π out of phase on the right. The death state for $N = 5$ on the other hand has an in-phase state on either side. These differences in the death island widths for even- and odd-numbered oscillators can be traced primarily to the behavior of the eigenvalues of the lowest permitted perturbation wave numbers. As N becomes large, the smallest perturbation mode for the N odd case gets closer to π and the death island boundaries of the two cases become indistinguishable. The size of the death island, for the odd case, decreases as N increases and finally asymptotes to the single ($N = $ even) island as $N \to \infty$. This asymmetry is intimately related to the nature of the coupling and as we will see later it disappears when we change the coupling to a non-local one.

1.3.2.2 Phase-Locked States

As mentioned briefly before, the spatial dependence of the coupling provides for a larger class of equilibrium states in a dispersively coupled system as compared to the globally coupled system. In particular plane wave states, which are characterized

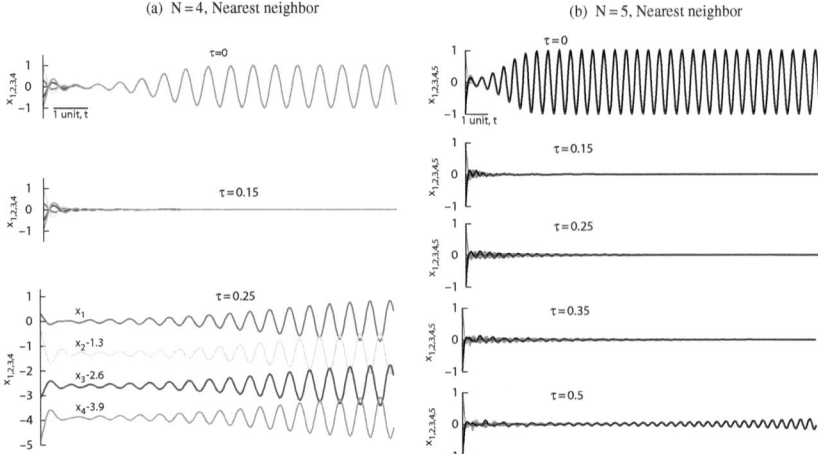

Fig. 1.10 (**a**) A nearest neighbor coupled network of four identical oscillators showing in-phase, death, and then anti-phase patterns as τ is increased, $K = 2$ and $\omega_0 = 10$. The initial conditions are $x_1 = 0.3$, $y_1 = 0$, $x_2 = 1$, $y_2 = 0$, $x_3 = -0.5$, $y_3 = 0$, $x_4 = -0.9$, and $y_4 = 0$. The time history vectors $Z_{1,2,3,4} = 0$. (**b**) A globally coupled network of five identical oscillators showing in-phase, death, and in-phase transitions as τ is increased. Time courses of the real parts of $Z_{1,2,3,4,5}$ are plotted for different τ at $K = 5$ (different than that in (**a**)) and $\omega_0 = 10$. Initial conditions are $x_1 = 1$, $y_1 = 0$, $x_2 = 0.45$, $y_2 = 0$, $x_3 = -0.2$, $y_3 = 0$, $x_4 = -0.61$, $y_4 = 0$, $x_5 = -0.97$, and $y_5 = 0$. The time history vectors $Z_{1,2,3,4,5} = 0$

by a frequency as well as a wave number, are one such possibility. Equation (1.44) admits plane wave solutions of the form

$$Z_j = Re^{i(jka+\omega t)}, \tag{1.49}$$

where a is the distance between any two adjacent oscillators and k is the wave number such that $-\pi \leq ka \leq \pi$. The values of ka are discrete due to the constraint imposed by the periodic boundary conditions, namely that $Z_{N+1} = Z_1$ and $Z_0 = Z_N$. This condition requires that we satisfy $e^{iNka} = 1$ which gives $Nka = 2m\pi$, $m = 0, 1, \ldots, N - 1$, that is,

$$ka = m\frac{2\pi}{N}, \quad m = 0, 1, \ldots, N - 1. \tag{1.50}$$

Thus, the various phase-locked states are now labeled by their characteristic wave number values. The wave numbers are further related to the frequencies of the states through a dispersion relation which can be obtained by substituting (1.49) in (1.44):

$$i\omega = 1 + i\omega_0 - R^2 + 2K\left[\cos(ka)e^{-i\omega\tau} - 1\right]. \tag{1.51}$$

Separating the real and imaginary parts of the above relation we get

$$\omega = \omega_0 - 2K\sin(\omega\tau)\cos(ka), \tag{1.52}$$
$$R^2 = 1 - 2K + 2K\cos(\omega\tau)\cos(ka). \tag{1.53}$$

It is interesting at this point to compare the above dispersion relation to the plane wave dispersion relation of the CGLE. We can obtain such a dispersion relation by substituting $Z(x,t) = R_{CGL} \exp(ikx - i\omega_{CGL}t)$ in (1.45) to get

$$\omega_{CGL} = \omega_0, \tag{1.54}$$
$$R^2_{CGL} = 1 - K(ka)^2, \tag{1.55}$$

where we have replaced K by its scaled value. Equations (1.54) and (1.55) can also be obtained from (1.52) and (1.53) by putting $\tau = 0$ and expanding the $\cos(ka)$ term in (1.53) by taking the long wavelength limit of $ka \ll 1$. The two sets of dispersion relations show interesting differences. First of all their domains of validity are different: (1.52) and (1.53) are valid for any arbitrary value of N whereas (1.54) and (1.55) are strictly valid only in the continuum limit (i.e., $N \to \infty$). The ka spectrum is therefore a continuous one for the CGLE whereas in our model they are discrete and also depend on the value of N. For $\tau = 0$, (1.52) and (1.54) become identical, but (1.53) reduces to

$$R^2 = 1 - 2K + 2K\cos(ka) = 1 - 4K\sin^2(ka/2). \tag{1.56}$$

Since $R^2 > 0$ is a necessary condition for a plane wave state to exist, we see that for a given value of K the domain of existence is considerably reduced in the case of the CGLE as compared to the discrete model equations. As an example, at $K = 1/4$, the discrete model allows all modes from 0 to π to exist, whereas the continuum model has an upper cutoff at $ka = 2.0$. For $K > 1/4$ one has cutoff regions in the discrete model as well that are defined by the expression

$$f_1 = \cos^{-1}(1 - 1/2K) < ka < f_2 = 2\pi - \cos^{-1}(1 - 1/2K). \tag{1.57}$$

From (1.57) it is clear that for $K > 1/4$ the anti-phase-locked state $(ka = \pi)$ is now no longer a permitted state. In the presence of time delay the existence region is defined by a more complex relation since it is now not only a function of ka (for a given value of K) but also depends on ω which is a solution of the transcendental equation (1.52). Thus, time delay can bring about interesting modifications such as enabling certain forbidden states (of the non-delayed system) to exist and in general reshaping the existence domain significantly. In addition, as we saw in the two oscillator model, the transcendental character of the dispersion relation can introduce additional branches of collective oscillations. A detailed analysis of some of these features are available in [35]. In Fig. 1.11(a) we have illustrated some of the salient findings of [35] by showing the existence regions of plane wave states for some special cases. To delineate the general existence regions which are now complicated functions of ka, K, and τ, one needs to have a simultaneous solution of (1.52) and (1.53). To appreciate the constraints imposed by (1.52) in Fig. 1.11(b) we have plotted the solution (ω vs. ka) for various values of τ and for a fixed value of ω_0 and K. For $\tau = 0$, the values of ka are constrained to be in the range $(|ka|) < \cos^{-1}(1 - 1/2K)$. At $K = 1$, the phase-locked patterns that have wave

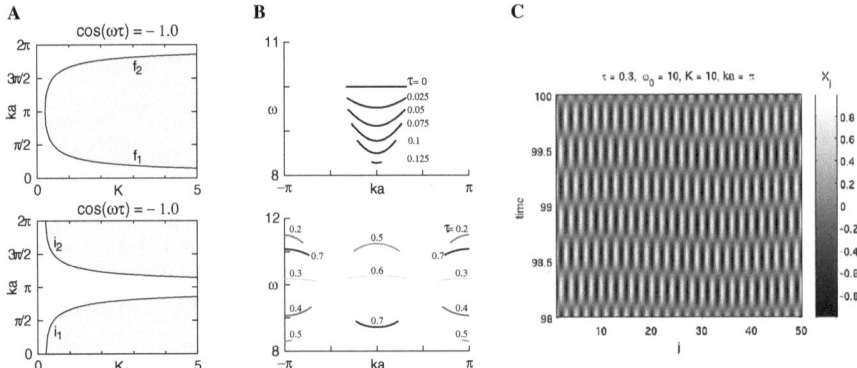

Fig. 1.11 (**a**) Allowed (*unshaded*) and forbidden (*shaded*) wave modes in the presence of time delay for two values of $\cos(\omega\tau)$. (**b**) Dispersion relation between allowed wave numbers and the corresponding frequency shown as τ is gradually increased, $K = 1$ and $\omega_0 = 10$. A range of τ values is forbidden. (**c**) A numerical example of out-of-phase state shown for $N = 50$ oscillators by plotting the level of the real component of Z_j as a function of the oscillator number j [35]

numbers less than $\pi/3$ are allowed, and all of them have an identical frequency. As τ is increased the frequency of oscillation decreases for small τ, and the dispersion relation acquires a nonlinear parabolic character. As τ is further increased, depending on the actual value of K, there are bands in τ values where no modes exist. The shrinking and disappearance of the dispersion curve at $ka = 0$ beyond $\tau = 0.125$ up to $\tau = 0.2$ in the top panel of Fig. 1.11(b) illustrate this phenomenon. One also notices from the bottom panel of Fig. 1.11(b) that at higher values of τ the dispersion curves become discontinuous and have bands of forbidden ka regions.

1.3.2.3 Stability of Phase-Locked States

So far we have only discussed the existence conditions for plane wave states of (1.44) in the parameter domain of the wave number, coupling strength, frequency, and time delay. We also need to know the stability of such states in order for them to be excited and sustained in the system. In this section we will carry out a linear stability analysis of the equilibrium phase-locked solutions discussed in the previous section. We let

$$Z_j(t) = \left[R_k e^{i\omega_k t} + u_j(t) \right] e^{i(jka)}, \tag{1.58}$$

where $k = 0, 1, \cdots, N - 1$, substiute it in (1.44), and carry out an order by order analysis in the perturbation amplitude u_j. In the lowest order we recover the dispersion relation (1.51). In the next order, where we retain terms that are linear in the perturbation amplitude, we get

$$\frac{\partial u_j(t)}{\partial t} = \left(1 + i\omega_0 - 2R_k^2 - 2K \right) u_j(t) - R_k^2 e^{2i\omega_k t} \bar{u}_j(t)$$

$$+ K \left[u_{j+1}(t - \tau) e^{ika} + u_{j-1}(t - \tau) e^{-ika} \right] \tag{1.59}$$

and, taking its complex conjugate,

$$\frac{\partial \bar{u}_j(t)}{\partial t} = \left(1 - i\omega_0 - 2R_k^2 - 2K\right) \bar{u}_j(t) - R_k^2 e^{-2i\omega_k t} u_j(t)$$

$$+ K\left[\bar{u}_{j+1}(t-\tau) e^{-ika} + \bar{u}_{j-1}(t-\tau) e^{ika}\right]. \tag{1.60}$$

We next multiply (1.59) and (1.60) term by term by $e^{i(jqa)}$ and make use of the identities

$$u_{j\pm 1}(t-\tau) e^{\pm ika} e^{i(jqa)} = u_{j\pm 1}(t-\tau) e^{i(j\pm 1)qa} e^{\pm i(k-q)a} \tag{1.61}$$

and

$$\bar{u}_{j\pm 1}(t-\tau) e^{\pm ika} e^{i(jqa)} = \bar{u}_{j\pm 1}(t-\tau) e^{i(j\pm 1)qa} e^{\pm i(k+q)a}, \tag{1.62}$$

and finally sum over $j = 0, 1, 2, \cdots, N-1$. Introducing adjoint amplitudes $w_q(t)$ and $\tilde{w}_q(t)$ by the definitions

$$\left[w_q(t), \tilde{w}_q(t)\right] = \sum_{j=0}^{N-1} \left[u_j(t), \bar{u}_j(t)\right] e^{i(jqa)}, \tag{1.63}$$

we can arrive at the following set of coupled equations:

$$\frac{dw_q(t)}{dt} = \left(1 + i\omega_0 - 2R_k^2 - 2K\right) w_q(t) - R_k^2 e^{2i\omega_k t} \tilde{w}_q(t)$$

$$+ 2K \cos\left[(k-q)a\right] w_q(t-\tau) \tag{1.64}$$

and

$$\frac{d\tilde{w}_q(t)}{dt} = \left(1 - i\omega_0 - 2R_k^2 - 2K\right) \tilde{w}_q(t) - R_k^2 e^{-2i\omega_k t} w_q(t)$$

$$+ 2K \cos\left[(k+q)a\right] \tilde{w}_q(t-\tau). \tag{1.65}$$

Now assuming the solutions to be of the form

$$\left[w_q(t), \tilde{w}_q(t)\right] = \left[c e^{i\omega_k t}, \tilde{c} e^{-i\omega_k t}\right] e^{\lambda t}, \tag{1.66}$$

one can, after some straightforward algebra, obtain the following eigenvalue equation:

$$\lambda^2 + (a_1 + a_2)\lambda + (a_1 a_2 - R^4) = 0, \tag{1.67}$$

where

$$a_1 = 2R^2 - 1 + 2K - i(\omega_0 - \omega) - 2K \cos{[(k - q)a]}e^{-(\lambda + i\omega)\tau}, \qquad (1.68)$$
$$a_2 = 2R^2 - 1 + 2K + i(\omega_0 - \omega) - 2K \cos{[(k + q)a]}e^{-(\lambda - i\omega)\tau}. \qquad (1.69)$$

The perturbation wave numbers q are a discrete set that obey the relation

$$qa = m\frac{2\pi}{N}, \quad m = 0, 1, \ldots, N - 1.$$

Thus, for any given plane wave pattern characterized by a given value of ka, one needs to examine the eigenvalues of (1.67) at each of the above permitted values of qa, which is a formidable task for any reasonably large value of N and one needs to carry out extensive numerical investigations. Some specific examples have been worked out in [35]. In general, time delay appears to expand the stability domain of plane wave states allowing for a richer spectrum of states to be sustained in the system as compared to the no-delay case. For example, the out of phase state (where each oscillator is π out of phase with its neighbor) which is always unstable in the absence of time delay can get stabilized for certain values of τ. A numerical simulation of such an out-of-phase state is shown in Fig. 1.11(c) where the level of the real part of Z_j is plotted as a function of the oscillator number j for $N = 50$ oscillators.

1.3.3 Non-Local Coupling

While global and local (nearest neighbor) coupling models have received much attention in the past, [1, 2, 39–41] there is now a growing interest in the collective dynamics of models with non-local couplings [42–48]. Non-local coupling implies a form of coupling in which the coupling extends over a wider domain than the local (nearest neighbor coupling) but with varying (usually diminishing) coupling strength. The coupling strength can fall off exponentially or in some cases even change sign with distance (for example, in a Mexican hat fashion). Non-local coupling can be relevant to a variety of applications such as in the modeling of Josephson junction arrays [49], chemical oscillators [47, 48, 50], neural networks for producing snail shell patterns, and ocular dominance stripes [51–53]. They can also arise in a large class of reaction diffusion systems under certain limiting assumptions for the diffusion strength and local kinetics such that the dynamics is governed by an equation which is a non-local generalization of the CGLE [50, 54]. Another interesting and unique feature of a non-locally coupled system is that it can sustain some unusual collective states in which the oscillators separate into two groups—one that is synchronized and phase-locked and the other desynchronized and incoherent [47]. Such a state of co-existence of coherence and incoherence does not occur in either globally or locally coupled systems and has been named as a 'chimera' state by Strogatz [55]. The nature and properties of this exotic collective state as well as its potential applications are still not fully explored or understood and therefore continue to offer exciting future possibilities. In this section we will explore the effect of time delay on the collective states of a non-locally coupled system, where in addition to looking at the amplitude death and phase-locked states we will also discuss the novel 'chimera' state.

1.3.3.1 Non-Local Model Equations

We begin by first extending our set of model equations from the previous section to include non-local time-delayed coupling. Considering again a closed chain of N identical limit cycle oscillators, we can generalize (1.44) to the following form:

$$\frac{\partial Z_j}{\partial t} = (1 + i\omega_0) Z_j(t) - |Z_j(t)|^2 Z_j(t) + KQ_j(t),\qquad(1.70)$$

where K is the coupling constant, $Z_j(t)$ are the complex amplitudes of the oscillators, $j = 0, 1, 2, \ldots, (N-1)$, and

$$Z_{j+nN}(t) = Z_j(t), \quad n = 0, \pm 1, \pm 2, \ldots\qquad(1.71)$$

The total coupling function acting on the jth oscillator, $Q_j(t)$, is given by

$$Q_j(t) = S_{e,o} \sum_{\{e,o\}} e^{-m\kappa a} \left\{ \left[Z_{j+m}(t - m\tau) - Z_j(t) \right] + \left[Z_{j-m}(t - m\tau) - Z_j(t) \right] \right\},$$

$$(1.72)$$

where a is the distance and τ is the time delay between two adjacent oscillators, respectively. The labels e and o indicate cases in which, respectively, N is even and N is odd. The two cases must be treated separately. The total coupling function is a sum over m for pairwise couplings of oscillators and excludes self-coupling. If we compare (1.70) (keeping in mind (1.72)) with the model set of equations for the locally coupled system (1.44), we notice the following essential differences. The coupling strength K is now weighted by a factor $\exp(-m\kappa a)$ which is distance dependent and therefore makes the coupling from distant oscillators progressively weaker. The time delay dependence in the argument of each Z_j is likewise a function of $m\tau$ which implies that the amount of time a signal takes to arrive at a given oscillator location increases linearly as the distance it has to travel from another oscillator. Physically this amounts to assuming a constant signal velocity v in the system such that $\tau = a/v$. The exponentially decaying weight factor that we have chosen for this model can be replaced by other functions to change the nature of coupling, but we will restrict ourselves to this function as it has been used in some past calculations [45, 47, 48] that were done in the absence of delay and hence provides a convenient benchmark for assessing delay effects. The exponential damping coefficient κ provides a measure of the amount of non-locality in the coupling with $\kappa = 0$ being the global coupling limit. In order that the coupling amplitude be an exponential function of the distance between coupled oscillators, rather than a truncated exponential function of that distance, it is necessary that the effective coupling range be less than half the length of the ring. If we denote the length of the ring by $2L$, then the minimum value of κ for which truncation does not occur is the value that yields the largest value of $\exp(-\kappa L)$ that can be considered negligible. It is convenient for this purpose to adopt the condition $\kappa L \geq 2\pi$, because $\exp(-2\pi) = 0.00187$. The choice $L = \pi$ is particularly convenient for the Fourier analysis of the discrete system because it yields the primitive basis set $e_k(m) = \exp(imka)$, where $a = (2\pi/N)$ and $m, k = 0, 1, 2, \ldots, (N - 1)$. For other choices of L, k and a have different values, but ka is invariant with respect to the choice of L.

The quantity $S_{\{e,o\}}$ is a normalization factor. By assigning the value 1 to each coupling term, $\left[Z_{j\pm m}(t - m\tau) - Z_j(t)\right]$, and requiring that the value of the associated total coupling function is 1, we obtain

$$S_{\{e,o\}} = \left(2 \sum_{\{e,o\}} e^{-m\kappa a}\right)^{-1}. \tag{1.73}$$

The correctness of this choice is demonstrated in the case that $N = 2$ and $\kappa = 0$, where one obtains (1.44) as intended. The form of the normalization factor makes it possible to express the total coupling function in the simplified form

$$Q_j(t) = S_{\{e,o\}} \sum_{\{e,o\}} e^{-m\kappa a} \left[Z_{j+m}(t - m\tau) + Z_{j-m}(t - m\tau)\right] - Z_j(t). \tag{1.74}$$

In the case in which N is even, the summation is

$$\sum_e = \sum_{m=1}^{\frac{1}{2}N} \left(1 - \frac{1}{2}\delta_{m,\frac{1}{2}N}\right). \tag{1.75}$$

The quantity in parentheses is introduced to account for the fact that the subscripts $j + \frac{1}{2}N$ and $j - \frac{1}{2}N$ denote the same oscillator.

In the case in which N is odd, the summation is

$$\sum_o = \sum_{m=1}^{\frac{1}{2}(N-1)}. \tag{1.76}$$

One obtains for S_e the result

$$S_e = \frac{1}{2}\left[\frac{1 - e^{-\kappa a}}{e^{-\kappa a} - \frac{1}{2}e^{-\frac{1}{2}N\kappa a} - \frac{1}{2}e^{-\left(\frac{1}{2}N+1\right)\kappa a}}\right]. \tag{1.77}$$

In the limit $\kappa \to 0$, one obtains $S_e = (N-1)^{-1}$. In that limit and for the case $N = 2$, one obtains the correct result, $S_e = 1$. One obtains for S_o the result

$$S_o = \frac{1}{2}\left[\frac{1 - e^{-\kappa a}}{e^{-\kappa a} - e^{-\frac{1}{2}(N+1)\kappa a}}\right]. \tag{1.78}$$

In the limit $\kappa \to 0$, one obtains $S_o = (N-1)^{-1}$.

It is interesting to also consider the continuum limit of (1.70), (1.71), and (1.72) in the spirit of what we did for the locally coupled system. This limit can be achieved by adopting the following procedure: We let the number of oscillators N increase without limit, i.e., $N \to \infty$, but keep the system length $2L$ unchanged. This implies that $a \to 0$ but $Na = 2L$. As before, the discrete variable that denotes the position of an oscillator, j, is replaced by the continuous variable x $(-L \leq x < L)$, so that the discrete oscillator amplitude, $Z_j(t)$, becomes the continuous oscillator amplitude, $Z(x, t)$. The discrete variable that denotes the distance between coupled oscillators, m, is replaced by the continuous variable $0 \leq y \leq L$, so that the discrete total coupling function $Q_j(t)$ becomes the continuous total coupling function $Q(x, t)$. Consider first the normalization factors $S_{\{e,o\}}$. As the continuous limit is approached, it is necessary that the range of the non-local coupling be much greater than the distance between adjacent oscillators, i.e., that $\kappa a \ll 1$, which ensures that $\exp(-\kappa a) \approx 1 - \kappa a$. The condition $\kappa L \geq 2\pi$, which is necessary to guarantee that the form of the damping function is an exponential function, rather than a truncated exponential function, ensures that $\frac{1}{2}N\kappa a \geq 2\pi$, which in turn ensures that $\exp\left(-\frac{1}{2}N\kappa a\right) \ll 1$. Accordingly, we conclude that in the continuous limit $S_{\{e,o\}} = \frac{1}{2}\kappa a$. Assuming that the delay time between adjacent oscillators is proportional to the distance between them, we obtain the following relation between these quantities and σ, the reciprocal of the speed of propagation of delay coupling in the continuous case: $\sigma = \tau/a$. In the continuous limit, t_d, the earlier time at which a coupling signal originates from position $x \pm y$ to reach position x at time t is $t_d = t - \sigma y$. The correspondence between summation in the discrete case and integration in the continuous case is

$$\sum_{\{e,o\}} a \to \int_0^L dy.$$

The condition $\kappa L \geq 2\pi$ permits to replace the upper limit of integration, L, by ∞. Thus, we obtain as the continuous limit of (1.70), (1.71) and (1.72)] the set of equations

$$\frac{\partial}{\partial t}Z(x, t) = (1 + i\omega_0)Z - |Z|^2 Z + KQ(t), \tag{1.79}$$

$$Z(x + 2Ln, t) = Z(x, t) \quad n = 0, \pm 1, \pm 2, \ldots, \tag{1.80}$$

and

$$Q(x,t) = \int_0^\infty dy \frac{1}{2} \kappa e^{-\kappa y} \left[Z(x+y, t-\sigma y) + Z(x-y, t-\sigma y) \right] dy. \quad (1.81)$$

Equation (1.79) is the non-local time-delayed generalization of the complex Ginzburg–Landau equation. One can also obtain a phase-reduced version of the above equation in the limit when the coupling between the oscillators is weak. We then let $Z = A \exp(i\phi(x,t))$ and treat the amplitude A to be a constant. In general the coupling constant K may be complex and can be written as $K(1+ia) = K' \exp(i\alpha)$ where a and α are real constants. Substituting for Z in (1.79) and separating the real and imaginary parts, we get from the imaginary part,

$$\frac{\partial \phi(x,t)}{\partial t} = \omega_0 + K' \int_{-L}^{L} dy \, G(x-x') \sin\left[\phi(x', t-\sigma \mid x-x' \mid) - \phi(x,t) + \alpha \right],$$
$$(1.82)$$

where $\alpha = \tan^{-1}(a)$ and we have chosen the system length to be $2L$. In the absence of time delay (setting $\sigma = 0$) equations (1.79) and (1.82) have recently been studied by a number of authors [47, 48, 50] in the context of chimera states. We will return to these equation later in this section after we have discussed some aspects of the death state and phase-locked states in the discrete non-local system.

1.3.3.2 Amplitude Death in the Non-Local System

A linear perturbation analysis about the origin $Z_j = 0$, carried out for (1.70) in the manner described in the previous section, yields the following eigenvalue equation for N odd

$$\lambda_j = 1 + i\omega_0 - K + 2KS_o \sum_{m=1}^{\frac{1}{2}(N-1)} \cos\left[\frac{2\pi}{N}(j-1)m\right] e^{-m(\kappa a + \lambda_j \tau)}, \quad (1.83)$$

and likewise for even N one can get

$$\lambda_j = 1 + i\omega_0 - K\left[1 + S_e e^{-N(\kappa a + \lambda_j \tau)/2} \cos(\pi(j-1)) \right]$$
$$+ 2KS_e \sum_{m=1}^{N/2} \cos(\frac{2\pi}{N}(j-1)m) e^{-m(\kappa a + \lambda_j \tau)}. \quad (1.84)$$

Notice the similarities and some essential differences between (1.83) (or (1.84)) and the individual factors of the eigenvalue equation (1.46) that we derived earlier for the locally coupled system. In addition to the weight factor $\exp(-\kappa a)$ we also see that there is a summation over all perturbation wave numbers $2\pi(j-1)/N$. One can therefore expect a more complex shape for the island structure compared to the two or at most four curves for the locally coupled system. The parameter κa

provides a measure of the degree of non-locality—a large κa corresponds to a highly localized interaction region and a small value of κa implies stronger non-locality. One can also expect therefore a dependence of the island complexity on the value of κa. These expectations are indeed borne out [36] when one constructs the stability islands in the $K - \tau$ parameter space by using the marginal stability curves of (1.83) or (1.84). For the odd N case the marginal curves are defined by the following two relations:

$$1 - K + 2KS_o \sum_{m=1}^{(N-1)/2} e^{-m\kappa a} \cos{(m\tau\beta)} \cos{(\frac{2\pi}{N}(j - 1)m)} = 0, \qquad (1.85)$$

$$\omega_0 + 2KS_o \sum_{m=1}^{(N-1)/2} e^{-m\kappa a} \sin{(m\tau\beta)} \cos{(\frac{2\pi}{N}(j - 1)m)} = \beta, \qquad (1.86)$$

where $\beta = \text{Im}(\lambda_j)$. To construct the marginal curves, i.e., to derive a relation between K and τ one needs to eliminate β from the above two equations. This is difficult to do analytically but can be accomplished numerically by rewriting the above equations in the following parametric form:

$$K = \frac{1}{1 - 2S_o \sum_{m=1}^{(N-1)/2} e^{-m\kappa a} \cos{(mx)} \cos{(my)}}, \qquad (1.87)$$

$$\tau = \frac{x}{\omega_0 + 2K(x)S_o \sum_{m=1}^{(N-1)/2} e^{-m\kappa a} \sin{(mx)} \cos{(my)}}, \qquad (1.88)$$

where $x = \beta\tau$ and $y = \frac{2\pi}{N}(j - 1)$. For a given value of κa and N, the idea is to evaluate K and τ numerically over a range of x values, e.g. $(-2\pi, 0)$, for a particular value of y and thereby eliminate β. The evaluation is repeated for each value of y. The stable region bounded by these curves then constitutes the death island. In Fig. 1.12(a) we have plotted the lower portions of the islands for two cases, namely $\kappa a = 1$, for which the non-locality is strong, and $\kappa a = 3$, which is close to being a locally coupled system. As can be seen the lower boundary of the strongly coupled case is quite complex and is made up of portions of several marginal stability curves arising from different mode number perturbations. By contrast, for the weakly non-local case the island region lies between just two curves as shown in Fig. 1.12(b). Another important difference from the local coupling case is that the island size is no longer invariant for even N but changes as a function of N.

1.3.3.3 Plane Wave States and their Stability

As in the case of the nearest neighbor coupled system, a non-locally coupled system can also sustain plane wave solutions. The plane wave form of the complex amplitude of the jth oscillator, including a perturbation of order ϵ, is

$$Z_j(t) = \left[1 + \epsilon a_j(t)\right] Z_j^0(t), \qquad (1.89)$$

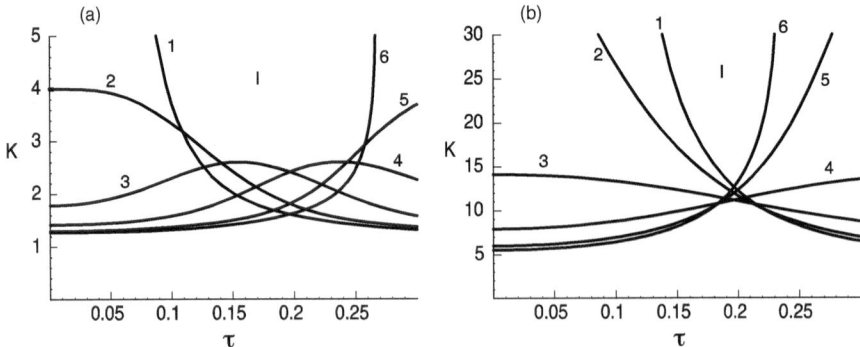

Fig. 1.12 (**a**) The lower portion of a death island for strong non-locality $\kappa a = 1, N = 10$, for small K. All the perturbation modes participate in defining the boundary where the curves are labeled by $j = 1, 2, 3...$ (**b**) A similar plot for weak non-locality, $\kappa a = 3, N = 10$ where only two perturbation modes $j = 1$ and $j = 6$ are seen to define the boundary

where

$$Z_j^0 (t) = \mathcal{R}_k e^{i(jka+\omega_k t)}, \tag{1.90}$$

$k = 0, 1, 2, \ldots, (N-1)$, $\mathcal{R}_k^2$ is required to be positive, and $a_j (t)$ are the complex amplitudes of the perturbation, which satisfy the periodicity conditions $a_{j+nN} (t) = a_j (t)$, $n = 0, \pm 1, \pm 2, \ldots$. In general, ω_k is complex, but for plane wave equilibria ω_k is required to be real. The quantities $\mathcal{R}_k^2$ and ω_k are determined simultaneously by solution of a complex dispersion relation obtained at $O(\epsilon^0)$ of a perturbation expansion in ϵ. A system of evolution equations for the set of functions $a_j (t)$, from which the linear stability of the equilibria is determined, is obtained at $O(\epsilon)$ of the perturbation expansion. The dispersion relation for plane wave solutions of (1.70) is given by the pair of equations

$$\omega_k = \omega_0 + K\Im\left\{\tilde{Q}\right\} \tag{1.91}$$

and

$$\mathcal{R}_k^2 = 1 + K\Re\left\{\tilde{Q}\right\}, \tag{1.92}$$

where

$$\tilde{Q} = S_{\{e,o\}} \sum_{\{e,o\}} e^{-m(\kappa a + i\omega_k \tau)} \left(e^{imka} + e^{-imka}\right) - 1 \tag{1.93}$$

and the symbols $\Re\{\}$, $\Im\{\}$ stand for the real and imaginary parts of the quantity within the braces. For a given value of k, one determines numerically the values of ω_k that satisfy (1.91). For each set of values of k and ω_k, one then determines from (1.92) the value of $\mathcal{R}_k^2$, which, for an acceptable set of values of k and ω_k, is required to be positive. In the case of vanishing time delay, the quantities $\tilde{Q}$ are independent of ω_k, a given value of k determines a single value of ω_k, and the solution of (1.92) and (1.91) for ω_k and $\mathcal{R}_k^2$ is considerably simplified. Having determined a set of equilibrium states $(\omega_k, \mathcal{R}_k)$, one needs to next determine their stability from an eigenvalue analysis. The derivation of such an eigenvalue equation is quite straightforward, if somewhat tedious, and is along the lines of the method followed for the nearest neighbor case. The analysis of such an equation for large N is however quite challenging even numerically and remains an open problem at the present time.

1.3.3.4 Time-Delayed Chimera States

Non-locally coupled oscillator systems can exhibit a remarkable class of patterns called *chimeras*, in which identical limit cycle oscillators separate sharply into two domains, one synchronized and phase-locked and the other desynchronized and drifting [47]. This peculiar mode, in which coherence and incoherence co-exist at the same time in a system of oscillators, was first noticed by Kuramoto and his coworkers [47, 48, 50] in their simulations of the complex Ginzburg–Landau equation (CGLE) with non-local coupling and was later named a *chimera* state by Abrams and Strogatz [55]. In Fig. 1.13(a) we show a typical plot of the chimera state in the absence of delay obtained from a numerical solution of (1.82) with $\sigma = 0$. One clearly sees two distinct regions—a central region where the phases of the oscillators are locked to each other and an outer region where they are randomly distributed. The central portion drifts with a certain fixed velocity while the incoherent part has no fixed velocity. A chimera is a stationary stable pattern that co-exists with a fully phase-locked coherent state and occurs in a limited parameter

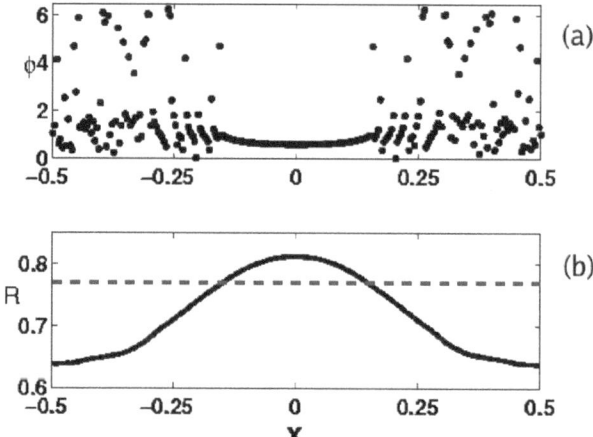

Fig. 1.13 Phase pattern for a typical chimera state. Here $K' = 1$, $\kappa = 4.0$, $\alpha = 1.45$, $\omega = 0$, and $N = 257$

regime defined by the coupling strength K' and the tuning parameter α. Access to such a state is also dependent on initial conditions and Kuramoto et al. [47] took special precautions to obtain such solutions numerically. Chimera solutions exist for both the CGLE and for the reduced phase-only version of the equation. For convenience we rewrite the phase-reduced NDCGL equation (1.82) once again here and set $K' = 1$ and drop the subscript on ω.

$$\frac{\partial}{\partial t}\phi(x,t) = \omega - \int_{-L}^{L} G(x - x')\sin\left[\phi(x,t) - \phi(x',t - \tau_{x,x'}) + \alpha\right] dx'. \quad (1.94)$$

The kernel $G(x - x')$ provides a non-local coupling among the oscillators over a finite spatial range of the order of κ^{-1}, which is chosen to be less than the system size. The coupling is time delayed through the argument of the sinusoidal interaction function, namely, the phase difference between two oscillators located at x and x' is calculated by taking into account the temporal delay for the interaction signal to travel the intervening geodesic (i.e., shortest) distance determined as $d_{x,x'} = \min \{|x - x'|, 2L - |x - x'|\}$. The time delay term is therefore taken to be of the form $\tau_{x,x'} = d_{x,x'}/v$ where v is the signal propagation speed.

The question we now ask is whether (1.94) has a chimera solution in the presence of finite time delay and what it looks like. This problem was addressed in [37] exploring both numerical solutions of (1.94) and some analytical insights obtained from the behavior of the order parameter. We first discuss the numerical results obtained by using the discretized version of (1.94) and employing a large number of oscillators (typically 257) for the simulation. The set of system parameters chosen for the simulations were $2L = 1.0$, $\alpha = 0.9$, $k^{-1} = 0.25$, $\omega = 1.1$, and $v = 0.09765625$ corresponding to a maximum delay time ($\tau_{\max}$) in the system of 5.12. Initially all the oscillators were given uniformly random phases (mirror symmetric) between 0 and 2π, and the equations were evolved long enough to get a time-stationary solution. Figure 1.14 provides a comprehensive summary of the evolution and final state of the time-delayed chimera. Panels (a) and (b) show a space–time plot of the simulation in the early stages of evolution (starting from random initial phases) and in the final stages of the formation of a *clustered chimera* state, respectively. Panel (c) shows a snapshot of the spatial distribution of the phases in the final stationary state. We see four coherent regions interspersed by incoherence and also find that the adjacent coherent regions are π out of phase with each other. Panel (d) is a blowup of the region between $x = -0.5$ and $x = -0.25$ giving an enlarged view of an incoherent region and portions of the adjacent coherent regions.

To gain a better understanding of the nature of this pattern and of the dynamics of its formation, it is instructive to adopt a generalized mean field approach and try to examine the behavior of an averaged quantity like an order parameter of the system. Such an approach and formalism was developed by Kuramoto and Battogtokh [47] to understand the formation of the non-delayed chimera state. For this, we first rewrite (1.94) in terms of a relative phase θ given by

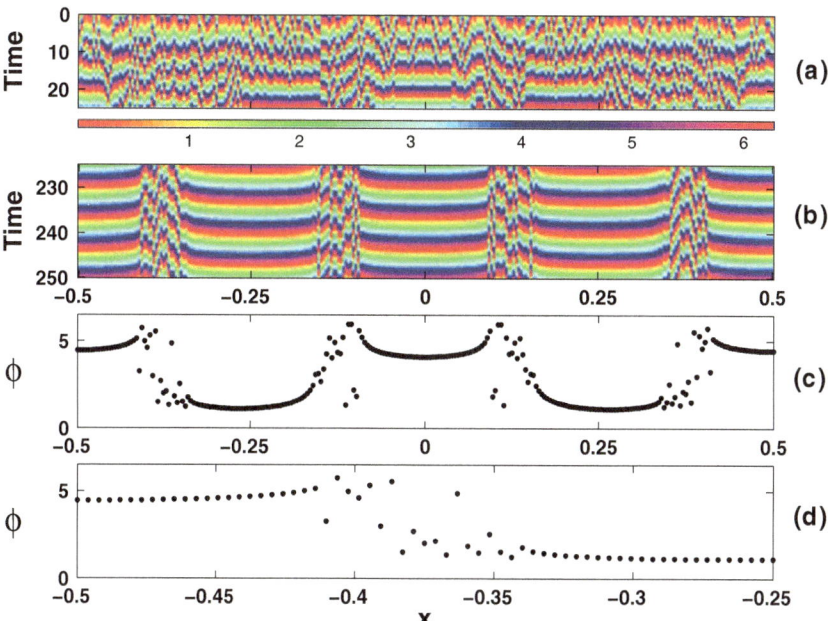

Fig. 1.14 (**a**) The space–time plot of the oscillator phases ϕ for the parameters $2L = 1.0$, $\kappa = 4.0$, $1/v = 10.24$, $\omega = 1.1$, and $\alpha = 0.9$ in the early stages of evolution from a random set of initial phases. Panel (**b**) shows a later time evolution and panel (**c**) gives a snapshot of the final stationary state. Panel (**d**) is a blowup of the region between $x = -0.5$ and $x = -0.25$ giving an enlarged view of an incoherent region and portions of the adjacent coherent regions

$$\theta(x,t) = \phi(x,t) - \Omega t, \tag{1.95}$$

where Ω represents a rotating frame in which the dynamics simplifies as much as possible and is the constant drift frequency of the phase-locked portions. In terms of θ (1.94) becomes

$$\frac{\partial}{\partial t}\theta(x,t) = \omega - \Omega - \int_{-L}^{L} G(x-x')\sin[\theta(x,t)-\theta(x',t-\tau_{x,x'})+\alpha+\Omega\tau_{x,x'}]dx'. \tag{1.96}$$

Following Kuramoto's approach [47], we now define a complex order parameter $Re^{i\Theta}$, in a manner analogous to what we had done for a globally coupled system in Sect. 1.3.1, as,

$$R(x,t)e^{i\Theta(x,t)} = \int_{-L}^{L} G(x-x')e^{i[\theta(x',t-\tau_{x,x'})-\Omega\tau_{x,x'}]}dx'. \tag{1.97}$$

The above order parameter differs from the usual definition for globally coupled systems in a number of significant ways: The spatial average of $e^{i\theta}$ is now weighted by the coupling kernel $G(x-x')$, the phase θ is evaluated in a time-delayed fashion, and the factor $e^{-i\Omega\tau_{x,x'}}$ adds a complex phase to the kernel $G(x - x')$. The latter two features provide a further generalization of Kuramoto's analysis carried out for a non-delayed system [47, 48, 55, 56].

In terms of R and Θ, (1.94) can be rewritten as

$$\frac{\partial}{\partial t}\theta(x, t) = \Delta - R(x, t)\sin[\theta(x, t) - \Theta(x, t) + \alpha], \qquad (1.98)$$

where $\Delta = \omega - \Omega$. Equation (1.98) is in the form of a single-phase oscillator equation driven by a force term which in this case is the mean field force. To obtain a stationary pattern (in a statistical sense) we require that R and Θ depend only on space and be independent of time. Under such a circumstance the oscillator population can be divided into two classes: those which are located such that $R(x) > |\Delta|$ can approach a fixed point solution ($\partial\theta(x, t)/\partial t = 0$) and the other oscillators that have $R(x) < |\Delta|$ would not be able to attain such an equilibrium solution. The oscillators approaching a fixed point in the rotating frame would have phase coherent oscillations at frequency Ω in the original frame whereas the other set of oscillators would drift around the phase circle and form the incoherent part.

One can easily solve (1.98) for the motion of the oscillator at each x, subject to the assumed time-independent values of $R(x)$ and $\Theta(x)$. The oscillators with $R(x) \geq |\Delta|$ asymptotically approach a stable fixed point θ^*, defined implicitly by

$$\Delta = R(x)\sin[\theta^* - \Theta(x) + \alpha]. \qquad (1.99)$$

The fact that they approach a fixed point in the rotating frame implies that they are phase-locked at frequency Ω in the original frame. On the other hand, the oscillators with $R(x) < |\Delta|$ drift around the phase circle monotonically. To be consistent with the assumed stationarity of the solution, these oscillators must distribute themselves according to an invariant probability density $\rho(\theta)$. And for the density to be invariant, the probability of finding an oscillator near a given value of θ must be inversely proportional to the velocity there. From (1.98), this condition becomes

$$\rho(\theta) = \frac{\sqrt{\Delta^2 - R^2}}{2\pi|\Delta - R\sin(\theta - \Theta + \alpha)|}, \qquad (1.100)$$

where the normalization constant has been chosen such that $\int_{-\pi}^{\pi}\rho(\theta)\,d\theta = 1$ and R, Θ, and θ are functions of x.

The resulting motions of both the locked and the drifting oscillators must be consistent with the assumed time-independent values of $R(x)$ and $\Theta(x)$. To calculate the contribution that the locked oscillators make to the order parameter (1.97), we note that

$$\sin{(\theta^* - \Theta + \alpha)} = \frac{\Delta}{R}, \tag{1.101}$$

$$\cos{(\theta^* - \Theta + \alpha)} = \pm \frac{\sqrt{R^2 - \Delta^2}}{R} \tag{1.102}$$

for any fixed point of (1.98). Taking plus sign for the stable fixed point, one can write

$$e^{i(\theta^* - \Theta + \alpha)} = \frac{\sqrt{R^2 - \Delta^2} + i\Delta}{R} \tag{1.103}$$

which implies that the locked oscillators contribute

$$\int dx' G(x - x') e^{i\theta^*(x')} = e^{-i\alpha} \int dx' G(x - x') e^{i[\Theta(x') - \Omega \tau_{x,x'}]} \left(\frac{\sqrt{R^2 - \Delta^2} + i\Delta}{R} \right) \tag{1.104}$$

to the order parameter (1.97). Here the integral is taken over the portion of the domain where $R(x') \geq \Delta$.

Next, to calculate the contribution from the drifting oscillators, following the prescription provided by Kuramoto [47, 48] for the undelayed case, we replace $e^{i\theta(x')}$ in (1.97) with its statistical average $\int_{-\pi}^{\pi} e^{i\theta} \rho(\theta) \, d\theta$. Using (1.100) and contour integration, one obtains

$$\int_{-\pi}^{\pi} e^{i\theta} \rho(\theta) d\theta = \frac{i}{R} \left(\Delta - \sqrt{\Delta^2 - R^2} \right). \tag{1.105}$$

The contribution of the drifting oscillators to the order parameter can therefore be written as

$$\int dx' G(x - x') e^{-i\Omega \tau_{x,x'}} \int_{-\pi}^{\pi} e^{i\theta} \rho(\theta) \, d\theta = ie^{-i\alpha} \int dx' G(x - x')$$
$$e^{i[\Theta(x') - \Omega \tau_{x,x'}]} \left(\frac{\Delta - \sqrt{\Delta^2 - R^2(x')}}{R(x')} \right), \tag{1.106}$$

where now the integral is over the complementary portion of the domain where $R(x') < |\Delta|$. We substitute these solutions of (1.98) for the two classes of oscillators into the integrand on the right-hand side of (1.97) and obtain the following functional self-consistency condition:

$$R(x) e^{i\Theta(x)} = e^{i\beta} \int_{-L}^{L} G(x - x') e^{i[\Theta(x') - \Omega \tau_{x,x'}]} \left(\frac{\Delta - \sqrt{\Delta^2 - R^2(x')}}{R(x')} \right) dx', \tag{1.107}$$

where $\beta = \pi/2 - \alpha$. For a chimera state to exist, R, Θ, and Δ must satisfy the above self-consistency condition. Note that we have three unknowns, and condition

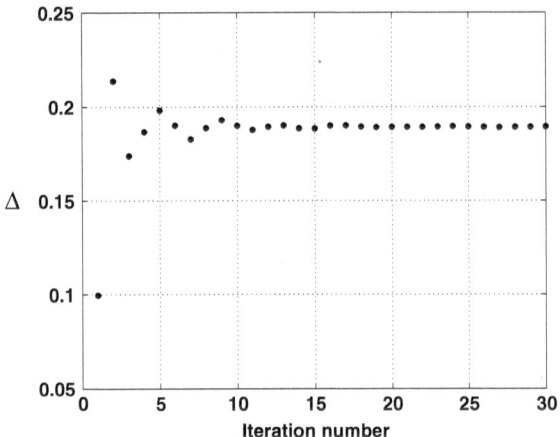

Fig. 1.15 Variation of Δ with the iteration number showing a rapid convergence in the numerical solution of the self-consistency equation (1.107). The system parameters are identical to those used in the direct solution of (1.94)

(1.107) provides only two equations when we separate its real and imaginary parts. A third condition can be obtained by exploiting the fact that the equation is invariant under any rigid rotation $\Theta(x) \rightarrow \Theta(x) + \Theta_0$. We can therefore specify the value of $\Theta(x)$ at any arbitrary chosen point, e.g., $\Theta(L) = 0$. In [37] (1.107) was solved numerically by following a three-step iterative procedure consisting of the following steps: Arbitrary but well-behaved initial guess functions were chosen for $R(x)$ and $\Theta(x)$ and the condition $\Theta(L) = 0$ was used in one of the equations of (1.107) to obtain a value for Δ. The initial profiles and the Δ value so obtained were then used to evaluate the right-hand side of (1.107) to generate new profiles for R and Θ. These were next used to generate a new value of Δ and the procedure was repeated until a convergence in the value of Δ and the functions R and Θ was obtained.

Figure 1.15 shows the rapid and excellent convergence in Δ to a unique value of $\Delta = 0.189$ for the solution of (1.107) with system parameters chosen identical to the ones that were used to obtain a clustered chimera state by a direct solution of (1.94). The converged spatial profiles of the order parameter (R and Θ) are shown in Fig. 1.16 and the converged value of Δ is marked in the upper panel by the horizontal line. The amplitude of the order parameter (R) shows a periodic spatial modulation—peaking at four symmetrically placed spatial locations. The corresponding phases of the order parameter are seen to be in anti-phase for adjacent peaks in R. In between the peaks R is seen to dip to very small values at certain locations such that $R(x) < |\Delta|$ which should correspond to the incoherent drifting parts of the chimera. To better appreciate the agreement between the direct solutions of (1.94) and the mean field solutions of (1.107) the results are plotted together in Fig. 1.17. As is clearly seen the measured order parameter (R and Θ) and Δ from the direct simulations of (1.94) match well with the results of solving (1.107). The spatial profile of the phases (ϕ) of the oscillators as obtained from the direct simulation of (1.94) is shown in the top panel of Fig. 1.17. One finds four coherent regions interspersed by incoherence as expected from the results of solving (1.107).

Fig. 1.16 Spatial profiles of the amplitude R and the phase Θ of the order parameter obtained by solving the self-consistency equation (1.107) by an iterative scheme. The horizontal line in the upper panel marks the converged value $\Delta = 0.189$

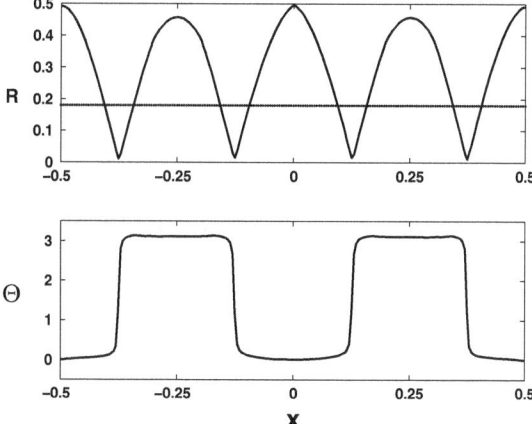

Also note that for the no-delay case there is only one peak of the order parameter as shown in Fig. 1.13(b).

It is appropriate at this juncture to point out some other general features of the clustered chimera states. Figures 1.16 and 1.17 show that both R and Θ are mirror symmetric (i.e., $R(x) = R(-x), \Theta(x) = \Theta(-x)$), a property that the original phase equation (1.94) also possesses. Equation (1.94) is also invariant under the

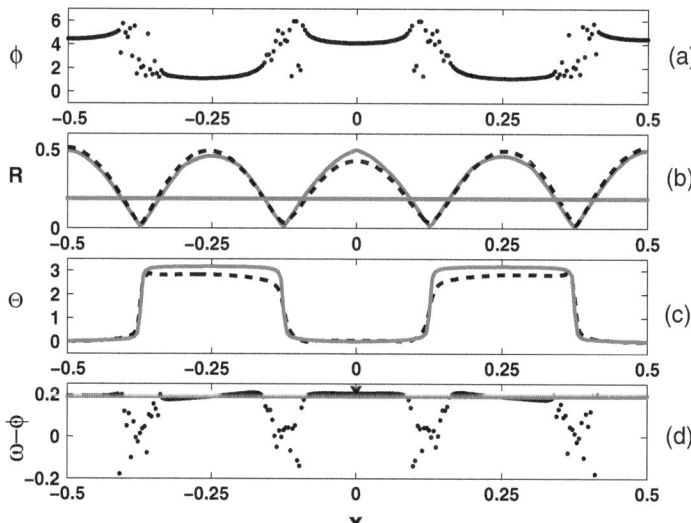

Fig. 1.17 (**a**) The phase pattern for a clustered chimera state as obtained by direct simulation of (1.94). The measured spatial profiles of the order parameter (R and Θ) from these simulations are shown in panels (**b**) and (**c**) as *dashed curves* and compared with the solutions from the self-consistency Equation (1.107) shown as *solid curves*. (**d**) $\omega - \dot{\phi}$ for the oscillators from a direct simulation of (1.94). The horizontal lines in (**b**) and (**d**) mark the converged value of $\Delta = 0.189$

transformation ($\phi(x, t) \rightarrow -\phi(x, t)$, $\omega \rightarrow -\omega$, $\alpha \rightarrow -\alpha$) and can have solutions with such a symmetry as well, namely, traveling wave solutions given by $\phi(x, t) = \Omega t + \pi q x/L$. In the numerical simulations with a change of the initial conditions keeping the same system parameters, one can get traveling wave solutions. For the non-delayed chimera case the co-existent stable state is that of a uniformly coherent state. The corresponding co-existent state in the time-delayed case is a traveling wave. In fact, there seems to be a clear correspondence between the number of clusters of the observed chimera state and the wave number q of the co-existent traveling wave solution. For the four-cluster chimera of Fig. 1.16 the co-existent traveling wave has $q = 2$ and similar results were observed in [37] for six-cluster ($q = 3$) and eight-cluster ($q = 4$) chimera solutions.

To summarize, we see that chimera type solutions do exist in a time-delayed system of non-locally coupled identical phase oscillators and that time delay leads to novel clustered states with a number of spatially disconnected regions of coherence with intervening regions of incoherence. The adjacent coherent regions of this clustered chimera state are found to be in anti-phase relation with respect to each other. These results are also well understood in terms of the behavior of a generalized order parameter for the system.

1.4 Summary and Perspectives

In the preceding sections we have shown how time delay can have subtle and sometimes profound effects on the collective dynamics of a coupled oscillator system. Various scenarios have been considered—starting from a simple two-oscillator model to a system of a large number of oscillators coupled in various ways. Time delay is introduced in the coupling mechanism and is seen to affect the existence, stability, and nature of the various collective states. Most of the basic effects associated with time delay are well demonstrated by the simple two-oscillator model. These include the phenomenon of time delay-induced amplitude death of identical oscillators, the existence of higher frequency oscillatory states, multistability and the co-existence of in-phase and anti-phase states, and multi-connectedness of death islands. For large N systems and in the presence of spatial dependence of the coupling mechanism, one has a wider variety of collective states such as traveling waves and the peculiar chimera state. Time delay affects the domain of existence of traveling waves as well as significantly altering their stability properties. The chimera state acquires a spatial modulation in the presence of time delay whose periodicity is closely linked to the co-existent stable traveling wave that the system can support. While the principal features of time-delayed dynamics, as displayed by the minimal model, are seen to be present in the N-oscillator models, their detailed analysis gets progressively difficult with increasing complexities of the coupling mechanisms and coupling topologies. Thus, as we found for the case of traveling wave states in nearest neighbor coupled as well as in non-locally coupled systems, the dispersion relations for their equilibrium states are extremely complicated and demand extensive numerical analysis. The eigenvalue equations for linear stability are even more complex and do not permit an easy determination of the stability domains in parameter space. Due to the transcendental nature of these equations, there is some debate about the practical feasibility of applying some of the standard stability analysis techniques and this is very much an open area of research for future investigations. We would also like to remark here that in our analysis we have restricted ourselves to a very simple model of time delay—either a single fixed discrete delay or a discrete set of delays for the non-local case. Alternative representations of time delay are possible, such as the one used by Atay [57] who showed that the parameter space of amplitude death for the coupled oscillator is enhanced when the oscillators are connected with time delays distributed over an interval rather than concentrated at a point. Distributed delays provide for a more realistic model for the description of larger physical systems where the delay parameter can be space- or time-dependent or in biological systems where memory effects are important. Another limitation of our treatment has been the adherence to a single and simple kind of a collective model—namely an array of Stuart–Landau oscillators. Networks of pulse coupled oscillators (also known as integrate and fire models) have been widely explored in the context of neuronal studies and provide a more realistic description of such systems. Such systems are also found to be quite sensitive to the presence of time-delayed coupling, e.g., in enhancing the onset of neural synchrony [58]. Another area that we have not discussed in this chapter is the response of the oscillator system to an external time-delayed stimulus. This is an active area of research today with important applications in neuroscience and control of neural disorders. The basic idea here is again the influence of a time-delayed feedback in enhancing or suppressing self-synchronization in an assembly of oscillators. We have also restricted ourselves to the study of time delay effects on stationary (equilibrium) states of the system. Their influence can extend to time-dependent states as well as chaotic dynamics, Hopf oscillations, etc. In fact, the nature of transition between chaotic and unsynchronized states is to date a poorly understood and open problem in the study of coupled oscillator systems. Time delay effects, which provide a sensitive probe by its subtle influence on the behavior of the order parameter, can prove useful in the further exploration of this problem [33].

References

1. A. T. Winfree. *The Geometry of Biological Time*. Springer-Verlag, New York, 1980.
2. Y. Kuramoto. *Chemical Oscillations, Waves and Turbulence*. Springer, Berlin, 1984.
3. S. H. Strogatz. From Kuramoto to Crawford: exploring the onset of synchronization in populations of coupled oscillators. *Physica D*, 143:1–20, 2000. and references therein.
4. K. Satoh. Computer experiment on the cooperative behavior of network of interacting nonlinear oscillators. *J. Phys. Soc. Japan*, 58:2010–2021, 1989.
5. G. B. Ermentrout. Oscillator death in populations of "all to all" coupled nonlinear oscillators. *Physica D*, 41:219–231, 1990.
6. A. A. Brailove and P. S. Linsay. An experimental study of a population of relaxation oscillators with a phase-repelling mean-field coupling. *Int. J. Bifurcation Chaos*, 6:1211–1253, 1996.

7. P. Hadley, M. R. Beasley, and K. Wiesenfeld. Phase locking of Josephson-junction series arrays. *Phys. Rev. B*, 38:8712–8719, 1988.
8. K. Wiesenfeld, P. Colet, and S. H. Strogatz. Synchronization transitions in a disordered Josephson series array. *Phys. Rev. Lett.*, 76:404–407, 1996.
9. P. M. Varangis, A. Gavrielides, T. Erneux, V. Kovanis, and L. F. Lester. Frequency entrainment in optically injected semiconductor lasers. *Phys. Rev. Lett.*, 78:2353–2356, 1997.
10. A. Hohl, A. Gavrielides, T. Erneux, and V. Kovanis. Localized synchronization in two coupled nonidentical semiconductor lasers. *Phys. Rev. Lett.*, 78:4745–4748, 1997.
11. G. Grüner and A. Zettl. Charge-density wave conduction – a novel collective transport phenomenon in solids. *Phys. Rep.*, 119:117–232, 1985.
12. J. Benford, H. Sze, W. Woo, R. R. Smith, and B. Harteneck. Phase locking of relativistic magnetrons. *Phys. Rev. Lett.*, 62:969–971, 1989.
13. I. Schreiber and M. Marek. Strange attractors in coupled reaction diffusion cells. *Physica*, 5D:258–272, 1982.
14. M. F. Crowley and I. R. Epstein. Experimental and theoretical studies of a coupled chemical oscillator: Phase death, multistability, and in-phase and out-of-phase entrainment. *J. Phys. Chem.*, 93:2496–2502, 1989.
15. M. Dolnik and I. R. Epstein. Coupled chaotic chemical oscillators. *Phys. Rev. E*, 54:3361–3368, 1996.
16. M. Kawato and R. Suzuki. Two coupled neural oscillators as a model of the circadian pacemaker. *J. Theor. Biol.*, 86:547–575, 1980.
17. Y. Kuramoto. Cooperative dynamics of oscillator community. *Prog. Theor. Phys. Suppl.*, 79:223–240, 1984.
18. H. Daido. Intrinsic fluctuations and a phase transition in a class of large populations of interacting oscillators. *J. Stat. Phys.*, 60:753–800, 1990.
19. S. Strogatz. *Sync: The Emerging Science of Spontaneous Order*. Hyperion, New York, 2003.
20. Y. Aizawa. Synergetic approach to the phenomena of mode-locking in nonlinear systems. *Prog. Theor. Phys.*, 56:703–716, 1976.
21. M. Shiino and M. Frankowicz. Synchronization of infinitely many coupled limit-cycle oscillators. *Phys. Lett. A*, 136:103–108, 1989.
22. M. Poliashenko and S. R. McKay. Chaos due to homoclinic orbits in two coupled oscillators with nonisochronism. *Phys. Rev. A*, 46:5271–5274, 1992.
23. J. L. Rogers and L. T. Wille. Phase transitions in nonlinear oscillator chains. *Phys. Rev. E*, 54:R2193–R2196, 1996.
24. D. G. Aronson, G. B. Ermentrout, and N. Kopell. Amplitude response of coupled oscillators. *Physica D*, 41:403–449, 1990.
25. R. E. Mirollo and S. H. Strogatz. Amplitude death in an array of limit-cycle oscillators. *J. Stat. Phys.*, 60:245–262, 1990.
26. K. Bar-Eli. On the stability of coupled chemical oscillators. *Physica D*, 14:242–252, 1985.
27. P. C. Matthews, R. E. Mirollo, and S. H. Strogatz. Dynamics of a large system of coupled nonlinear oscillators. *Physica D*, 52:293–331, 1991.
28. H. G. Schuster and P. Wagner. Mutual entrainment of two limit cycle oscillators with time delayed coupling. *Prog. Theor. Phys.*, 81:939–945, 1989.
29. E. Niebur, H. G. Schuster, and D. Kammen. Collective frequencies and metastability in networks of limit-cycle oscillators with time delay. *Phys. Rev. Lett.*, 67:2753–2756, 1991.
30. Y. Nakamura, F. Tominaga, and T. Munakata. Clustering behavior of time-delayed nearest-neighbor coupled oscillators. *Phys. Rev. E*, 49:4849–4856, 1994.
31. S. Kim, S. H. Park, and C. S. Ryu. Multistability in coupled oscillator systems with time delay. *Phys. Rev. Lett.*, 79:2911–2914, 1997.
32. D. V. Ramana Reddy, A. Sen, and G. L. Johnston. Time delay induced death in coupled limit cycle oscillators. *Phys. Rev. Lett.*, 80:5109–5112, 1998.
33. D. V. Ramana Reddy, A. Sen, and G. L. Johnston. Time delay effects on coupled limit cycle oscillators at Hopf bifurcation. *Physica D*, 129:15–34, 1999.

34. D. V. Ramana Reddy, A. Sen, and G. L. Johnston. Experimental evidence of time-delay induced death in coupled limit-cycle-oscillators. *Phys. Rev. Lett.*, 85:3381–3384, 2000.
35. R. Dodla, A. Sen, and G. L. Johnston. Phase-locked patterns and amplitude death in a ring of delay-coupled limit cycle oscillators. *Phys. Rev. E*, 69:056217, 2004.
36. M. P. Mehta, A. Sen, and G. L. Johnston. Amplitude death states of a ring of nonlocally-coupled limit cycle oscillators with time delay. *Proceedings of the National Conference on Nonlinear Systems and Dynamics*, Chennai, Feb.5-8, 2006, 2006.
37. G. C. Sethia, A. Sen, and F. M. Atay. Clustered chimera states in delay-coupled oscillator systems. *Phys. Rev. Lett.*, 100:144102, 2008.
38. P. C. Matthews and S. H. Strogatz. Phase diagram for the collective behavior of limit cycle oscillators. *Phys. Rev. Lett.*, 65:1701–1704, 1990.
39. S. Peles and K. Wiesenfeld. Synchronization law for a van der Pol array. *Phys. Rev. E*, 68:026220, 2003.
40. L. L. Bonilla, C. J. Perez Vicente, F. Ritort, and J. Soler. Exactly solvable phase oscillator models with synchronization dynamics. *Phys. Rev. Lett.*, 81:3643–3646, 1998.
41. H. Daido. Multibranch entrainment and scaling in large populations of coupled oscillators. *Phys. Rev. Lett.*, 77:1406–1409, 1996.
42. G.B. Ermentrout. The behavior of rings of coupled oscillators. *J. Math. Biol.*, 23:55–74, 1985.
43. P. C. Bressloff, S. Coombes, and B. de Souza. Dynamics of a ring of pulse-coupled oscillators: Group theoretic approach. *Phys. Rev. Lett.*, 79:2791–2794, 1997.
44. M. Barahona and L.M. Pecora. Synchronization in small-world systems. *Phys. Rev. Lett.*, 89:054101, 2002.
45. Y. Kuramoto. Scaling behavior of oscillators with non-local interaction. *Prog. Theor. Phys.*, 94:321–330, 1995.
46. Y. Kuramoto and H. Nakao. Origin of power-law spatial correlations in distributed oscillators and maps with nonlocal coupling. *Phys. Rev. Lett.*, 76:4352–4355, 1996.
47. Y. Kuramoto and D. Battogtokh. Coexistence of coherence and incoherence in nonlocally coupled phase oscillators. *Nonlinear Phenomena in Complex Systems*, 5:380–385, 2002.
48. Y. Kuramoto. *Nonlinear Dynamics and Chaos:Where Do We Go from Here?*, p 209. Institute of Physics, Bristol, UK, 2003.
49. J. R. Phillips, H. S. J. van der Zant, J. White, and T. P. Orlando. Influence of induced magnetic-fields on the static properties of josephson-junction arrays. *Phys. Rev. B*, 47:5219–5229, 1993.
50. S. I. Shima and Y. Kuramoto. Rotating spiral waves with phase-randomized core in nonlocally coupled oscillators. *Phys. Rev. E*, 69:036213, 2004.
51. J.D. Murray. *Mathematical Biology*. Springer, New York, 1989.
52. B. Ermentrout, J. Campbell, and G. Oster. A model for shell patterns based on neural activity. *Veliger*, 28:369–388, 1986.
53. N.V. Swindale. A model for the formation of ocular dominance stripes. *Proc. R. Soc. London B*, 208:243–264, 1980.
54. D. Tanaka and Y. Kuramoto. Complex Ginzburg-Landau equation with nonlocal coupling. *Phys. Rev. E*, 68:026219, 2003.
55. D. M. Abrams and S. H. Strogatz. Chimera states for coupled oscillators. *Phys. Rev. Lett.*, 93:174102, 2004.
56. D. M. Abrams and S. H. Strogatz. Chimera states in rings of nonlocally coupled oscillators. *Int. J. Bifurcation Chaos*, 16:21–37, 2006.
57. F. M. Atay. Distributed delays facilitate amplitude death of coupled oscillators. *Phys. Rev. Lett.*, 91:094101, 2003.
58. M. Dhamala, V. K. Jirsa, and M. Ding. Enhancement of neural synchrony by time delay. *Phys. Rev. Lett.*, 92:74104, 2004.

Chapter 2
Delay-Induced Stability: From Oscillators to Networks

Fatihcan M. Atay

2.1 Introduction

To those who have ever dealt with delay equations, the instabilities and oscillatory behavior caused by delays are all too well known. What perhaps not so familiar is that the delays could also have the opposite effect, namely that they could suppress oscillations and stabilize equilibria which would be unstable in the absence of delays. Research in this area is relatively sparse, as stabilization is not a typical or generic effect of time delays and one usually has to hit small regions in parameter spaces to observe it, which of course differ from system to system. The source of delays can also vary; ranging from feedback delays in control systems to communication delays in networks. It is therefore an interesting question to determine the general characteristics of delay-induced stability without reference to a particular system. One way to go about this is to focus on categories of dynamics rather than categories of systems. For instance, oscillatory behavior can often be connected to a Hopf bifurcation of an equilibrium solution under the variation of some parameter, and such local bifurcations share common qualities expressed in terms of the behavior on a low-dimensional center manifold. Hence, an analysis of stabilization near a generic Hopf bifurcation would yield general results applicable to any system near a Hopf instability and serve as a useful guide for understanding the behavior of oscillatory systems under time delays.

In this chapter, we present a systematic analysis of delay-induced stability for oscillatory systems whose oscillations result from a supercritical Hopf bifurcation of an equilibrium point. The main mathematical tool we use is averaging theory for delay differential equations, which is briefly introduced in Sect. 2.2 together with the main notation. Section 2.3 considers systems under delayed-feedback control, also considered in Chaps. 3 and 4. The goal here, however, is to obtain generic results near the bifurcation point. Section 2.4 extends the analysis to networks of systems with coupling delays. In this case, the suppression of the oscillations in

F. M. Atay (✉)
Max Planck Institute for Mathematics in the Sciences, D-04103 Leipzig, Germany
e-mail: atay@member.ams.org

F.M. Atay (ed.), *Complex Time-Delay Systems,* Understanding Complex Systems,
DOI 10.1007/978-3-642-02329-3_2, © Springer-Verlag Berlin Heidelberg 2010

the network corresponds to the phenomenon often referred to as *amplitude death* or *oscillator death*, as described in Chap. 1. The goal in our case is to character- ize stability in general delay equations and under general coupling conditions, and in particular to determine the role of the network topology, i.e., of the connection structure, on stability. Moreover, the results can be used to discover additional novel phenomena, for instance, the *partial amplitude death* of the network when the cou- pled oscillators are not identical. Section 2.5 studies diffusively coupled networks and uses tools of spectral graph theory to study delay-induced stability, where the largest eigenvalue of the graph Laplacian plays a prominent role. Finally Sect. 2.6 treats discrete-time networks. Here, we actually drop the assumption of being near a bifurcation, although several results turn out to be analogous to the continuous-time case.

2.2 A Brief Synopsis of Averaging Theory

We briefly describe the averaging theory for delay differential equations and intro- duce some notation in the process. The essential development here is based on Hale [1, 2]. Averaging theory yields rigorous results that are particularly useful for study- ing weakly nonlinear oscillatory systems, which is the case near a Hopf bifurcation. In fact, averaging can be used to obtain normal forms for Hopf bifurcation [3], setting an appropriate background for the problem of this chapter.

Consider the differential equation

$$\dot{x}(t) = \mathcal{F}(x_t; \alpha), \tag{2.1}$$

where $x(t) \in \mathbb{R}^n$, and $\alpha \in \mathbb{R}$ is a parameter. The usual notation x_t denotes the values of the system state over a time window of finite length τ, that is, $x_t(\theta) = x(t + \theta) \in \mathbb{R}^n$, $\theta \in [-\tau, 0]$, and $x_t \in \mathcal{C}$, where $\mathcal{C} := C([-\tau, 0], \mathbb{R}^n)$ denotes the Banach space of continuous functions over the interval $[-\tau, 0]$ equipped with the supremum norm. It is assumed that $\mathcal{F} : \mathcal{C} \times \mathbb{R} \to \mathbb{R}^n$ is twice continuously differentiable in its arguments and $\mathcal{F}(0; \alpha) = 0$ for all α. Assume further that the origin undergoes a supercritical Hopf bifurcation at $\alpha = 0$. Hence, for small positive α the origin is unstable and there exists a small amplitude limit cycle. To study the behavior near the origin, it is convenient to scale the variables $x \mapsto \varepsilon x$ and $\alpha \mapsto \varepsilon \alpha$, where ε is a small positive parameter. This transforms (2.1) into a weakly nonlinear system of the form

$$\dot{x}(t) = Lx_t + \varepsilon f(x_t; \varepsilon), \tag{2.2}$$

where $L : \mathcal{C} \to \mathbb{R}^n$ is a linear operator and f is a C^2 function with $f(0; \varepsilon) = 0$ for all ε.

Equation (2.2) is a perturbation of the linear equation

$$\dot{x}(t) = Lx_t, \tag{2.3}$$

which can also be expressed as a Riemann–Stieltjes integral

$$\dot{x}(t) = \int_{-\tau}^{0} d\eta(\theta)x(t+\theta) \tag{2.4}$$

for some $n \times n$ matrix η whose components are functions of bounded variation on $[-\tau, 0]$. By the assumption of Hopf bifurcation, the linear equation (2.4) has a pair of complex conjugate eigenvalues $\pm i\omega$, $\omega > 0$. We assume all other eigenvalues have negative real parts. Hence (2.4) has a two-dimensional center subspace. The adjoint equation corresponding to (2.4) is

$$\dot{z}(t) = -\int_{-\tau}^{0} d\eta^{\top}(\theta)z(t-\theta), \tag{2.5}$$

where $z_t(\theta) = z(t+\theta)$ for $\theta \in [0, \tau]$. Thus $z_t \in C^* := C([0, \tau], \mathbb{R}^n)$. Equations (2.4) and (2.5) have the same eigenvalues. The spaces C and C^* are related by the bilinear form

$$(\psi, \varphi) = \psi^{\top}(0)\varphi(0) - \int_{-\tau}^{0}\int_{0}^{\theta} \psi^{\top}(\zeta - \theta)d\eta(\theta)\varphi(\zeta)\,d\zeta$$

defined for $\psi \in C^*$ and $\varphi \in C$. If Φ is an $n \times 2$ matrix whose columns span the eigenspace of (2.4) corresponding to the eigenvalue $\pm i\omega$, then

$$\Phi(\theta) = \Phi(0)e^{\omega J\theta}, \quad \theta \in [-\tau, 0],$$

where

$$J = \begin{bmatrix} 0 & -1 \\ 1 & 0 \end{bmatrix}.$$

A analogous statement holds for the $n \times 2$ matrix Ψ whose columns span the eigenspace corresponding to $\pm i\omega$ for the adjoint equation. Moreover, it is possible to choose Φ and Ψ such that $(\Psi, \Phi) = I$.

The above ideas form the basis of the analysis of the perturbed equation (2.2). Hence, if $\bar{x}_t \in C$ and $y(t) \in \mathbb{R}^2$ are defined by

$$x_t = \Phi y(t) + \bar{x}, \quad y(t) = (\Psi, x_t),$$

then y satisfies

$$\dot{y}(t) = \omega J y(t) + \varepsilon \Psi^{\top}(0)h(\Phi y(t) + \bar{x}; \varepsilon). \tag{2.6}$$

By a change of variables $y = \exp(\omega Jt)u$, one arrives at the slowly varying system

$$\dot{u}(t) = \varepsilon e^{-\omega Jt} \Psi^\top(0) f(\Phi e^{\omega Jt} u(t) + \bar{x}; \varepsilon).$$

The corresponding averaged equation is defined as

$$\dot{u}(t) = \varepsilon \bar{f}(u), \tag{2.7}$$

where

$$\bar{f}(u) = \lim_{T \to \infty} \frac{1}{T} \int_0^T e^{-\omega Jt} \Psi^\top(0) f(\Phi e^{\omega Jt} u; 0) \, dt.$$

The qualitative behavior of the infinite-dimensional system (2.1) that is relevant for the present problem can be studied on the two-dimensional ODE (2.7). More precisely, under the assumptions of the above paragraphs, one has the following result.

Theorem 2.1 (Averaging Theorem, [2]) *Suppose that equation (2.7) has a fixed point u_0 and the Jacobian $D\bar{f}(u_0)$ has no eigenvalues with zero real part. Then there exists $\varepsilon_0 > 0$ such that for each $\varepsilon \in [0, \varepsilon_0]$ there exists an almost periodic solution $x^*(\varepsilon)$ of (2.2) which has the same stability type as the fixed point u_0 of (2.7). Furthermore, $x^*(0) = 0$, and x^* is unique in a neighborhood of $\Phi e^{\omega Jt} u_0$ and $\varepsilon = 0$.*

2.3 Stability by Delayed Feedback

Suppose now that a feedback function g acts on system (2.2),

$$\dot{x}(t) = Lx_t + \varepsilon f(x_t; \varepsilon) + \varepsilon \kappa g(x_t; \varepsilon) \tag{2.8}$$

with $\kappa \in \mathbb{R}$ denoting the feedback gain. The feedback is taken to be of the same order of magnitude as the perturbation f, so it is also scaled by ε. By assumption the zero solution of the uncontrolled system ($\kappa = 0$) is unstable for small $\varepsilon > 0$. We wish to study the effect of the feedback control for nonzero κ.

Equation (2.8) is a perturbation of a linear system; hence, the tools of the previous section can be applied. Since we are interested only in the stability of the zero solution, it suffices to work with the linearizations of the functions f and g. To this end, we introduce $n \times n$ matrices F and G with elements of bounded variation on $[-\tau, 0]$, such that

$$[D_1 f(0;0)]\phi = \int_{-\tau}^0 dF(\theta)\phi(\theta),$$

$$[D_1 g(0;0)]\phi = \int_{-\tau}^0 dG(\theta)\phi(\theta)$$

for all $\phi \in C$. (D_1 denotes the partial derivative with respect to the first variable.) The averaged equation is a two-dimensional linear ODE, which, in the present case, can be written as a complex scalar equation

$$\dot{z} = \varepsilon(q + \kappa p)z, \tag{2.9}$$

where $p, q \in \mathbb{C}$ are defined by

$$p = \operatorname{tr}\left[\Psi^\top(0) \int_{-\tau}^0 dG(\theta)\Phi(\theta)\right] + i\operatorname{tr}\left[J\Psi^\top(0) \int_{-\tau}^0 dG(\theta)\Phi(\theta)\right],$$

$$q = \operatorname{tr}\left[\Psi^\top(0) \int_{-\tau}^0 dF(\theta)\Phi(\theta)\right] + i\operatorname{tr}\left[J\Psi^\top(0) \int_{-\tau}^0 dF(\theta)\Phi(\theta)\right].$$

where tr denotes the matrix trace. We refer the interested reader to [4] for the details of the calculations, and state the stability result based on the averaged equation (2.9).

Theorem 2.2 Let $\kappa \in \mathbb{R}$. There exists $\varepsilon_0 > 0$ such that for $\varepsilon \in (0, \varepsilon_0)$ the zero solution of (2.8) is asymptotically stable (resp., unstable) if

$$\operatorname{Re}(q + \kappa p) < 0 \quad (resp., > 0).$$

The proof is based on the averaging theory of the previous section: For the averaged equation, the eigenvalues of the Jacobian at the origin have real parts equal to $\operatorname{Re}(q + \kappa p)$, whose sign determines stability. On the other hand, by Theorem 2.1, there exists a solution $x^*(t) = 0 + \mathcal{O}(\varepsilon)$ of the original equation (2.8) having the same stability type as the averaged equation. By uniqueness of x^*, one has $x^*(t) \equiv 0$ since 0 is always a solution of (2.8). Hence, the stability of the zero solution of (2.8) is determined by the sign of $q + \kappa p$.

Note that, in the statement of Theorem 2.2, q depends only on the controlled system and p depends only on the feedback. Hence, $\kappa \operatorname{Re}(p)$ is the quantity that characterizes the effect of feedback for a given oscillator. In particular, the feedback has no effect on stability if $\operatorname{Re}(p) = 0$. We can equivalently characterize this case as

$$\cos \zeta = 0, \tag{2.10}$$

where $\zeta = \arg(p)$ can be viewed as a projection angle. While at first (2.10) appears to describe a singular case, it does indeed arise often when the feedback depends on incomplete state information, as we shall see below. More generally, Theorem 2.2 shows that the feedback can be stabilizing or destabilizing depending on whether $\kappa \operatorname{Re}(p)$ is negative or positive, respectively.

To study the effect of delays on $\operatorname{Re}(p)$, we fix $\kappa = 1$ and consider a linear feedback of the form

$$g(x_t;\varepsilon) = \int_{-\tau}^{0} Cx(t+\theta)\,dh(\theta), \tag{2.11}$$

where $C \in \mathbb{R}^{n\times n}$ is a structure matrix and $h:[-\tau,0] \to \mathbb{R}$ is a scalar delay distribution. In this case, it can be calculated that [4]

$$\operatorname{Re}(p) = \alpha\operatorname{tr}(\hat{C}) + \beta\operatorname{tr}(\hat{C}J), \tag{2.12}$$

where $\hat{C} = \Psi^{\top}(0)C\Phi(0)$ and

$$\alpha = \int_{-\tau}^{0} \cos\theta\,dh(\theta), \quad \beta = \int_{-\tau}^{0} \sin\theta\,dh(\theta). \tag{2.13}$$

For undelayed feedback, h is a Dirac delta at 0; so $\alpha = 1$ and $\beta = 0$, yielding $p = \operatorname{tr}(\hat{C})$. Hence if $\operatorname{tr}(\hat{C}) = 0$ the undelayed feedback cannot stabilize the origin; however, if $\beta\operatorname{tr}(\hat{C}J) < 0$ then applying the same feedback signal with some time delay can achieve stabilization. In fact, from (2.12) it follows that delayed feedback is more stabilizing than undelayed feedback if

$$(1-\alpha)\operatorname{tr}(\hat{C}) > \beta\operatorname{tr}(\hat{C}J). \tag{2.14}$$

A particular application area is when only partial system state is available for feedback. We illustrate with an example.

Example Consider the classical van der Pol oscillator with linear feedback

$$\ddot{y} + \varepsilon(y^2 - 1)\dot{y} + y = \varepsilon\left[\kappa_1 y(t-\tau) + \kappa_2 \dot{y}(t-\tau)\right], \quad 0 < \varepsilon \ll 1. \tag{2.15}$$

With $x = (y, -\dot{y})$, the linear equation around the origin is

$$\dot{x}(t) = Jx(t) + \varepsilon\begin{pmatrix} 0 & 0 \\ 0 & 1 \end{pmatrix}x(t) + \varepsilon\begin{pmatrix} 0 & 0 \\ -\kappa_1 & \kappa_2 \end{pmatrix}x(t-\tau),$$

which has the same form as (2.6). Carrying out averaging as above, we obtain $q = 1$ and

$$\hat{C} = C = \begin{pmatrix} 0 & 0 \\ -\kappa_1 & \kappa_2 \end{pmatrix},$$

giving $\text{Re}(p) = \alpha \kappa_2 + \beta \kappa_1$. From (2.13) we have $\alpha = \cos \tau$ and $\beta = -\sin \tau$. If the feedback is instantaneous (i.e., without delays), then $\beta = 0$ and $\text{Re}(p)$ depend only on the velocity feedback κ_2. In this case, the origin cannot be stabilized if $\kappa_2 = 0$, that is, if velocity information is not used in the feedback. By contrast, if the feedback is delayed, then using only position information can yield stability. In this latter case, Theorem 2.2 gives the condition for stability as $\kappa_1 \sin \tau > 1$. Hence, for instance, the choices $\tau = \pi/2$ and $\kappa_1 = 1.5$ will stabilize the origin, whereas stability is not possible for any feedback gain if $\tau = 0$.

Remark. The form of the feedback function g in (2.8) or (2.11) is of course more general than the example above and allows, e.g., multiple delays or a combination of delayed and instantaneous feedback. For instance, the "non-invasive" form of delayed feedback used in some applications (notably in chaos control, see Chap. 4 in this volume) falls in this category, where, instead of (2.15), one would have, for instance,

$$\ddot{y} + \varepsilon(y^2 - 1)\dot{y} + y = \varepsilon \kappa_1 \left[y(t - \tau) - y(t) \right].$$

It is immediate by the above arguments that in this example the presence of the term $y(t)$ in the feedback has no effect on stability for small ε. Indeed, one can calculate $\text{Re}(p) = -\kappa_1 \sin \tau$ as before. Hence, as far as stability is concerned, direct and difference feedback schemes are equivalent on the order $\mathcal{O}(\varepsilon)$. The distinction becomes important, however, when several oscillators are coupled, as we will see in Sect. 2.4.

The suppression of limit-cycle oscillations in the van der Pol system was studied in [5]. The rigorous analysis presented here points to the essence of some novel uses of delayed feedback control in engineering, for instance, in vibration control of mechanical systems [6]. Examples of earlier work on the engineering use of deliberate delays in controllers can be found in [7–10]. A further example is the classical control problem of balancing an inverted pendulum, which is treated in [11] using delayed position feedback. Although here we have focused on stability of equilibria, and hence essentially on linear systems, the delayed feedback actually offers finer control on the behavior of oscillations. For instance, the amplitude [12] and period [13] of limit-cycle oscillations can be changed by nonlinear delayed feedback schemes.

We finally comment on the role of distributed delays. While most mathematical models and analytical results concern discrete delays, distributed delays also arise in applications. The foregoing theory applies to arbitrary delays, which makes it possible to make a comparison regarding the stabilizing effect near a Hopf bifurcation. Such an analysis is given in [4], where it is shown that discrete delays constitute a local extremum among all delay distributions having the same mean value. More precisely, if the delayed feedback is stabilizing (resp, destabilizing), then the discrete delay is locally the most stabilizing (resp., destabilizing) distribution, and the stabilizing or destabilizing effect of the feedback decreases as the variance of the delay distribution increases.

2.4 Amplitude Death in Networks of Oscillators

The spontaneous cessation of oscillations in coupled systems is a dynamical phenomenon of both theoretical and practical interest. The terms *amplitude death* and
oscillator death have been used to describe this quenching effect, early observations of which can be traced to Lord Rayleigh's experiments on acoustics [14],
or to the more recent results from chemical oscillators [15]. There are various
causes of amplitude death; for example, the differences in the intrinsic oscillation
frequencies of the units can suppress oscillations in the coupled system [16, 17].
Indeed, it is known that death is not possible in identical limit-cycle oscillations
[18], provided that the information transmission is instantaneous (see Fig. 1.1 in
Chap. 1). However, in the presence of time delays even identical oscillators can
exhibit amplitude death [19]. In this section, we extend our analysis to networked
systems to give a general treatment of amplitude death and its causes for oscillators near Hopf bifurcation. The new parameters here will be descriptors of connection structure of the oscillators, that is, the network topology. We shall in particular identify the relevant network parameters responsible for the suppression of
oscillations.

Consider N oscillators of the form (2.2), indexed by $k = 1, \ldots, N$, and coupled
in a general manner:

$$\dot{x}^k(t) = L_k x_t^k + \varepsilon f_k(x_t^k; \varepsilon) + \varepsilon \kappa g_k(x_t^1, \ldots, x_t^N; \varepsilon), \quad k = 1, \ldots, N, \qquad (2.16)$$

with the function f_k denoting the individual dynamics of the kth oscillator, and g_k
describing the effect of all other oscillators on the kth one. The parameter $\kappa \in$
$\mathbb{R}$ denotes the coupling strength, $x_t^k \in \mathcal{C} = C([-\tau, 0], \mathbb{R}^n)$, and f_k and g_k are
assumed to have continuous second derivatives and vanish at the origin for all ε.
Consequently, the trivial solution is an equilibrium of (2.16) for all ε. As before,
we study oscillators arising from Hopf bifurcations. Thus, for each k, the isolated
system

$$\dot{x}^k(t) = L_k x_t^k + \varepsilon f_k(x_t^k; \varepsilon) \qquad (2.17)$$

obtained from (2.16) by setting $\kappa = 0$, is assumed to undergo a supercritical Hopf
bifurcation. That is, the unperturbed problem

$$\dot{x}^k(t) = L_k x_t^k,$$

has a pair of eigenvalues $\pm i\omega_k \neq 0$ and all others eigenvalues have negative real
parts, the origin is unstable for the perturbed system (2.17), and there exists a small
amplitude limit cycle for $0 < \varepsilon \ll 1$. Amplitude death refers to the possibility
that for some nonzero coupling strength κ the trivial solution becomes stable in the
coupled system (2.16), although it is unstable in the isolated oscillator (2.17).

We again use averaging theory to study the problem for small ε and obtain a
similar result, given in Theorem 2.3. To introduce some notation, let F_k and G_{kl} be

$N \times N$ matrices corresponding to the linear parts of f_k and g_k, that is, matrices whose elements are of bounded variation on $[-\tau, 0]$, such that

$$[D_1 f_k(0;0)]\phi = \int_{-\tau}^{0} dF_k(\theta)\phi(\theta),$$

$$[D_l g_k(0,\ldots,0;0)]\phi = \int_{-\tau}^{0} dG_{kl}(\theta)\phi(\theta)$$

for all $\phi \in \mathcal{C}$. Define $P = [p_{kl}] \in \mathbb{C}^{N \times N}$ by

$$p_{kl} = \mathrm{tr}\left[\Psi_k^\top(0) \int_{-\tau}^{0} dG_{kl}(\theta)\Phi_l(\theta)\right] + i\mathrm{tr}\left[J\Psi_k^\top(0) \int_{-\tau}^{0} dG_{kl}(\theta)\Phi_l(\theta)\right]$$

if $\omega_k = \omega_l$, otherwise set $p_{kl} = 0$. Define the diagonal matrix $Q = \mathrm{diag}\{q_1,\ldots,q_N\} \in \mathbb{C}^{N \times N}$ by

$$q_k = \mathrm{tr}\left[\Psi_k^\top(0) \int_{-\tau}^{0} dF_k(\theta)\Phi_k(\theta)\right] + i\mathrm{tr}\left[J\Psi_k^\top(0) \int_{-\tau}^{0} dF_k(\theta)\Phi_k(\theta)\right].$$

Similar to (2.9) we end up with a linear ODE

$$\dot{z} = (Q + \kappa P)z, \quad z \in \mathbb{C}^N,$$

and deduce the following counterpart of Theorem 2.2.

Theorem 2.3 *Let $\kappa \in \mathbb{R}$. There exists $\varepsilon_0 > 0$ such that for $\varepsilon \in (0, \varepsilon_0)$ the origin is asymptotically stable if all eigenvalues of $Q + \kappa P$ have negative real parts, and unstable if there is an eigenvalue with positive real part.*

Proof Similar to the proof of Theorem 2.2. We now work on an N-fold Cartesian product of the space $\mathcal{C}$. Since the unperturbed system ($\varepsilon = 0$) is decoupled, block diagonal matrices of Φ_i and Ψ_i, defined as in Sect. 2.2, form basis for the appropriate eigenspaces.

Remark The stability of the uncoupled system is determined by the sign of $\mathrm{Re}(q_i)$ since Q is a diagonal matrix. Hence, $\mathrm{Re}(q_i) > 0 \,\forall i$ by our assumption that the origin is unstable for the isolated oscillators.

As P is an $N \times N$ matrix, one cannot in general expect to move all eigenvalues of $Q + \kappa P$ to the left-hand complex plane by varying only the scalar quantity κ. The particular cases when such a κ exists constitute amplitude death phenomena. As corollaries of Theorem 2.3, we can now recover several characteristics of amplitude death in the present general setting of coupled delay differential equations.

Death by frequency mismatch. Suppose that the individual oscillator frequencies $\omega_1, \ldots, \omega_N$ are all distinct. Then by definition P is a diagonal matrix.[1] Now if Re(p_{ii}) are nonzero and have the same sign for all i (e.g., when coupling conditions are identical or similar), then stability is possible for sufficiently strong coupling. In fact, since $Q + \kappa P$ is in this case diagonal, it suffices to choose $|\kappa| > \max_i |\text{Re}(q_i)/\text{Re}(p_{ii})|$ with the correct sign to ensure that $q_i + \kappa p_{ii}$ has negative real part for all i. This behavior corresponds to the oscillation suppression by frequency mismatch studied in [16, 17]. Nevertheless, it can be observed only under the correct coupling conditions, since all Re(p_{ii}) need to be nonzero and have the same sign.

Partial amplitude death. If the frequencies $\omega_1, \ldots, \omega_N$ are distinct and the p_{ii} are not identical, then there may exist values of κ for which Re$(q_i + \kappa p_{ii})$ is negative only for some i. Since the averaged equations are decoupled in this case, one sees that only some of the oscillators are suppressed while the others continue to exhibit their limit-cycle behavior. This phenomenon has been termed *partial amplitude death* [20]. We illustrate with an example.

Consider the pair of coupled van der Pol oscillators

$$\ddot{y}_1 + \varepsilon(y_1^2 - \alpha_1)\dot{y}_1 + \omega_1^2 y_1 = \varepsilon\kappa(\dot{y}_2(t - \tau) - \dot{y}_1(t)),$$
$$\ddot{y}_2 + \varepsilon(y_2^2 - \alpha_2)\dot{y}_2 + \omega_2^2 y_2 = \varepsilon\kappa(\dot{y}_1(t - \tau) - \dot{y}_2(t)) \tag{2.18}$$

with non-identical frequencies and amplitudes. That is, $\alpha_1, \alpha_2 > 0$, $\alpha_1 \neq \alpha_2$, and $\omega_1 \neq \omega_2$. In the absence of coupling, the ith oscillator can be shown to have an attracting limit-cycle solution given by $y_i = 2\sqrt{\alpha_i} \cos \omega_i t + \mathcal{O}(\varepsilon)$. Figure 2.1 shows the behavior of (2.18) the coupled system as the coupling strength κ is varied. It is seen that amplitude death occurs for $\kappa > 1$, where both oscillators are quenched. Furthermore, en route to amplitude death, a partial death occurs for $0.5 < \kappa < 1$, where only the second oscillator is damped while the first one continues to exhibit high-amplitude oscillations. In the partial death range the amplitude of the second oscillator is almost (but not exactly) zero, whereas in the full amplitude death range both amplitudes are exactly zero as the origin of the full system (2.18) becomes asymptotically stable.

Direct coupling. This case refers to the instance when the coupling function g_i has no explicit dependence on x_t^i, or more generally $p_{ii} = 0$, for all i. It is a corollary of Theorem 2.3 that amplitude death is not possible under direct coupling. To see this, note that Re$(\text{tr}(Q + \kappa P)) = \sum_{i=1}^{N} \text{Re}(q_i) > 0$. Since trace equals sum of eigenvalues, $Q + \kappa P$ has an eigenvalue with positive real part.

As opposed to direct coupling, the oscillators can also interact via the so-called *diffusive coupling,* where the function g_i depends on x_t^i as well as some of the other oscillators x_t^j in the network in a way that aims to reduce the differences $|x_t^i - x_t^j|$, much like a diffusion process. This type of coupling is of particular importance in

---

[1] The underlying reason is that in this case the system essentially decouples on the order $\mathcal{O}(\varepsilon)$; see [20].

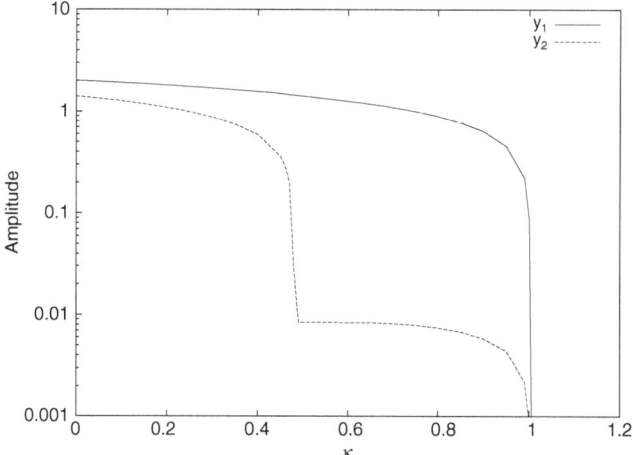

Fig. 2.1 Amplitude of oscillations in the coupled van der Pol system (2.18) with non-identical amplitudes and frequencies. Partial amplitude death occurs for coupling strength in the range $0.5 < \kappa < 1$, where one of the oscillators is suppressed and the other one continues to exhibit high-amplitude oscillations. The region $\kappa > 1$ corresponds to the full amplitude death of the whole system. Note that the vertical scale is logarithmic. The parameter values are $\alpha_1 = 1$, $\alpha_2 = 0.5$, $\omega_1 = 0.1$, $\omega_2 = 1$, and $\varepsilon = 0.1$

network studies because of its nature to facilitate synchronization. The dependence of g_i on its arguments defines a graph, where there is a link from vertex j to i if and only if g_i depends on x_t^j. One can then study the relation of dynamics to the coupling topology in terms of the properties of the underlying graph. This will be the topic of next section, which treats delay-induced stability in identical oscillators.

2.5 Diffusively Coupled Networks

In this section, we consider the coupled system

$$\dot{x}^k(t) = Lx_t^k + \varepsilon f(x_t^k;\varepsilon) + \varepsilon \frac{\kappa}{d_k} \sum_{j=1}^{N} a_{kj} g(x_t^k, x_{t-\tau_0}^j), \quad k = 1, \dots, N, \qquad (2.19)$$

which is a special case of (2.16) where the units are identical and the coupling function is

$$g_k(x_1^t, \dots, x_N^t; \varepsilon) = \frac{1}{d_k} \sum_{j=1}^{N} a_{kj} g(x_t^k, x_t^j{}_{\tau_0}).$$

The assumptions on L and f are analogous to Sect. 2.4; in particular, there exists a pair of purely imaginary and nonzero eigenvalues $\pm i\omega$ for the linear problem

$\dot{x} = Lx_t$, all other eigenvalues having negative real parts. Without loss of generality, $\omega = 1$ by a rescaling of time. The numbers a_{kj} are such that $a_{kj} = a_{jk} = 1$ if the kth and jth units have direct influence on each other; otherwise $a_{kj} = 0$. The symmetric array $A = [a_{kj}]$ forms the adjacency matrix of the underlying graph structure. The *degree* of the kth vertex in this graph is $d_k = \sum_{j=1}^{N} a_{kj}$, that is, the number of neighbors of the kth unit. Disregarding the trivial case of isolated vertices, $d_k > 0$ for all k. The function $g{:}\mathcal{C} \times \mathcal{C} \to \mathbb{R}^n$ is twice continuously differentiable and satisfies a generalized diffusion condition

$$g(\phi, \phi) = 0 \quad \forall \phi \in \mathcal{C}. \tag{2.20}$$

Finally, $\tau_0 \geq 0$ denotes the signal transmission delay along the links of the network.

By Theorem 2.3, the eigenvalues of the matrix $Q + \kappa P$ determine the stability of the zero solution of (2.19) for small ε. Since the oscillators are now identical, $Q = qI$ is a scalar matrix, with $\mathrm{Re}(q) > 0$, as the origin is assumed to be unstable for the isolated oscillator. Therefore, the stability condition is that

$$\kappa\,\mathrm{Re}(\rho_i) < -\mathrm{Re}(q)$$

for all eigenvalues ρ_i of the matrix P. It turns out [21] that P has identical diagonal elements and can be written as

$$P = p_{11}(I - e^{i\tau_0} D^{-1} A), \tag{2.21}$$

where $D = \mathrm{diag}\{d_1, \dots, d_N\}$ is the diagonal matrix of vertex degrees. The expression in parenthesis above is reminiscent of the *graph Laplacian* Δ, defined by

$$\Delta = I - D^{-1} A, \tag{2.22}$$

(in fact, the two expressions coincide for $\tau_0 = 0$), and one can rightly guess that the stability properties of the network would be related to the eigenvalues of Δ. The latter are a topic of spectral graph theory (see e.g., [22]), from which we briefly recall the relevant notions.

Let $\mathcal{G}$ be a simple and nontrivial graph (i.e., $\mathcal{G}$ contains at least one edge and no self-connections) on N vertices with Laplacian matrix Δ defined by (2.22). Although Δ is in general not symmetric, it can be shown to be a self-adjoint and positive semidefinite operator with respect to a certain inner product on $\mathbb{R}^N$. Consequently, the eigenvalues λ_k of Δ are real and nonnegative, and its eigenvectors $\{v_N, \dots, v_N\}$ form a complete orthonormal basis for $\mathbb{R}^N$. The smallest eigenvalue is zero and corresponds to the eigenvector $(1, 1, \dots, 1)$. For a graph without isolated vertices, the largest eigenvalue $\lambda_{\max}$ of Δ satisfies

$$\frac{N}{N-1} \leq \lambda_{\max} \leq 2. \tag{2.23}$$

Furthermore, $\lambda_{\max} = N/(N-1)$ if and only if $\mathcal{G}$ is a complete graph of N vertices, and $\lambda_{\max} = 2$ if and only if $\mathcal{G}$ has a nontrivial bipartite component. Here, a *complete graph* refers to one where every vertex is connected to every other vertex, and a *bipartite graph* is one whose vertex set can be divided into two parts V_1 and V_2 such that every edge has one end in V_1 and one in V_2. Common examples of bipartite graphs include open chains, grids, cycles with an even number of vertices, and all trees.

With this background, the stability result can be stated as follows.

Theorem 2.4 ([21]) *Suppose $p_{11} \neq 0$, let $\zeta = \arg(p_{11})$, and define $H \in \mathbb{R}$ by*

$$H = (\cos \zeta - \cos(\zeta + \tau_0))(\cos \zeta + (\lambda_{\max} - 1)\cos(\zeta + \tau_0)). \qquad (2.24)$$

If $H > 0$, then there exist $\kappa \in \mathbb{R}$ and $\varepsilon_0 > 0$ such that the zero solution of (2.19) is asymptotically stable for all $\varepsilon \in (0, \varepsilon_0)$. If $H < 0$ and $\kappa \in \mathbb{R}$, then there exists $\varepsilon_0 > 0$ such that the zero solution of (2.19) is unstable for all $\varepsilon \in (0, \varepsilon_0)$.

Note that if $\tau_0 = 0$, then H cannot be positive and the stability condition of the theorem is not satisfied. This is related to the fact observed in [18] that amplitude death does not occur for identical oscillators in the absence of time delays.

The quantity $\zeta = \arg(p_{11})$ appearing in (2.24) is analogous to the one defined in (2.10): If $\cos \zeta = 0$, then $H = -(\lambda_{\max} - 1)\sin^2(\tau_0) \leq 0$, because $\lambda_{\max} > 1$ by (2.23). In this case, the network stability condition of the theorem is not satisfied, similar to the condition (2.10) for the failure of feedback stabilization for a single oscillator.

Note that ζ does not depend on the connection structure of the network or on the delay τ_0. Indeed, the only contribution of the network structure comes through the largest eigenvalue $\lambda_{\max}$ of the Laplacian. In other words, $\lambda_{\max}$ completely characterizes the effect of network topology on stability. Thus, networks having the same value of $\lambda_{\max}$ have the same stability characteristics. These include, in particular, all bipartite network configurations, for which $\lambda_{\max}$ is always equal to 2. In fact, for small ε one can prove a monotone dependence of stability on $\lambda_{\max}$ in the following sense.

Corollary 2.5 ([21]) *Consider two graphs $\mathcal{G}_a$ and $\mathcal{G}_b$, with the corresponding Laplacians Δ_a, Δ_b having largest eigenvalues $\lambda_{\max}^a \geq \lambda_{\max}^b$, respectively. If, for a fixed pair of values of ζ and τ_0, asymptotic stability is possible in $\mathcal{G}_a$, then it is also possible in $\mathcal{G}_b$.*

As a direct implication that follows by (2.23), we have that bipartite graphs have the worst stability characteristics and complete graphs the best.

Directed networks and multiple delays. In a more general setting, one can consider the system

$$\dot{x}^k(t) = Lx_t^k + \varepsilon f(x_t^k; \varepsilon) + \varepsilon \frac{\kappa}{d_k} \sum_{j=1}^{N} a_{kj} g(x_t^k, x_{t-\tau_{kj}}^j), \quad k = 1, \ldots, N, \qquad (2.25)$$

which allows different values of the delay τ_{kj} between different pairs of nodes k and j. Furthermore, the requirement of interaction symmetry may be relaxed, so that $a_{kj} \neq a_{jk}$ and $\tau_{kj} \neq \tau_{jk}$ in general. The number d_k refers then to the in-degree of vertex k. Carrying out a similar analysis, it turns out that the matrix P now takes the form

$$P = p_{11}(I - D^{-1}A_\tau),$$

where $A_\tau = [\hat{a}_{kj}]$ is a delay-adjacency matrix with components

$$\hat{a}_{kj} = a_{kj} \exp(i\tau_{kj}),$$

which generalizes the expression (2.21) to the multiple-delay case. One can then express a stability condition in terms of the eigenvalues of the delay-Laplacian $\Delta_\tau = I - D^{-1}A_\tau$. For details, the reader is referred to [21].

2.6 Discrete-Time Systems

A discrete-time analogue of (2.19) is the coupled map network

$$x^k(t) = f(x^k(t)) + \frac{\kappa}{d_k} \sum_{j=1}^{N} a_{kj} g(x^k(t), x^j(t - \tau)), \quad k = 1, \ldots, N, \tag{2.26}$$

with $a_{kj} = a_{jk}, t \in \mathbb{Z}$ and the delay τ a nonnegative integer. The interaction function g satisfies the diffusion condition (2.20), so that any fixed point x^* of f yields a spatially uniform equilibrium solution $X^* = (x^*, \ldots, x^*)$ of (2.26). Our aim is to relate the stability of x^* as a fixed point of f to the stability of X^* as an equilibrium solution of the network (2.26). As in the previous sections, we are particularly interested in the case when x^* is unstable but X^* is stable. In the present section, however, it is possible to relax the assumption of being near a bifurcation and obtain sharp conditions without reference to a small parameter ε.

We note that, in the framework of condition (2.20), the interaction function g can have the form of, e.g., linear diffusion,

$$g(x, y) = y - x, \tag{2.27}$$

or can assume some other form such as

$$g(x, y) = f(y) - f(x), \tag{2.28}$$

which arises in the paradigm of coupled map lattices [23]. The constant κ can be viewed as a diffusion coefficient or a global coupling strength.

For simplicity, we restrict ourselves to scalar maps $f: \mathbb{R} \to \mathbb{R}$ and $g: \mathbb{R} \times \mathbb{R} \to \mathbb{R}$, which are assumed to be differentiable at x^* and (x^*, x^*), respectively. Upon linearizing about X^*, we obtain

$$u^k(t+1) = bu^k(t) + \frac{c}{d_i} \sum_{j=1}^{n} a_{kj}[u^j(t-\tau) - u^k(t)], \quad i = 1, \ldots n, \qquad (2.29)$$

where

$$b = f'(x^*) \quad \text{and} \quad c = \kappa D_2 g(x^*, x^*) = -\kappa D_1 g(x^*, x^*).$$

The asymptotic stability of the zero solution of the linear equation (2.29) yields the exponential stability of X^* in (2.26). If (2.29) is written in vector form,

$$U(t+1) = (b-c)U(t) + cD^{-1}AU(t-\tau), \qquad (2.30)$$

with $U = (u_1, \ldots, u_n)$, then the matrix $D^{-1}A = I - \Delta$ displays how the Laplacian Δ comes to the forefront in the analysis.

The exact conditions for the stability of the fixed point of (2.26) are derived in [24]. Rather than going into details here, we summarize some of the highlights. To begin with, if x^* is an unstable fixed point of f, then stability of the network can be achieved only for odd delays. In particular, if the delay is zero, stabilization of identical maps is not possible, which is the analog of the continuous-time result on amplitude death mentioned in Sect. 2.5. In a similar fashion, the largest eigenvalue $\lambda_{\max}$ of the Laplacian turns out to completely characterize the role of the network topology on stability. Furthermore, the stability has a monotone dependence on $\lambda_{\max}$, and we can cite a similar result to Corollary 2.5. Consequently, complete graphs yield the largest stability regions and bipartite graphs the smallest, as in Sect. 2.5.

Corollary 2.6 ([24]) *Consider two graphs $\mathcal{G}_a$ and $\mathcal{G}_b$, with the corresponding Laplacians Δ_a, Δ_b having largest eigenvalues $\lambda_{\max}^a \geq \lambda_{\max}^b$, respectively. Let $\tau \in \mathbb{Z}^+$. If the zero solution of (2.29) is asymptotically stable under the connection topology of $\mathcal{G}_a$, then it is also asymptotically stable for $\mathcal{G}_b$.*

Furthermore, discrete-time systems have the following monotonicity property with respect to delays.

Proposition 2.7 ([24]) *Let τ_1 and τ_2 be positive integers, $\tau_1 < \tau_2$, which are both odd or both even. If the zero solution of (2.29) is asymptotically stable for $\tau = \tau_2$, then it is also asymptotically stable for $\tau = \tau_1$.*

Stabilization of chaotic networks. The stability criteria given in [24] reveal that stabilization of an unstable fixed point x^* may be achieved for some range of the parameters when $b = f'(x^*) < -1$, but not when $b > 1$. For the nonlinear system, the implication is that only flip-bifurcations can be stabilized in the network. In a

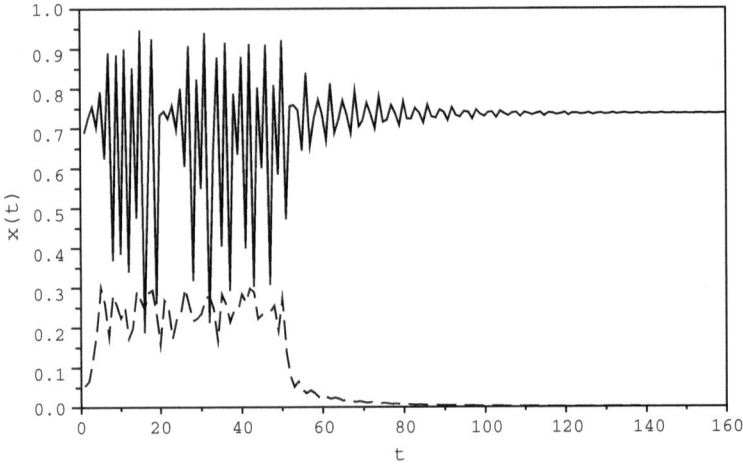

Fig. 2.2 Time evolution of a randomly selected unit (*solid line*) in a diffusively coupled network of 30 logistic maps forming a complete graph. The coupling is turned on at $t = 50$, after which the chaotic dynamics is replaced by a transient to a stable equilibrium solution. The *dotted line* below is the standard deviation over the vertices of the network, showing that each vertex asymptotically approaches the same equilibrium value

sense this is comparable to the stabilization of Hopf bifurcation discussed in the previous sections for continuous systems, since both bifurcations give rise to periodic solutions from an equilibrium point. The flip bifurcation is also the underlying mechanism for the well-known period-doubling route to chaos. Hence, the presence of delays may suppress chaos in the network in this case.

As an example, we consider a network of coupled logistic maps, with the individual dynamics given by $f(x) = \rho x(1-x)$ and the interaction function given by (2.28). For $\rho = 3.8, f$ is chaotic. Starting from random initial conditions, we simulate the system (2.26) without coupling ($\kappa = 0$) for 50 time steps, and afterward turn on the delayed coupling with parameters $\kappa = 0.5$ and $\tau = 1$. Figure 2.2 shows that chaos is rapidly suppressed in the coupled network and the system tends to the positive fixed point of the logistic map.

2.7 Concluding Remarks

The role of the largest eigenvalue of the Laplacian matrix in the stability of undirected networks has been first formulated in [21] for continuous-time systems and in [24] for discrete time. The generalization to directed networks with multiple delays is also indicated in [21], but remains mostly open for the discrete-time case. In another direction, the extension to weighted networks is straightforward when the edges carry nonnegative weights, but is more challenging to characterize in the presence of both positive and negative weights. Further complications ensue when the network structure itself is dynamic, that is, the edges or their weights vary in time according to some rule, or change randomly due to failure or disease, or in response to the dynamics on the nodes.

Although stability is interesting in its own right, it is certainly not the only effect of delays on the collective behavior of coupled systems. For instance, one can also observe different types of synchronization in coupled oscillators. In addition to the usual in-phase synchronization in the undelayed case, anti-phase synchronization can be induced by time delays, and in fact both types of synchronized solutions can stably co-exist at certain parameter regions for positive delay values [25]. Beyond synchronization there are many other dynamical regimes, such as cluster formation, and even more waiting to be discovered and studied. Examples abound in applications from diverse fields and form a rich source of problems.

References

1. J. K. Hale. Linear functional-differential equations with constant coefficients. *Contrib. Diff. Eqs.*, 2:291–319, 1963.
2. J. K. Hale. Averaging methods for differential equations with retarded arguments and a small parameter. *J. Diff. Eqs.*, 2:57–73, 1966.
3. S.-N. Chow and J. Mallet-Paret. Integral averaging and bifurcation. *J. Diff. Eqs.*, 26:112–159, 1977.
4. F. M. Atay. Delayed feedback control near Hopf bifurcation. *Discrete Continuous Dyn. Syst., S*, 1(2):197–205, 2008.
5. F. M. Atay. Van der Pol's oscillator under delayed feedback. *J. Sound Vib.*, 218(2):333–339, 1998.
6. N. Olgac and B. T. Holm-Hansen. A novel active vibration absorption technique: delayed resonator. *J. Sound Vib.*, 176(1):93–104, 1994.
7. I. H. Suh and Z. Bien. Proportional minus delay controller. *IEEE Trans. Automat. Contr.*, AC-24:370–372, 1979.
8. I. H. Suh and Z. Bien. Use of time-delay actions in the controller design. *IEEE Trans. Automat. Contr.*, AC-25:600–603, 1980.
9. N. Shanmugathasan and R. D. Johnston. Exploitation of time delays for improved process control. *Internat. J. Contr.*, 48(3):1137–1152, 1988.
10. W. H. Kwon, G. W. Lee, and S. W. Kim. Performance improvement using time delays in multivariable controller design. *Internat. J. Contr.*, 52(60):1455–1473, 1990.
11. F. M. Atay. Balancing the inverted pendulum using position feedback. *Appl. Math. Lett.*, 12(5):51–56, 1999.
12. F. M. Atay. Delayed-feedback control of oscillations in non-linear planar systems. *Internat. J. Contr.*, 75(5):297–304, 2002.
13. F. M. Atay. Oscillation control in delayed feedback systems. In: F. Colonius and L. Grüne, (eds), *Dynamics, Bifurcations and Control*, volume 273 of *Lecture Notes in Control and Information Sciences*, pp. 103–116. Springer-Verlag, Berlin, 2002.
14. J. W. S. Rayleigh. *The Theory of Sound*. Dover, New York, 2nd ed., 1945. Originally published in 1877–1878.
15. K. Bar-Eli. On the stability of coupled chemical oscillators. *Physica D*, 14:242–252, 1985.
16. M.Shiino and M. Frankowicz. Synchronization of infinitely many coupled limit-cycle type oscillators. *Phys. Lett. A*, 136(3):103–108, 1989.
17. R. E. Mirollo and S. H. Strogatz. Amplitude death in an array of limit-cycle oscillators. *J. Statist. Phys.*, 60(1–2):245–262, 1990.

18. D. G. Aronson, G. B. Ermentrout, and N. Kopell. Amplitude response of coupled oscillators. *Physica D*, 41:403–449, 1990.
19. D. V. Ramana Reddy, A. Sen, and G. L. Johnston. Time delay induced death in coupled limit cycle oscillators. *Phys. Rev. Lett.*, 80(23):5109–5112, 1998.
20. F. M. Atay. Total and partial amplitude death in networks of diffusively coupled oscillators. *Physica D*, 183(1–2):1–18, 2003.
21. F. M. Atay. Oscillator death in coupled functional differential equations near Hopf bifurcation. *J. Diff. Eqs.*, 221(1):190–209, 2006.
22. F. R. K. Chung. *Spectral Graph Theory*. American Mathematical Society, Providence, 1997.
23. K. Kaneko (ed). *Theory and applications of coupled map lattices*. Wiley, New York, 1993.
24. F. M. Atay and Ö. Karabacak. Stability of coupled map networks with delays. *SIAM J. Appl. Dyn. Syst.*, 5(3):508–527, 2006.
25. F. M. Atay. Synchronization and amplitude death in coupled limit cycle oscillators with time delays. *Lect. Notes Contr. Inf. Sci.*, 388:383–389, 2009.

Chapter 3
Delay Effects on Output Feedback Control of Dynamical Systems

Silviu-Iulian Niculescu, Wim Michiels, Keqin Gu, and Chaouki T. Abdallah

3.1 Introduction

The existence of time delays at the actuating input in a feedback control system is usually known to cause *instability* or poor performance for the corresponding closed-loop schemes (see, for instance [10, 18, 20, 23] and the references therein).

Consider the following class of linear, time-invariant dynamical systems:

$$\begin{cases} \dot{x}(t) = Ax(t) + bu(t) \\ y(t) = c^T x(t), \end{cases} \qquad (3.1)$$

where $x \in \mathbb{R}^n$ denotes the *state*, u and y are real scalar-valued functions denoting the *input* and the *output*, respectively, and $A \in \mathbb{R}^{n \times n}$, $b, c \in \mathbb{R}^{n \times 1}$. The transfer function associated with (3.1) is given by

$$H_{yu}(s) := c^T (sI_n - A)^{-1} b, \qquad (3.2)$$

which is a strictly proper ratio of two coprime polynomials:

$$H_{yu}(s) = \frac{P(s)}{Q(s)}, \text{ where } deg(P) < deg(Q).$$

In the language of control feedback theory, the system above is called a *SISO* (single-input single-output) system.

Consider now the following control law:

$$u(t) = -ky(t - \tau), \qquad (3.3)$$

S.-I. Niculescu (✉)
Laboratoire des Signaux et Systèmes (L2S), CNRS-Supélec, 91190, Gif-sur-Yvette, France
e-mail: Silviu.Niculescu@lss.supelec.fr

F.M. Atay (ed.), *Complex Time-Delay Systems,* Understanding Complex Systems,
DOI 10.1007/978-3-642-02329-3_3, © Springer-Verlag Berlin Heidelberg 2010

where the quantity k defines the corresponding *gain*, and τ denotes the *delay* in the output feedback. The presence of a delay in a system may have several causes, mainly due to *transport* and *propagation* (control over networks, teleoperation, and telemanipulation, to cite a few examples). For further discussions and related classifications of the delays present in engineering and other fields, see, for instance, [10, 18, 20, 23].

The control problem can be simply formulated as *finding a controller gain k such that the system* (3.1) *under the control law* (3.3) *becomes exponentially stable in closed loop.* In other words, the problem is reduced to find a gain k such that the roots of the characteristic equation (see, for instance, [10] and the references therein),

$$Q(s) + P(s)ke^{-s\tau} = 0, \tag{3.4}$$

are located in $\mathbb{C}_-$. Indeed, a system is *exponentially stable* if *all* the roots of its characteristic equation are located in $\mathbb{C}_-$ (roots with negative real part). It is important to point out that, in our case, the characteristic equation (3.4) is *transcendental*, that is, it has an infinite number of roots, a fact leading to further complications in closed-loop stability analysis. It is quite clear that the complication is essentially due to the presence of delay. Indeed, the characteristic equation of the closed-loop system free of delays ($\tau = 0$) reduces to a polynomial $P(s) + kQ(s)$ with a *finite* number of roots. Further discussions on the behavior of the characteristic roots and related continuity properties for small delay values can be found in [7] (see also, [2, 23]).

3.1.1 Existing Methodologies

There exists several methods and techniques in handling such a control problem, and most existing approaches consider the control problem for the case free of delay as a starting point due to the simplicity of the characteristic equation in such a case (polynomial type).

Thus, a classical (2 steps) procedure to handle such a control problem can be summarized as follows:

- first, derive a stabilizing control law $u(t) = -ky(t)$ for the system free of delays (setting the delay to zero) and
- next, under the assumption that a stabilizing gain k exists, find an upper bound for the delay value τ_m such that the corresponding closed-loop system is asymptotically stable for all delays in the interval $[0, \tau_m)$.

If the "real" input delay $\tau \notin [0, \tau_m)$, it is necessary to go back to the previous step in order to find a "better" gain k. In the case when $\tau_m = +\infty$ (or $\tau_m < +\infty$), we will have *delay-independent* (*delay-dependent*) closed-loop stability.

However, such a method does not give any answer concerning the case when the system cannot be *stabilized* by a *static output feedback* (see, for instance, the case of an oscillator) or it does not give any information on the behavior for *large delay values*, that is, the existence of delay intervals achieving closed-loop stability. In conclusion, the method above simply helps to put in evidence the degree of robustness of the closed-loop system with respect to the delay (bounds on the delay margin such that closed-loop stability is guaranteed for any delay smaller than the corresponding bound). In other words, the *destabilizing effect* of the *delay* was considered.

The method mentioned above can be illustrated by the following simple system: $H_{yu}(s) = 1/s$ (an integrator) subject to the control law $u(t) = -ky(t-\tau)$. The stability of the closed-loop system is given by the roots location of the quasipolynomial: $s + ke^{-s\tau}$. Since for $\tau = 0$, all positive gains k stabilize the system, the problem is to see how the delay affects the location of the roots.

As seen in Fig. 3.1, if the delay is increased from 0, the closed-loop stability is guaranteed for all delays $\tau \geq 0$, satisfying $k\tau < \pi/2$. The closed-loop system has two complex conjugate roots on the imaginary axis when $\tau = \pi/2$ (oscillatory behavior), and the closed-loop system becomes unstable for larger delays.

Several procedures and algorithms for computing the optimal delay margin τ_m are available in the existing literature as discussed in [10, 23]. Most of the existing approaches make use of tools and methods of robust control theory such as matrix pencils and 2D methods, μ-analysis and frequency-sweeping tests, pseudo-delay technique and parameter-dependent polynomials tests, etc. For the sake of brevity, we do not discuss such methods here.

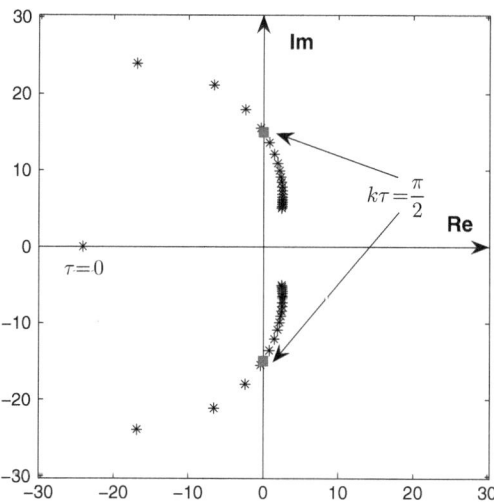

Fig. 3.1 Rightmost characteristic roots of the closed-loop quasipolynomial $s + ke^{-s\tau}$ for various values of $\tau \in [0, 0.4)$ with $k = \sqrt{2}e^{3\pi/4}$

3.1.2 Problem Formulation and Related Remarks

This chapter addresses the opposite problem, namely *characterizing* the *situations* where a *delay* has a *stabilizing effect*. In other words, we consider the situation where the delay-free feedback system is unstable, but becomes asymptotically stable due to the presence of an appropriate delay and gain in the actuating input.

The *stabilizing delay effect* problem is defined as follows:

Problem 3.1 (Delay stabilizing effect) Find explicit conditions on the pair (k, τ), such that the controller (3.3) stabilizes the system (3.2), but the closed-loop system would be unstable if the delay τ is set to zero.

As we will see below, the conditions derived lead to an explicit construction of the controller. Furthermore, for each stabilizing delay–gain pair, we may define a stabilizing delay interval, which can be treated as a *robustness measure* of the corresponding control law if the delay is uncertain.

The interest in solving Problem 3.1 is twofold:

- first, the resulting design procedure is rather simple and the controller is easy to implement;
- second, it allows us to *explore* the *potential* of using such a *controller* (that is, using delay as a *design parameter*, defining thus a "wait-and-act" strategy) in situations where it is not easy to design or implement a controller without delay (see, for instance, the congestion controllers in high-speed networks [15, 16, 24], etc.).

Some discussions on the stabilizing effect of delays in control systems have been considered in [1, 4, 25], but without any attempt to treat the problem in its most general setting. Nyquist criterion was used in [1] to prove that a pair (gain, delay) may stabilize second-order oscillatory systems. A different approach was proposed in [4], where upper and lower bounds on the delay are given such that the closed-loop system is stable, under the assumption that the system is stable for some known nominal delay values. Next, the paper [25] addressed the general static delayed output feedback problem, and some *existence results* (delay-independent, delay-dependent, instability persistence) were derived, but without any *explicit construction* of the controllers. More specifically, [25] compares the stability of the closed-loop schemes with and without delays in the corresponding control laws. Finally in [22] the stabilizing effect of a variation of time delays was investigated, motivated by machine tool vibration applications.

It is important to note that the approach proposed in this chapter may also be used for a different control problem: the characterization of *all stabilizing proportional controllers* for a dead-time SISO plant, and we believe that our method gives a simpler answer to that problem than existing results in the literature (see, for instance [26, 29] and the references therein, for a different frequency-domain approach).

Although only strictly proper SISO systems are considered above, most of the ideas still work for more general SISO systems, such as a restricted class of (not necessarily strictly) proper systems, or systems with internal delays in addition to the feedback input delay.

3.1.3 Methodology and Approach

The problem discussed here will be handled using frequency-domain methods. First, we shall analyze the location of roots in the complex plane of two appropriately defined polynomials that depend on the gain parameter.

Next, we analyze the sensitivity of the roots in terms of delays and derive *necessary and sufficient conditions* on the delay values for the asymptotic stability of the closed-loop system. Furthermore, an explicit construction of the controller will be given in the following form: for any gain satisfying some assumptions, a delay interval guaranteeing stability will be computed.

Several cases will be treated, and a complete characterization of the stabilizability using a gain–delay pair will be given. Illustrative examples complete the presentation.

The chapter is organized as follows: the main (existence) results as well as the resulting controller design procedure are presented in Sect. 2. Illustrative examples are given in Sect. 3. Some concluding remarks complete the chapter.

3.2 Main Results

In this section, we consider Problem 3.1. In order to prove our main results (stabilizability using a gain–delay pair), some notation is needed.

3.2.1 Notation

The characteristic equation of the closed-loop system is given by

$$H(s; k, \tau) = Q(s) + kP(s)e^{-s\tau}.$$

Two quantities will play a major role in the stabilizability study:

- card($\mathcal{U}_+$), where $\mathcal{U}_+$ is the set of roots of $H(s; k, 0) = Q(s) + kP(s)$, located in the *closed right half plane*, and card($\cdot$) denotes the cardinality (number of elements).
- card($\mathcal{S}_+$), where $\mathcal{S}_+$ the set of real strictly positive roots of the polynomial

$$F(\omega; k) = |\, Q(j\omega)\, |^2 - k^2 \,|\, P(j\omega)\, |^2 . \tag{3.5}$$

Both quantities depend on the gain and independent of the delay τ. We now clarify this dependence.

The quantity card($\mathcal{U}_+$) Its characterization as a function of the gain corresponds to the static (un-delayed) output feedback stabilizability problem. The difficulty of this problem is well known (see, for instance [28] and the references therein). However, in the SISO system case, the problem is reduced to a one-parameter problem, which is relatively easy. Indeed, there exist several methods to solve it. They include (standard) graphical tests (root-locus, Nyquist) and computation of the real roots of an appropriate set of polynomials. In addition to these standard methods, we may cite two interesting approaches [3, 14] based on *generalized eigenvalues computation* of some appropriate matrix pencils defined by the corresponding Hurwitz [3] and Hermite [14] matrices. The approach below is inspired by Chen's characterization [3] for systems without delay.

Introduce the following Hurwitz matrix associated with the denominator polynomial $Q(s) = \sum_{i=0}^{n} q_i s^{n-i}$ of the transfer function:

$$
H(Q) = \begin{bmatrix} q_1 & q_3 & q_5 & \cdots & q_{2n-1} \\ q_0 & q_2 & q_4 & \cdots & q_{2n-2} \\ 0 & q_1 & q_3 & \cdots & q_{2n-3} \\ 0 & q_0 & q_2 & \cdots & q_{2n-4} \\ \vdots & & \ddots & & \vdots \\ 0 & 0 & 0 & \cdots & q_n \end{bmatrix} \in \mathbf{R}^{n \times n}, \tag{3.6}
$$

where the coefficients $q_l = 0$, for all $l > n$. Next, corresponding to the numerator polynomial $P(s)$ of the transfer function, we construct $H(P)$ as a $n \times n$ matrix using the same procedure as (3.6) with the understanding that $p_l = 0$ for all $l > m$. The following result is a slight generalization of Theorem 2.1 by Chen [3]:

Lemma 3.2 *Let $\lambda_1 < \lambda_2 < \ldots \lambda_h$, with $h \leq n$ be the real eigenvalues of the matrix pencil $\Sigma(\lambda) = \mathbf{det}(\lambda H(P) + H(Q))$. Then the system (3.2) cannot be stabilized by the controller $u(t) = -ky(t)$ for any $k = \lambda_i$, $i = 1, 2, \ldots, h$. Furthermore, if there are r unstable closed-loop roots $(0 \leq r \leq n)$ for $k = k^*$, $k^* \in (\lambda_i, \lambda_{i+1})$, then there are r unstable closed-loop roots for any gain $k \in (\lambda_i, \lambda_{i+1})$. In other words, card($\mathcal{U}_+$) remains constant as k varies within each interval $(\lambda_i, \lambda_{i+1})$. The same holds for the intervals $(-\infty, \lambda_1)$ and (λ_h, ∞).*

Proof First, we need to show that as k varies, there are closed-loop roots on the imaginary axis if and only if $k = \lambda_i$, $i = 1, 2, ..., h$. The proof follows the same steps as those proposed by Chen in [3], and, therefore, will be omitted.

The above implies that for any gain $k \in (\lambda_i, \lambda_{i+1})$, the corresponding closed-loop system has no roots crossing the imaginary axis. Based on the *continuous dependence* of the roots of the polynomial $Q(s) + kP(s)$ on the parameter k, if there exists a k^* in the interval $(\lambda_i, \lambda_{i+1})$ such that $Q(s) + k^*P(s)$ has exactly r unstable roots, then the property is valid for any $k \in (\lambda_i, \lambda_{i+1})$ since the roots cannot jump from $\mathbf{C}_-$ to $\mathbf{C}_+$ or from $\mathbf{C}_+$ to $\mathbf{C}_-$, without crossing the imaginary axis. $\qquad\square$

Thus by computing the generalized eigenvalues of the matrix pencil $\Sigma(\lambda)$, yielding the critical gain values, and computing $\mathcal{U}_+$ for intermediate gain values, the function $k \mapsto \text{card}(\mathcal{U}_+)(k)$ is completely determined.

The quantity card($\mathcal{S}_+$) Without any loss of generality assume that $P(0) \neq 0$.[1] Its dependence on k is expressed in the following proposition:

Proposition 3.3 *Assume that* card($\mathcal{S}_+$) *changes at a gain value* k^*. *Then there exists a frequency* $\omega^* \geq 0$, *such that for* $\omega = \omega^*$:

$$|Q(j\omega)|^2 - k^{*2}|P(j\omega)|^2 = 0 \tag{3.7}$$

and

$$|P(j\omega)|^2 \frac{d}{d\omega}|Q(j\omega)|^2 - |Q(j\omega)|^2 \frac{d}{d\omega}|P(j\omega)|^2 = 0. \tag{3.8}$$

Proof For any k, F cannot have any real root ω satisfying $P(j\omega) = 0$. Otherwise, $Q(j\omega) = 0$, and P and Q are not coprime. The roots of F therefore coincide with the roots of

$$G(\omega; k) := \frac{|Q(j\omega)|^2}{|P(j\omega)|^2} - k^2. \tag{3.9}$$

A change of card($\mathcal{S}_+$) at $k = k^*$ implies that $G(\omega; k^*)$ has a root with multiplicity larger than one at some frequency ω^*, i.e.,

$$G(\omega^*; k^*) = G'(\omega^*; k^*) = 0.$$

This leads to (3.7) and (3.8). □

Proposition 3.3 allows one to compute systematically the behavior of card($\mathcal{S}^+$) as a function of the gain. First, one has to determine the real roots of the polynomial (3.8). Then the critical values of the gain k follow from (3.7). The characterization is complete when computing $\mathcal{S}_+$ for intermediate gain values, which again corresponds to finding the roots of a polynomial.

Remark 3.4 From the symmetry of the function (3.9), it follows that the pairs $(\omega, k) = (0, \pm \frac{Q(0)}{P(0)})$ always satisfy (3.7) and (3.8) and furthermore, at these k-values, card($\mathcal{S}_+$) always changes by 1. It also follows from (3.9) that at other k-values, card($\mathcal{S}_+$) can only change by 2.

---

[1] If not, then $Q(0) \neq 0$, and the results below still holds by defining the dependence in terms of $\frac{1}{k}$, etc.

3.2.2 Stabilizability in the Delay Parameter

Using the preliminary results presented above (dependence of card(S_+) and card(U_+) on the gain k), we can give a constructive solution to the problem considered in this chapter. Thus, for a given value of the gain k we derive conditions for the existence of stabilizing delay values.

We make the following assumptions:

Assumption 1 *Let the gain $k \in \mathbb{R}$ be such that*

1. *all the roots of F are simple,*
2. $0 \notin U_+$,
3. $\text{card}(U_+) \neq 0$.

Notice that the first condition is satisfied for almost all k. The second condition is *necessary* for stabilization because it excludes a characteristic root at zero, the latter being *invariant* with respect to delay changes. The third assumption excludes the trivial case where the system is asymptotically stable for $\tau = 0$.

Following [6], a crucial result is the following:

Theorem 3.5 *The characteristic equation has a root $j\omega$ for some delay value τ_0 if and only if*

$$\omega \in S_+. \tag{3.10}$$

Furthermore, for any ω satisfying (3.10), the set of corresponding delay values is given by[2]

$$T_\omega = \left\{ \frac{1}{\omega} \left[-j\text{Log}\left(-\frac{kP(j\omega)}{Q(j\omega)} \right) + 2\pi l \right] \geq 0, \ l \in \mathbb{Z} \right\}. \tag{3.11}$$

When increasing the delay, the corresponding crossing direction of characteristic roots is toward instability (stability) when $F'(\omega) > 0$ (< 0).

Proof Substituting $s = j\omega$ in the characteristic equation yields $H(j\omega; \tau) = 0$, or equivalently

$$Q(j\omega) = -kP(j\omega)e^{-j\omega\tau}. \tag{3.12}$$

By equating the modulus and argument of the left- and right-hand sides, (3.10) and (3.11) follow.

For an $\omega_0 \in S_0$ and $\tau_0 \in T_{\omega_0}$ the continuity of the characteristic roots with respect to the delay implies the existence of a root function $r(\tau)$ satisfying $r(\tau_0) = j\omega_0$ and

---

[2] Here, "Log" denotes the principal value of the logarithm. Consequently for $|z| = 1$, $\text{Log}(z) = j \arg(z)$ with $\arg(z) \in (-\pi, \pi)$.

$$H(r(\tau);\tau) = 0.$$

Inspired by [5, 6], taking the derivative of this expression with respect to τ at $\tau = \tau_0$ yields

$$r'(\tau_0)^{-1} = -\frac{Q'(j\omega_0)}{j\omega_0 Q(j\omega_0)} + \frac{kP'(j\omega_0)}{j\omega_0 kP(j\omega_0)} - \frac{\tau_0}{j\omega_0}. \qquad (3.13)$$

Since $\omega_0 \in S_+$, we have $| Q(j\omega_0) |^2 = k^2 | P(j\omega_0) |^2$ and, therefore,

$$r'(0)^{-1} = -\frac{1}{j\omega_0 | Q(j\omega_0) |^2} \left[Q'(j\omega_0)Q(-j\omega_0) - k^2 P'(j\omega_0)P(-j\omega_0) \right] - \frac{\tau_0}{j\omega_0}. \qquad (3.14)$$

The roots will cross the imaginary axis toward stability (instability) if $\Re(r'(0)) < 0$ (> 0). Then, (3.14) leads to (same steps as in [6]):

$$\Re(r'(0)^{-1}) = -\frac{1}{\omega_0 | Q(j\omega_0) |^2} \Im \left[Q'(j\omega_0)Q(-j\omega_0) - k^2 P'(j\omega_0)P(-j\omega_0) \right]$$

$$= \frac{1}{2\omega_0 | Q(j\omega_0) |^2} F'(\omega_0),$$

which confirms that the *sign* of F' will give the crossing direction.   □

Remark 3.6 Theorem 3.5 is based on the observation that with the delay as parameter, $H(j\omega; k, \tau) = 0$ can be seen as an equation in two independent variables, $j\omega \in j\mathbb{R}$ and $z := \exp(-j\omega\tau)$ on the unit circle of the complex plane. Taking the modulus in (3.12) allows us to eliminate the second variable. As seen in [23], the first variable may also be eliminated via matrix pencil techniques, leading to an alternative characterization of the stability regions.

Theorem 3.5, combined with the continuous dependence of the characteristic roots with respect to the delay, allows us to completely characterize the stability/instability regions in the delay parameter. Indeed, the set

$$\mathcal{T} = \bigcup_{\omega \in S_+} \mathcal{T}_\omega$$

partitions the delay space ($\mathbb{R}^+$) into intervals in which the number of roots in the open right half plane is constant. Moreover, we have the following result:

Remark 3.7 (Crossing direction) Assume that S_+ is not empty and denote its elements in descending order by $\omega_1 > \omega_2 > \dots$. Since $\lim_{\omega \to \infty} F(\omega) = +\infty$ and the roots $\{\omega_1, \omega_2, \dots\}$ of F are simple, the sign of F' at these roots alternates, with

$F'(\omega_1) > 0$. As a consequence, as the delay monotonically increases, all root crossings for delay values $\mathcal{T}_{\omega_1}$ are toward instability, while all root crossings for delay values $\mathcal{T}_{\omega_2}$ are toward stability, etc.

Taking into account the number of unstable roots for $\tau = 0$ and the crossing direction in the points of $\mathcal{T}$, the number of unstable roots for each delay value may be determined.

Based on Theorem 3.5 and its underlying ideas, we now focus on the delay *stabilization* problem. First, we analyze various cases for which the delay stabilization problem has no solution, then present the simplest case (in terms of card($\mathcal{S}_+$) and card($\mathcal{U}_+$)) for which the delay stabilization problem has a positive answer. Finally, as a consequence of all the cases treated, we give the *necessary and sufficient conditions* such that a pair (k, τ) can stabilize the SISO system (3.2). We have the following results:

Proposition 3.8 *Assume that* card($\mathcal{U}_+$) *is an odd number. Then the delay stabilization problem has no solution.*

Proof By contradiction. Assume that the closed-loop system is asymptotically stable for some delay value τ_s. Because the number of roots in the closed right half plane changes *from odd to even* when increasing the delay from zero (number of closed right half plane roots equal to card($\mathcal{U}_+$)) to τ_s (number of closed half plane roots equal to zero), a characteristic root at *zero* must occur for some $\tau_0 \in [0, \tau_s]$. But $H(0; k, \tau_0) = 0$ implies $H(0; k, \tau) = 0, \forall \tau \geq 0$, which contradicts the asymptotic stability at $\tau = \tau_s$. $\square$

Remark 3.9 The above result is relatively simple and proves the existence of a strictly positive root of the characteristic equation for all delay values if, for the delay-free system, such a root exists, and if the number of roots in the open right half plane is odd. Similar results, with slightly different formulations, and similar (or different) proofs have already been proposed in the literature (see, for instance [9, 19, 27] to cite only a few).

Proposition 3.10 *If either* card($\mathcal{S}_+$) $= 0$ *or* card($\mathcal{S}_+$) $= 1$, *then the delay stabilizing problem has no solution.*

Proof When card($\mathcal{S}_+$) $= 0$, characteristic roots cannot cross the imaginary axis as the delay is varied, and the instability for $\tau = 0$ persists for all delay values. On the other hand, when card($\mathcal{S}_+$) $= 1$, there is one crossing frequency and from Remark 3.7 the crossing direction is always toward instability as the delay is increased. Combining this fact with the instability for $\tau = 0$ yields the statement of the proposition. $\square$

Notice that the condition card($\mathcal{S}_+$) $= 0$ corresponds to the *delay-independent hyperbolicity* property (fixed number of unstable roots for all positive delay values), as defined in [11] (see also [12]). For the remaining cases, one needs to count the roots crossing the imaginary axis toward stability/instability and to define the corresponding delay intervals (see also [23], Chaps. 4 and 7).

Thus, it follows that the first case when the delay has a stabilizing effect may appear if $\text{card}(\mathcal{U}_+) = 2$, and $\text{card}(\mathcal{S}_+) \geq 2$, and if the first crossing is toward stability. Furthermore, as proved in the sequel, such a condition becomes also *necessary* if $\text{card}(\mathcal{S}_+) \leq 3$. As we shall see in the next section devoted to illustrative examples, such a case may appear with a second-order system.

Proposition 3.11 *Assume that* $\text{card}(\mathcal{S}_+) = 2$ *or* $\text{card}(\mathcal{S}_+) = 3$. *Then the delay stabilizing problem has a solution if and only if*

1. $\text{card }(\mathcal{U}_+) = 2$,
2. $\tau_- < \tau_+$,

where

$$\tau_- = \min \bigcup\nolimits_{\omega \in \mathcal{S}_+, F'(\omega) < 0} \mathcal{T}_\omega,$$
$$\tau_+ = \min \bigcup\nolimits_{\omega \in \mathcal{S}_+, F'(\omega) > 0} \mathcal{T}_\omega \setminus \{0\}.$$

If stabilizable, all delay values $\tau \in (\tau_-, \tau_+)$ *are stabilizing.*

Proof "$\Rightarrow$" Consider first the case where $\text{card}(\mathcal{S}_+) = 2$. Let $\mathcal{S}_+ = \{\omega_1, \omega_2\}$, with $\omega_1 > \omega_2$. Then from Remark 3.7 we have $F'(\omega_1) > 0$ (associated with crossings for $\mathcal{T}_{\omega_1}$ toward instability when increasing the delay) and $F'(\omega_2) < 0$ (associated with crossings for $\mathcal{T}_{\omega_2}$ toward stability). The set $\mathcal{T}_{\omega_1}$ consists of delay values, equally spaced with $2\pi/\omega_1$, whereas the elements of $\mathcal{T}_{\omega_2}$ are equally spaced with $2\pi/\omega_2 > 2\pi/\omega_1$. As a consequence, between two stability crossings, an instability crossing must occur, i.e., the number of unstable roots in the closed right half plane cannot be reduced by more than two while increasing the delay. Thus, the occurrence of a stabilizing delay value necessarily implies that the number of roots in the closed right half plane is two for $\tau = 0$ (i.e., $\text{card}(\mathcal{U}_+) = 2$) and that the first[3] crossing is toward stability, mathematically expressed by $\tau_- < \tau_+$.

If $\text{card}(\mathcal{S}_+) = 3$ then $\mathcal{S}_+ = \{\omega_1, \omega_2, \omega_3\}$ with $\omega_1 > \omega_2 > \omega_3$ and $F'(\omega_1) > 0, F'(\omega_2) < 0, F'(\omega_3) > 0$. Compared to the previous case, there is one additional crossing frequency ω_3 where additional crossings toward *in*stability occur, and the argument remains the same.

"$\Leftarrow$" The condition $\tau_- < \tau_+$ implies that the first crossing is toward stability when the delay in increased from zero. Since $\text{card}(\mathcal{U}_+) = 2$, the closed-loop system is asymptotically stable for any $\tau \in (\tau_-, \tau_+)$. $\square$

Remark 3.12 The importance of this result lies in the fact that, in order to check stabilizability in the delay parameter, one only has to investigate the *first* root crossing of the imaginary axis as the delay increases from zero. This is particularly useful when one determines stabilizability by numerically computing the rightmost characteristic roots as a function of the delay. After the first root crossing, one can then stop the computations.

In the case when $\text{card}(\mathcal{S}_+) = 2$, the set of *all* stabilizing delay values can be expressed analytically:

---

[3] When $0 \in \mathcal{T}_{\omega_1}$ the crossing at $\tau = 0$ is not counted since $\text{card}(\mathcal{U}_+)$ does not change.

Corollary 3.13 *Assume that the conditions of Proposition 3.11 are satisfied, and in addition* $\mathrm{card}(\mathcal{S}_+) = 2$. *Then all the stabilizing delay values are defined by* $\tau \in (\underline{\tau_l}, \overline{\tau_l})$, $l = 0, 1, 2, \ldots, l_m$, *where*

$$\underline{\tau_l} = \tau_- + \frac{2\pi l}{\omega_-}, \overline{\tau_l} = \tau_+ + \frac{2\pi l}{\omega_+},$$

$\mathcal{S}_+ = \{\omega_+, \omega_-\}$ *with* $\omega_+ > \omega_-$, *and* l_m *is the largest integer for which* $\underline{\tau_l} < \overline{\tau_l}$, *which can be expressed as*

$$l_m = \max_{l \in \mathbb{Z}} \left\{ l < \frac{\omega_+ \omega_-}{2\pi} \cdot \frac{\tau_+ - \tau_-}{\omega_+ - \omega_-} \right\}. \tag{3.15}$$

Proposition 3.14 *Assume that* $\mathrm{card}(\mathcal{S}_+) = 2n$ *or* $\mathrm{card}(\mathcal{S}_+) = 2n + 1$, *with* $n \geq 1$. *Assume further that* $\mathrm{card}(\mathcal{U}_+) > 2n$. *Then the delay stabilizing problem has no solution.*

Proof Let $\mathcal{S}_+ = \{\omega_1, \omega_2, \ldots\}$ with $\omega_1 > \omega_2 > \ldots$. From Remark 3.7 we have the alternating sequence: $F'(\omega_1) > 0, F'(\omega_2) < 0, F'(\omega_3) > 0, \ldots$. Consider the pair (ω_1, ω_2). By the same arguments used in the proof of Proposition 3.11, there must be an element of $\mathcal{T}_{\omega_1}$ between two elements of $\mathcal{T}_{\omega_2}$. When the delay is increased from zero, the root crossings at $j\omega_1$ and $j\omega_2$ can therefore not contribute to reduction of closed half plane roots with more than two. When $n > 1$ the same argument can be used for the pairs $(\omega_3, \omega_4), \ldots, (\omega_{2n-1}, \omega_{2n})$. Thus taking into account the root crossings at $j\omega_1, j\omega_2, \ldots, j\omega_{2n}$ no more than $2n$ unstable roots can be shifted to the left half plane and the proof is complete for $\mathrm{card}(\mathcal{S}_+) = 2n$. In the case $\mathrm{card}(\mathcal{S}_+) = 2n + 1$ the argument remains the same since $F'(\omega_{2n+1}) > 0$. ☐

Define now the following quantities:

$$n_+(\tau) = \sum_{\omega \in \mathcal{S}_+, \, F'(\omega) > 0} \mathrm{card}\left\{ \mathcal{T}_\omega \cap (0, \tau] \right\}, \tag{3.16}$$

$$n_-(\tau) = \sum_{\omega \in \mathcal{S}_+, \, F'(\omega) < 0} \mathrm{card}\left\{ \mathcal{T}_\omega \cap [0, \tau] \right\}, \tag{3.17}$$

for some positive $\tau > 0$. Furthermore, introduce the sets $\mathcal{T}^+$ and $\mathcal{T}^-$, which partition $\mathcal{T}$ according to the sign of the derivative F' evaluated at the corresponding crossing frequency, that is,

$$\mathcal{T}^+ = \bigcup_{\omega \in \mathcal{S}_+, \, F'(\omega) > 0} \mathcal{T}_\omega \setminus \{0\},$$

$$\mathcal{T}^- = \bigcup_{\omega \in \mathcal{S}_+, \, F'(\omega) < 0} \mathcal{T}_\omega.$$

Based on the conditions and notations above, we conclude with the following result:

Proposition 3.15 *For a given gain k, the stabilizing control problem has a solution of the form $u(t) = -ky(t - \tau)$ if and only if the following conditions hold simultaneously:*

(i) *card($\mathcal{U}_+(k)$) is a strictly positive even integer, which satisfies the inequality card($\mathcal{U}_+(k)$) $\leq$ card($\mathcal{S}_+(k)$) and*

(ii) *there exists at least one delay value $\hat{\tau} \in \mathcal{T}$ such that the following equality is verified:*

$$2n_-(\hat{\tau}) = 2n_+(\hat{\tau}) + \text{card}(\mathcal{U}_+(k)). \tag{3.18}$$

Then all delay values $\tau \in (\hat{\tau}, \hat{\tau}_+)$, with

$$\hat{\tau}_+ = \min\left\{\mathcal{T}^+ \cap (\hat{\tau}, +\infty)\right\}, \tag{3.19}$$

guarantee the closed-loop asymptotic stability.

Proof While condition (i) is clear, condition (3.18) in (ii) simply characterizes the existence of crossings such that there are no more unstable rightmost roots for delays $\tau = \hat{\tau} + \varepsilon$, for sufficiently small $\varepsilon > 0$, and the definition of the delay interval follows straightforwardly.   □

3.2.3 Controller Design

The main results of the previous section are displayed in Table 3.1. Recall that card($\mathcal{U}_+$) and card($\mathcal{S}_+$) depend *only* on the gain k. The first quantity can be efficiently determined as a function of k by computing the generalized eigenvalues of a matrix pencil (Proposition 3.2), the second quantity by computing the roots of a polynomial (Proposition 3.3).

Table 3.1 Output feedback stabilizability conditions when using the delay as controller parameter. Necessary and sufficient conditions are given by Proposition 3.15. In case of card($\mathcal{U}_+$) = 2 and card($\mathcal{S}_+$) ∈ {2,3}, Proposition 3.11 can be applied

	0	1	2	3	4	5	6	7	8	9	card($\mathcal{S}_+$)
1		✓	✓	✓	✓	✓	✓	✓	✓	✓	
2		✓	$\tau_- < \tau_+$								
3		✓	✓	✓	✓	✓	✓	✓	✓	✓	
4		✓	✓	✓	✓						
5		✓	✓	✓	✓	✓	✓	✓	✓	✓	
6		✓	✓	✓	✓	✓	✓				
7		✓	✓	✓	✓	✓	✓	✓	✓	✓	
8		✓	✓	✓	✓	✓	✓	✓	✓		
card($\mathcal{U}_+$)											

The procedure to derive a stabilizing pair (k, τ) (if any) can be summarized as follows:

- first, compute card($\mathcal{S}_+$) and card($\mathcal{U}_+$) as functions of the gain parameter k, then select possible gain intervals $(\underline{k}, \overline{k})$ such that condition (i) of Proposition 3.15 holds;
- second, for a given k, search for stabilizing delay values. In the special case where card($\mathcal{S}_+$) = 2 or card($\mathcal{S}_+$) = 3 and card($\mathcal{U}_+$) = 2, Proposition 3.11 can be applied, i.e., it is sufficient to investigate only whether the *first* root crossing of the imaginary axis is toward stability as the delay increases from zero. In general, a more complete characterization of stability/instability regions becomes necessary, as we shall illustrate in the next section. However, according to condition (ii) of Proposition 3.15, this is a systematic task. Indeed, one has to compute the set $\mathcal{T}$ and next the partition $\mathcal{T}^+$ and $\mathcal{T}^-$, which give the root crossings (function of the delay values) toward instability and stability, respectively. Condition (3.18) together with (3.19) will define the corresponding stabilizing delay intervals.

Remark 3.16 (SISO dead-time systems stabilization) As mentioned in the Introduction, our method may also be used for solving a different control problem, namely *finding* all *the stabilizing (proportional) controllers for a SISO dead-time plant.*

Such a problem was considered by [26, 29] using a Pontryagin approach. More precisely, [26] addresses the control of first-order system with a time delay in both cases (stable and unstable delay-free systems), and [29] deals with some robustness issues in terms of delays for the closed-loop system under the assumption that the delay-free system can be stabilized by a proportional controller.

Solving such problems is outside the scope of the chapter. However, our method and the design procedure may still be applied. More precisely, we can easily compute the gain, and the corresponding delay values for which the number of roots in the open right half plane changes, then find, by "duality," all the cases when for a *given delay* value τ_0, there exists at least one gain interval $(\underline{k_0}, \overline{k_0})$ such that the closed-loop dead-time system becomes asymptotically stable. Furthermore, our method allows us to explicitly compute a delay interval $(\underline{\tau}, \overline{\tau})$ including τ_0 for any given gain value $k \in (\underline{k_0}, \overline{k_0})$.

3.3 Illustrative Examples

In the sequel, we shall consider several examples to illustrate the theoretical results derived above. Thus, we present first a second-order system that represents the simplest transfer function for which a delay stabilizing effect could appear. Next, we focus on some general stabilization conditions for a chain of oscillators, and finally we present a sixth-order system with more than one stabilizing delay interval.

3.3.1 Second-Order System

Consider the following second-order system:

$$\frac{P(s)}{Q(s)} = \frac{1}{s^2 - \alpha s + 2},$$

(3.20)

where $\alpha > 0$ is a real parameter. The polynomial $Q(s) = s^2 - \alpha s + 2$ is unstable and for all $k \in \mathbf{R}$, the polynomial $Q(s) + kP(s)$ has at least one unstable root. Furthermore, if $\alpha = 0$, then (3.20) corresponds to an oscillator (the characteristic equation has two roots on the imaginary axis).

With the controller (3.3), the characteristic equation of the closed-loop system is given by

$$s^2 - \alpha s + 2 + ke^{-s\tau} = 0$$

(3.21)

and the polynomial $F(\omega; k)$ by

$$F(\omega) = |Q(j\omega)|^2 - |P(j\omega)|^2 = (2 - \omega^2)^2 + \alpha^2 \omega^2 - k^2$$
$$= \omega^4 - (4 - \alpha^2)\omega^2 + (4 - k^2).$$

(3.22)

For $\alpha > 2$ the quantities $\text{card}(\mathcal{S}_+)$ and $\text{card}(\mathcal{U}_+)$ are functions of the gain k as displayed in the following table:

k	< -2	$\in (-2, 2)$	> 2
$\text{card}(\mathcal{S}_+)$	1	0	1
$\text{card}(\mathcal{U}_+)$	1	2	2

while for $\alpha < 2$ we have

k	< -2	$\in (-2, -k^*)$	$\in (-k^*, k^*)$	$\in (k^*, 2)$	> 2
$\text{card}(\mathcal{S}_+)$	1	2	0	2	1
$\text{card}(\mathcal{U}_+)$	1	2	2	2	2

where

$$k^* = 2\sqrt{1 - \left(1 - \frac{\alpha^2}{4}\right)^2}.$$

According to the results of the previous section, summarized in Table 3.1, a necessary condition for asymptotic stability of the closed-loop system is therefore given by

$$\alpha < 2, \ |k| \in (k^*, 2).$$

(3.23)

Furthermore, for a gain satisfying (3.23) the existence of a stability region in the delay parameter is determined by the condition $\tau^- < \tau^+$. Summarizing, we have

Proposition 3.17 *The system (3.20) can be stabilized with a controller of the form* $u(t) = -ky(t - \tau)$ *if and only if the pair* (α, k) *satisfies*

$$\alpha \in [0,2), \quad |k| \in \left(2\sqrt{1 - \left(1 - \frac{\alpha^2}{4}\right)^2}, 2 \right), \tag{3.24}$$

and in addition $\tau_- < \tau_+$, *where*

$$\tau_{\pm} = \frac{1}{\omega_{\pm}}\left[\mathrm{Log}\left(\frac{\omega_{\pm}^2 - 2}{k} - j\frac{\alpha\omega_{\pm}}{k} \right) + (1 + \mathrm{sign}\, k)\pi \right], \alpha \neq 0$$

$$\begin{cases} \tau_+ = (3 + \mathrm{sign}\, k)\frac{\pi}{2\omega_+} \\ \tau_- = (1 + \mathrm{sign}\, k)\frac{\pi}{2\omega_-} \end{cases}, \qquad \alpha = 0 \tag{3.25}$$

and

$$\omega_{\pm} = \sqrt{1 - \frac{\alpha^2}{4}} \cdot \sqrt{2\left(1 \pm \sqrt{1 - \frac{1 - \frac{k^2}{4}}{\left(1 - \frac{\alpha^2}{4}\right)^2}} \right)}. \tag{3.26}$$

A stabilizing controller is then defined by the gain k and $\tau \in (\tau_-, \tau_+)$.

Remark 3.18 (Stabilizing oscillations) If $\alpha = 0$ and $k \in (-2, 0)$, we recover the results proposed in [1, 4, 25]:

$$\tau_- = 0, \quad \tau_+ = \frac{\pi}{\sqrt{2 + |k|}}.$$

Furthermore, the number of delay intervals is given by

$$\max_{l \in \mathbf{Z}} \left\{ l \leq \frac{1}{2} \cdot \frac{1}{\sqrt{\frac{2+|k|}{2-|k|}} - 1} \right\}.$$

Roughly speaking, the smaller the gain is, the fewer the stabilizing delay intervals there are, as displayed in the graphical representation in [1], etc.

3.3.2 Stabilizing a Chain of Oscillators

Consider now a chain of n oscillators characterized by the frequencies: $0 < \omega_1 < \ldots < \omega_n$ with the following input–output transfer function:

$$H_{yu}(s) = \frac{1}{\displaystyle\prod_{i=1}^{n} (s^2 + \omega_i^2)}. \tag{3.27}$$

A state space representation for such a transfer function is

$$\begin{cases} \prod_{i=1}^{n} \left(\dfrac{d^2}{dt^2} + \omega_i^2 \right) :x(t) = u(t) \\ y(t) = x(t), \end{cases}$$

where, the operator $(d^2/d\theta^2 + \omega^2):x(\theta)$, for some real variable θ is defined by

$$(d^2/d\theta^2 + \omega^2):x(\theta) = \ddot{x}(\theta) + \omega^2 x(\theta).$$

The use of the control feedback $u(t) = -ky(t - \tau)$ will lead to the following characteristic equation associated with the closed-loop system:

$$\prod_{i=1}^{n} (s^2 + \omega_i^2) + ke^{-s\tau} = 0. \tag{3.28}$$

Simple computations lead to

$$\text{card}(\mathcal{U}_+) = 2n \text{ and card}(\mathcal{S}_+) = 2n \tag{3.29}$$

for sufficiently small k, and according to the results of the previous section, summarized in Table 3.1, there is no "objection" in stabilizing such a system using small gain values.

The main idea of the way to control such a chain of oscillators by using a delayed output feedback can be easily understood from the single oscillator case. Indeed, assume that we have only one oscillator subject to a delayed output feedback. Then (3.28) becomes

$$h(s;k): = s^2 + \omega^2 + ke^{-s\tau} = 0. \tag{3.30}$$

Consider now the dependence of the roots with respect to the gain parameter, and let us define such dependence as $r = r(k)$ (see, for example, [21]). Motivated by the validity of the conditions (3.29) for small gain values, we look for the behavior of the roots around $k = 0$. It is easy to see that, for $k = 0$, (3.30) has one pair of imaginary roots $r = \pm j\omega$. To investigate the effect of roots moving away from the imaginary axis for small gain values, we need to differentiate $h(r(k);k) = 0$ with respect to k at the point $k = 0$. Simple computations lead to the following equality:

$$\frac{d}{dk} r(0) = \frac{\sin(\omega\tau) + j\cos(\omega\tau)}{2\omega}.$$

When choosing τ such that $\sin(\omega\tau) \neq 0$, it follows that

$$\Re\left(\frac{d}{dk} r(0) \right) \neq 0,$$

that is, the roots on the imaginary axis can be shifted in $\mathbb{C}_-$ with a suitable change of k. Such an idea was exploited by [17], and their result can be summarized as follows.

Proposition 3.19 *Given a set of frequencies* $0 < \omega_1 < \omega_2 < \ldots < \omega_n$, *if there exists* $\tau > 0$ *such that*

$$(-1)^\nu \sin(\omega_\nu \tau) > 0, \text{ for all } \nu = 1, 2, \ldots, n, \tag{3.31}$$

then for sufficiently small $k > 0$, *the delayed output feedback control law* $u(t) = -ky(t - \tau)$ *stabilizes the chain of oscillators (3.27).*

As a consequence, we have

Corollary 3.20 *If the frequencies* $\{\omega_i\}_{i=\overline{1,n}}$ *are rationally independent, then there always exists a pair* (k, τ) *such that the delayed output feedback control law* $u(t) = -ky(t - \tau)$ *stabilizes the chain of oscillators (3.27).*

Proof The result follows straightforwardly from (3.31) and Kronecker's theorem [13]. □

Remark 3.21 If $n > 1$, then for any fixed gain k the controlled chain of oscillators (3.27) is unstable for sufficiently small values of τ. This follows from a sensitivity analysis of the roots of (3.28) with respect to the delay τ at $\tau = 0_+$. An indication of this property is given by the fact that condition (3.31) is always violated for small delay values.

3.3.3 Multiple Crossing Frequencies Toward (in)Stability

We study the stabilization of the system

$$\frac{P(s)}{Q(s)} = \frac{1}{s^6 + p_1 s^5 + p_2 s^4 + p_3 s^3 + p_4 s^2 + p_5 s + p_6}, \tag{3.32}$$

where

$$p_1 = -6.0000000e - 04, \ p_2 = 1.4081634e + 00, \quad p_3 = -5.6326533e - 04,$$
$$p_4 = 4.3481891e - 01, \quad p_5 = -8.6963771e - 05, \ p_6 = 2.6655565e - 02$$

using a controller of the form (3.3). The uncontrolled system has six strictly unstable poles,

$$
\begin{aligned}
s_{1,2} &= 9.99999988e - 05 \pm 2.85714287e - 01j \\
s_{3,4} &= 1.00000009e - 04 \pm 5.71428578e - 01j \\
s_{5,6} &= 9.99999921e - 04 \pm 9.99999988e - 01j
\end{aligned} \tag{3.33}
$$

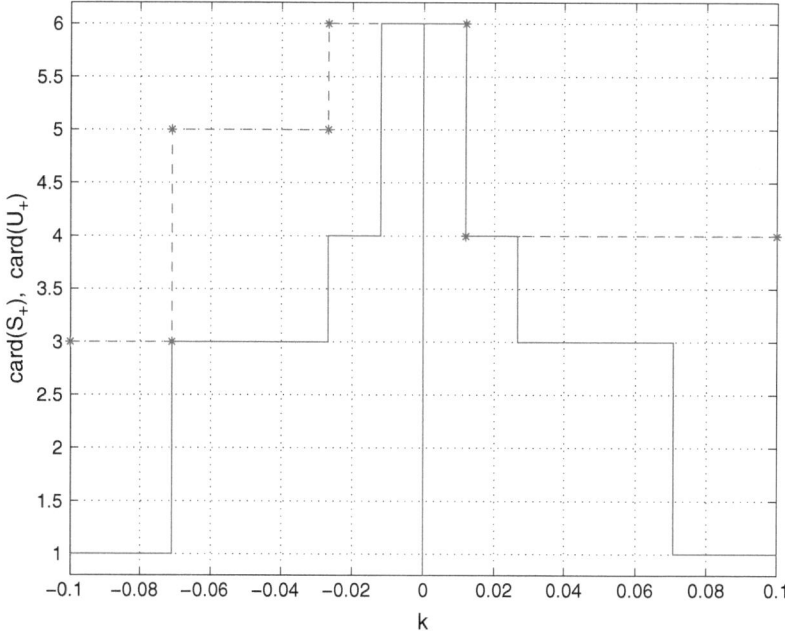

Fig. 3.2 card(S_+) (*solid line*) and card(U_+) (*dashed line*) as a function of the gain k

and, hence, it cannot be stabilized with static and delay-free output feedback. Note from (3.33) that (3.32) corresponds to a perturbed chain of three oscillators.

As a first step in the controller design, we compute card(U_+) and card(S_+) as a function of the gain, resulting in Fig. 3.2. The information given in Table 3.1 then helps to choose possible values of the gain. We take $k = 0.0025$, where card(S_+) = card(U_+) = 6.

Second, we characterize stability regions in the delay parameter. The set S^+ is given by

$$\omega_1 = 1.0019959e + 00, \ \omega_2 = 9.9795792e - 01, \ \omega_3 = 5.8408171e - 01,$$
$$\omega_4 = 5.5740265e - 01, \ \omega_5 = 3.0572050e - 01, \ \omega_6 = 2.6663916e - 01.$$

By computing the set T, using (3.11), and taking into account the crossing direction, we arrive at a complete characterization of stability regions in the delay, displayed in Table 3.2. This is in accordance with Fig. 3.3, where we show the rightmost characteristic roots of the closed-loop system as a function of the delay, computed with the software package DDE-BIFTOOL [8].

To summarize, the system (3.32) and (3.3) is asymptotically stable for

$$\begin{cases} k = 0.0025, \\ \tau \in (11.802168, 12.490817) \ \cup \ (15.788569, 16.121915) \\ \qquad\qquad\qquad\qquad \cup \ (35.366543, 37.573495). \end{cases}$$

Table 3.2 Characterization of stability regions in the delay parameter for the system (3.32) and (3.3) with $k = 0.0025$

Elements of $\mathcal{T} = \bigcup_{i=1}^{6} \mathcal{T}_{\omega_i}$	Crossing frequency	Number of unstable roots changes to
		6
1.2048745e-02	ω_4	4
3.1964843e+00	ω_2	2
5.3645410e+00	ω_3	4
6.2201470e+00	ω_1	6
9.4925266e+00	ω_2	4
1.1284305e+01	ω_4	2
1.1802168e+01	ω_6	0
1.2490817e+01	ω_1	2
1.5788569e+01	ω_2	0
1.6121915e+01	ω_3	2
1.8761486e+01	ω_1	4
2.0536234e+01	ω_5	6
2.2084611e+01	ω_2	4
2.2556560e+01	ω_4	2
2.5032156e+01	ω_1	4
2.6879289e+01	ω_3	6
2.8380653e+01	ω_2	4
3.1302825e+01	ω_1	6
3.3828816e+01	ω_4	4
3.4676696e+01	ω_2	2
3.5366543e+01	ω_6	0
3.7573495e+01	ω_1	2
3.7636663e+01	ω_3	4
$\vdots$	$\vdots$	$\vdots$

Notice that stability can be achieved by increasing the delay after having three pairs of unstable roots. Notice also that it is generally not sufficient to investigate only the first root crossing of the imaginary axis.

3.4 Concluding Remarks

This chapter was devoted to the stabilization problem of a class of dynamical systems with a single input, a single output subject to output (or input) delayed feedback. More precisely we considered the problem where the delay in the control law may have a *stabilizing* effect, that is, the closed-loop stability is guaranteed precisely due to the existence of delay. Various conditions have been derived using a frequency-domain approach. Some illustrative examples were also proposed.

Acknowledgments This chapter presents results of the Belgian Programme on Interuniversity Poles of Attraction, initiated by the Belgian State, Prime MinisterâĂŹs Office for Science, Technology and Culture, and of the Optimization in Engineering Centre OPTEC of the K.U.Leuven. The research of Silviu-Iulian Niculescu is partially supported by PICS CNRS-US cooperative

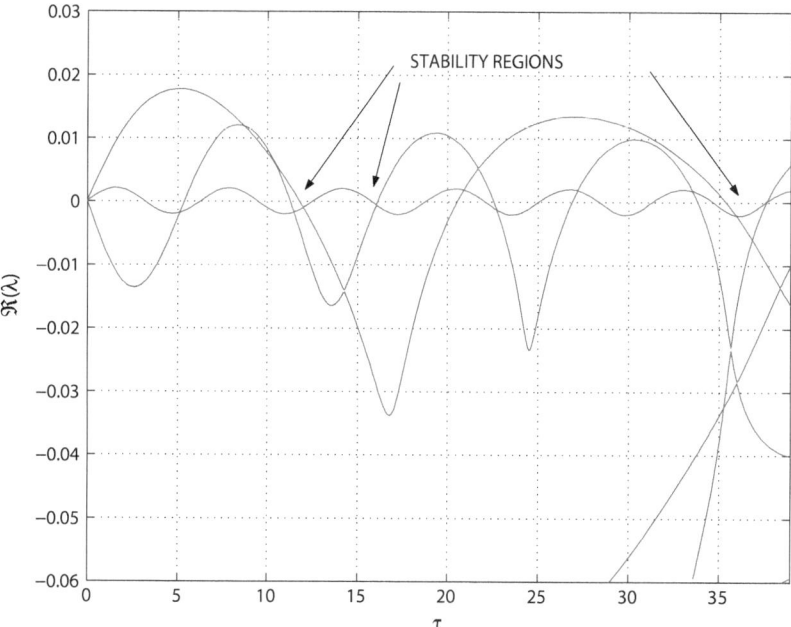

Fig. 3.3 Rightmost eigenvalues of the closed-loop system (3.32) and (3.3) as a function of the delay τ; $k = 0.0025$

project "Delays in interconnected systems: Analysis and applications" (2005–2007) and by PAI Tournesol (French-Belgium bilateral program): "Distributed delays in dynamical systems: Analysis and applications" (2006–2007). The research of Wim Michiels is partially supported by PAI Tournesol (French-Belgium bilateral program): "Distributed delays in dynamical systems: Analysis and applications" (2006–2007). The research of Keqin Gu and Chaouki T. Abdallah is partially supported by NSF-CNRS INT-0129062.

References

1. C. Abdallah, P. Dorato, J. Benitez-Read, and R. Byrne. Delayed positive feedback can stabilize oscillatory systems. *Proc. American Contr. Conf.*, 3106–3107, 1993.
2. R. E. Bellman, and K. L. Cooke. *Differential-Difference Equations.* Academic Press, New York, 1963.
3. J. Chen, Static output feedback stabilization for SISO systems and related problems: solutions via generalized eigenvalues. *Contr. - Theory Adv. Tech.*, 10: 2233–2244, 1995.
4. J. Chiasson and C.T. Abdallah, Robust stability of time delay systems: Theory. In: *Proc. 3rd IFAC Wshop on Time Delay Syst.* Santa Fe, NM (December 2001) 125–130.
5. K. L. Cooke, and Z. Grossman, Discrete delay, distributed delay and stability switches. *J. Math. Anal. Appl.*, 86: 592–627, 1982.
6. K. L. Cooke, and P. van den Driessche, On zeroes of some transcendental equations. *Funkcialaj Ekvacioj*, 29: 77–90, 1986.
7. R. Datko, A procedure for determination of the exponential stability of certain differential-difference equations. *Quart. Appl. Math.* 36, 279–292, 1978.

8. K. Engelborghs, T. Luzyanina, and G. Samaey, *DDE-BIFTOOL v. 2.00: a Matlab package for bifurcation analysis of delay differential equation*, T.W. Report 330, Department of Computer Science, K.U. Leuven, 2001.

9. K. Gopalsamy, *Stability and Oscillations in Delay Differential Equations of Population Dynamics*. Kluwer Academic Publishers, Math. Its Appl. Series, 74, 1992.

10. K. Gu, V.L. Kharitonov, and J. Chen, *Stability of Time-Delay Systems*. Birkhauser, Boston, 2003.

11. J. K. Hale, E. F. Infante, and F. S. -P. Tsen, Stability in linear delay equations. *J. Math. Anal. Appl.*, 105: 533–555, 1985.

12. J. K. Hale and S. M. Verduyn Lunel, *Introduction to Functional Differential Equations*. Applied Math. Sciences, 99, Springer-Verlag, New York, 1993.

13. G. Hardy and E. Wright, *An Introduction to the Theory of Numbers*. Oxford Univ. Press: Oxford, UK, 1968.

14. U. Helmke and B. D. O. Anderson, Hermitian pencils and output feedback stabilization of scalar systems. *Int. J. Contr.*, 56: 857–876, 1992.

15. R. Izmailov, Analysis and optimization of feedback control algorithms for data transfers in high-speed networks. *SIAM J. Contr. Optimiz.* 34: 1767–1780, 1996.

16. F.P. Kelly, Mathematical modelling of the Internet, In: B. ENGQUIST, W. SCHMID (eds) *Mathematics unlimited - 2001 and beyond*, Springer-Verlag, Berlin, 685–702, 2001.

17. V.L. Kharitonov, S.-I. Niculescu, J. Moreno, and W. Michiels, Static output feedback stabilization: Necessary conditions for multiple delay controllers. *IEEE Trans. Automat. Contr.*, 50 82–86, 2005.

18. V. B. Kolmanovskii, and A. D. Myshkis, *Appl. Theory Funct. Diff. Eqs.*, Kluwer, Dordrecht, Netherlands, 1992.

19. E. Malakhovski, and L. Mirkin, Stability analysis of a class quasi-polynomials with a single delay, *Proc. 5th IFAC Wshop Time Delay Systems*, Leuven, Belgium, September 2004.

20. M. Malek-Zavarei and M. Jamshidi, *Time Delay Systems: Analysis, Optimization and Applications*. North-Holland Systems and Control Series, 9, Amsterdam, 1987.

21. W. Michiels, S.-I. Niculescu, and L. Moreau, Using delays and time-varying gains to improve the static output feedback stabilizability of linear systems: A comparison. *IMA J. Math. Contr. Inf.*, 21: 393–418, 2004.

22. W. Michiels, V. Van Assche, and S.-I. Niculescu, Stabilization of time-delay systems with a controlled time-varying delay and applications. *IEEE Trans. Automat. Contr.*, 50(4): 493–505, 2005.

23. S.-I. Niculescu, *Delay effects on stability. A robust control approach*. Springer-Verlag, Heidelberg, LNCIS, vol. 269, 2001.

24. S.-I. Niculescu, On delay robustness of a simple control algorithm in high-speed networks. *Automatica*, 38: 885–889, 2002.

25. S.-I. Niculescu and C. T. Abdallah, Delay effects on static output feedback stabilization. *Proc. 39th IEEE Conf. Dec. Contr.*, Sydney, Australia, December 2000.

26. G.J. Silva, A. Datta, and S.P. Bhattacharrya. New results on the synthesis of PID controllers. *IEEE Trans. Automat. Contr.*, 47: 241–252, 2002.

27. G. Stépán, *Retarded Dynamical Systems: Stability and Characteristic Function* (Research Notes in Math. Series, 210, Longman Scientific, UK, 1989).

28. V. Syrmos, C. T. Abdallah, P. Dorato, and K. Grigoriadis, Static output feedback: A survey. *Automatica*, 33: 125–137, 1997.

29. H. Xu, A. Datta, and S.P. Bhattacharyya, PID stabilization of LTI plants with time-delay. *Proc. 42nd IEEE Conf. Dec. Contr.*, Mauii, Hawaii (2003) 4038–4043.

Chapter 4
Time-Delayed Feedback Control: From Simple Models to Lasers and Neural Systems

Eckehard Schöll, Philipp Hövel, Valentin Flunkert, and Markus A. Dahlem

4.1 Introduction

Over the past decade control of unstable states has evolved into a central issue in applied nonlinear science [1]. This field has various aspects comprising stabilization of unstable periodic orbits embedded in a deterministic chaotic attractor, which is generally referred to as *chaos control*, stabilization of unstable fixed points (steady states), or control of the coherence and timescales of stochastic motion. Various methods of control, going well beyond the classical control theory [2–4], have been developed since the ground-breaking work of Ott, Grebogi, and Yorke [5] in which they demonstrated that small time dependent changes in the control parameters of a nonlinear system can turn a previously chaotic trajectory into a stable periodic motion. One scheme where the control force is constructed from time-delayed signals [6] has turned out to be very robust and universal to apply and easy to implement experimentally. It has been used in a large variety of systems in physics, chemistry, biology, medicine, and engineering [1, 7, 8], in purely temporal dynamics as well as in spatially extended systems [9–25]. Moreover, it has recently been shown to be applicable also to noise-induced oscillations and patterns [26–29]. This is an interesting observation in the context of ongoing research on the constructive influence of noise in nonlinear systems [30–35].

In time-delayed feedback control (*time-delay autosynchronization* or TDAS) the control signal is built from the difference $s(t) - s(t - \tau)$ between the present and an earlier value of an appropriate system variable s. It is *non-invasive* since the control forces vanish if the target state (a periodic state of period τ or a steady state) is reached. Thus the unstable states themselves of the uncontrolled system are not changed, but only their neighborhood is adjusted such that neighboring trajectories converge to it, i.e., the control forces act only if the system deviates from the state to be stabilized. Involving no numerically expensive computations, time-delayed

E. Schöll (✉)

Institut für Theoretische Physik, Technische Universität Berlin, Hardenbergstraße 36, 10623 Berlin, Germany

e-mail: schoell@physik.tu-berlin.de

F.M. Atay (ed.), *Complex Time-Delay Systems,* Understanding Complex Systems, DOI 10.1007/978-3-642-02329-3_4, © Springer-Verlag Berlin Heidelberg 2010

feedback control is capable of controlling systems with very fast dynamics still in real-time mode [36–38]. Moreover, detailed knowledge of the target state is not required.

An extension to multiple time delays has been proposed by Socolar et al. [39], who considered multiple delays in form of an infinite series (ETDAS) or an average of N past iterates (N time-delay autosynchronization or NTDAS) [40] or coupling matrices (generalized ETDAS or GETDAS) [41]. Analytical insight into those schemes has been obtained by several theoretical studies, e.g., [42–54] as well as by numerical bifurcation analysis, e.g., [55, 56]. Time-delayed feedback can also stabilize fixed points using single [48, 49, 57] or multiple delay times [50, 58, 59]. The efficiency of these schemes can be improved by deterministic or stochastic modulation of the time delay [60].

Recent work has focused, on the one hand, on basic aspects like developing novel control schemes and gaining analytical insights, and on the other hand, on applications to optical and electronic systems, including laser diodes, electronic circuits, and semiconductor nanostructures [18, 61, 62], to chemical and electrochemical reaction systems [15, 16, 63–70], and to biological and medical systems, including the suppression of synchronization as therapeutic tools for neural diseases like Parkinson and epilepsy [71, 72], and control of cardiac dynamics [73]. In particular, networks of oscillatory or excitable elements, e.g., neural networks or coupled laser arrays, have been considered, where time delays naturally arise through signal propagation and processing times [74–82]. Systems composed of a small number of coupled oscillatory or excitable elements (lasers or neurons) can be conceived as network motifs of larger networks. Time-delayed feedback control schemes with different couplings of the control force have been applied to various models of nonlinear semiconductor oscillators, e.g., impact ionization-driven Hall instability [83], and semiconductor nanostructures described by an N-shaped [14, 84, 85], S-shaped [9, 11, 12, 24, 86], or Z-shaped [13] current-field characteristics. In semiconductor nanostructures complex chaotic spatio-temporal field and current patterns arise in the form of traveling field domains (for the N type) and breathing or spiking current filaments (for the S and Z types), which can be stabilized by time-delayed feedback control.

Time-delayed feedback control has also been applied to purely noise-induced oscillations and patterns in a regime where the deterministic system rests in a steady state, and in this way both the coherence and the mean frequency of the oscillations have been controlled in various nonlinear systems [26–28, 87–91], including chemical systems [92], neural systems [93, 94], laser diodes [95], and semiconductor nanostructures of N type [96–99] and Z type [100–102]. The control of deterministic and stochastic spatio-temporal patterns in semiconductor nanostructures by time-delayed feedback is reviewed elsewhere [62].

In this review we focus on simple models, for which some analytical results can be obtained in addition to computer simulations, and apply them to a selection of systems ranging from semiconductor lasers to neurosystems. We will show that time-delayed feedback control methods have a wider range of applicability than previously assumed, when applied to unstable steady states and to unstable periodic orbits, using generic normal forms. In the case of unstable periodic orbits the often invoked *odd number limitation*, which had been believed to impose serious restrictions for a long time, has recently been refuted [52]. Further, we will discuss applications to lasers and coupled neural systems in the framework of the Lang–Kobayashi laser model and the FitzHugh–Nagumo neuron model. We will demonstrate the suppression and enhancement of synchronization by time-delayed feedback, and point out some complex scenarios of synchronized in-phase or antiphase oscillations, bursting patterns, or amplitude death, induced by delayed coupling in combination with delayed feedback in simple network motifs.

4.2 Time-Delayed Feedback Control of Generic Systems

In this section we review basic properties of time-delayed feedback control, using simple normal form models which are representative of a large class of nonlinear dynamic systems [48–50, 52, 54].

4.2.1 Stabilization of Unstable Steady States

Time-delayed feedback methods, which have originally been used to control unstable periodic orbits [6], provide also a tool to stabilize unstable steady states [57, 103, 58, 59, 48–50, 60]. We present a numerical and analytical investigations of the feedback scheme using the Lambert function and discuss the extension to multiple time feedback control (ETDAS).

Other methods to control unstable steady states use the derivative of the current state as source of a control force [104]. It can be shown, however, that this *derivative control* is sensitive to high-frequency oscillations [105] and thus not robust in the presence of noise. Another control scheme calculates the difference of the current state to a low-pass filtered version [106].

Here we consider a general dynamic system given by a vector field $\mathbf{f}$ [48]:

$$\dot{\mathbf{x}} = \mathbf{f}(\mathbf{x}) \qquad\qquad (4.1)$$

with an unstable fixed point $\mathbf{x}^* \in \mathbb{R}^n$ given by $\mathbf{f}(\mathbf{x}^*) = 0$. The stability of this fixed point is obtained by linearizing the vector field around $\mathbf{x}^*$. Without loss of generality, let us assume $\mathbf{x}^* = 0$. In the following we will consider the generic case of an unstable focus for which the linearized equations in center manifold coordinates x, y can be written as

$$\dot{x} = \lambda x + \omega y \qquad\qquad (4.2)$$
$$\dot{y} = -\omega x + \lambda y,$$

where λ and ω are positive real numbers. They may be viewed as parameters governing the distance from the instability threshold, e.g., a Hopf bifurcation of system

(4.1), and the intrinsic eigenfrequency, respectively. For notational convenience, (4.2) can be rewritten as

$$\dot{\mathbf{x}}(t) = \mathbf{A}\mathbf{x}(t). \tag{4.3}$$

Alternatively, the components of $\mathbf{x}(t)$ can be understood as real and imaginary parts of a complex variable $z(t) = x(t) + iy(t)$ so that (4.2) reads $\dot{z}(t) = (\lambda + i\omega)z(t)$. The eigenvalues Λ_0 of the matrix $\mathbf{A}$ are given by $\Lambda_0 = \lambda \pm i\omega$, so that for $\lambda > 0$ and $\omega \neq 0$ the fixed point is indeed an unstable focus.

We shall now apply time-delayed feedback control [6] in order to stabilize this fixed point:

$$\dot{\mathbf{x}}(t) = \mathbf{A}\mathbf{x}(t) - \mathbf{F}(t), \tag{4.4}$$

where $\mathbf{F}$ denotes the control force given by

$$\mathbf{F}(t) = K[\mathbf{x}(t) - \mathbf{x}(t - \tau)], \tag{4.5}$$

with the feedback gain $K \in \mathbb{R}$ and the time delay $\tau > 0$. In components this yields

$$\dot{x}(t) = \lambda\,x(t) + \omega\,y(t) - K[x(t) - x(t - \tau)] \tag{4.6}$$
$$\dot{y}(t) = -\omega\,x(t) + \lambda\,y(t) - K[y(t) - y(t - \tau)].$$

The goal of the control method is to change the sign of the real part of the eigenvalue. Figure 4.1 depicts a schematic diagram of the time-delayed feedback loop. The red color shows the extension of the original Pyragas control including multiple delays (ETDAS) which will be discussed later on.

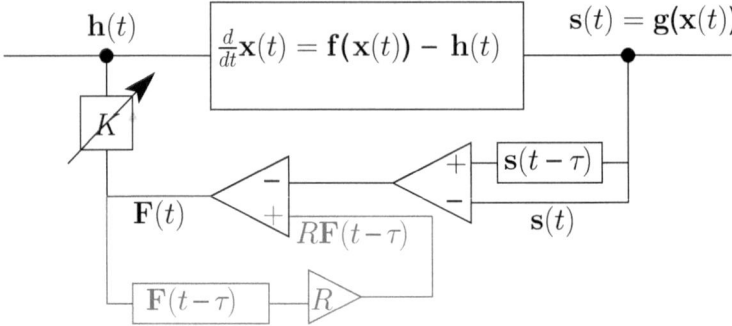

Fig. 4.1 Diagram of the time-delay autosynchronization method. $\mathbf{x}(t)$ denotes the state of the system at time t, $\mathbf{s}(t)$ is the control signal, i.e., some component of $\mathbf{x}(t)$ measured by $\mathbf{g}(\mathbf{x}(t))$, and $\mathbf{F}(t)$ is the control force. The real constants τ, K, and R denote the time delay, the feedback gain, and the memory parameter, respectively. The function $\mathbf{h}(t)$ describes the coupling of $\mathbf{F}$ to the dynamical system $\mathbf{x}$. The extension of the original time-delayed feedback [6] as introduced by Socolar et al. (see [39], ETDAS) is shown in *red color*

Since the control force applied to the ith component of the system involves only the same component, this control scheme is called *diagonal coupling* [11], which is suitable for an analytical treatment. Note that the feedback term vanishes if the unstable steady state is stabilized since $x^*(t - \tau) = x^*(t)$ and $y^*(t - \tau) = y^*(t)$ for all t, indicating the non-invasiveness of the TDAS method.

Figure 4.2 depicts the dynamics of the controlled unstable focus ($\lambda = 0.5$ and $\omega = \pi$) in the (x, y) plane for different values of the feedback gain K. Panels (a) through (d) correspond to increasing K. The time delay of the TDAS control scheme is chosen as $\tau = 1$ in all panels. Panel (a) displays the case of the absence of control, i.e., $K = 0$, and shows that the system is an unstable focus exhibiting undamped oscillations on a timescale $T_0 \equiv 2\pi/\omega = 2$. It can be seen from panel (b) that increasing K reduces the instability. The system diverges more slowly to infinity indicated by the tighter spiral. Further increase of K stops the unstable behavior completely and produces periodic motion, i.e., a center [see panel (c)]. The amplitude of the orbit depends on the initial conditions, which are chosen as $x = 0.01$ and $y = 0.01$. For even larger feedback gains, the trajectory becomes an inward spiral and thus approaches the fixed point, i.e., the focus. Hence the TDAS control scheme is successful.

An exponential ansatz for $x(t)$ and $y(t)$ in (4.6), i.e., $x(t) \sim \exp(\Lambda t)$ and $y(t) \sim \exp(\Lambda t)$, reveals how the control force modifies the eigenvalues of the system. The characteristic equation becomes

$$[\Lambda + K\left(1 - e^{-\Lambda \tau}\right) - \lambda]^2 + \omega^2 = 0, \tag{4.7}$$

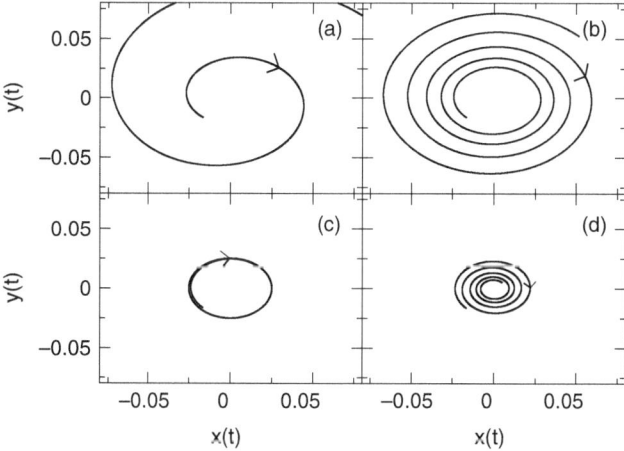

Fig. 4.2 Control of an unstable focus with $\lambda = 0.5$ and $\omega = \pi$ in the configuration space for different values of the feedback gain K. Panels (a), (b), (c), and (d) correspond to $K = 0, 0.2, 0.25,$ and 0.3, respectively. The time delay τ of the TDAS control scheme is chosen as 1 corresponding to $\tau = T_0/2 = \pi/\omega$ [48]

so that the complex eigenvalues Λ are given in the presence of a control force by the implicit equation

$$\lambda \pm i\omega = \Lambda + K\left(1 - e^{-\Lambda\tau}\right). \tag{4.8}$$

Using the Lambert function W, which is defined as the inverse function of $g(z) = ze^z$ for complex z [51, 107–111], (4.8) can be solved analytically

$$\Lambda\tau = W\left(K\tau e^{-(\lambda\pm i\omega)\tau+K\tau}\right) + (\lambda\pm i\omega)\tau - K\tau. \tag{4.9}$$

Panel (a) of Fig. 4.3 shows the dependence of the largest real part of the complex eigenvalues Λ upon the time delay τ according to (4.8) and (4.9) for $\lambda = 0.5$ and $\omega = \pi$. The solid curve corresponds to a feedback gain of $K = 0.3$, the dashed curve to $K = 0.25$, and the dotted curve to $K = 0.2$. All curves start at $\mathrm{Re}(\Lambda) = \lambda$ for $\tau = 0$, i.e., when no control is applied to system. For increasing time delay, the real part $\mathrm{Re}(\Lambda)$ decreases. It can be seen in the case of $K = 0.3$ that there exist values of the time delay for which $\mathrm{Re}(\Lambda)$ becomes negative, and thus the control is successful. The curve for $K = 0.25$ shows the threshold case where $\mathrm{Re}(\Lambda)$ becomes zero for $\tau = 1$, but does not change sign. The TDAS control scheme generates an infinite number of additional eigenmodes. The corresponding eigenvalues are the solutions of the transcendental equation (4.8). The real parts of the eigenvalues all originate from $-\infty$ for $\tau = 0$. Some of these lower eigenvalues are displayed for $K = 0.3$. The different branches of the eigenvalue spectrum originate from the multiple-leaf structure of the complex Lambert function. The real part of each eigenvalue branch exhibits a typical nonmonotonic dependence upon τ which leads to crossover of different branches resulting in an oscillatory modulation of the largest real part as a function of τ. Such behavior of the eigenvalue spectrum appears to be quite general and has been found for various delayed feedback coupling schemes, including the Floquet spectrum of unstable periodic orbits [11, 86] and applications to noise-induced motion where the fixed point is stable [26].

The notch at $\tau = 1$ corresponds to Fig. 4.2, so that at this value of τ the solid, dashed, and dotted curves correspond to panels (d), (c), and (b) of Fig. 4.2, respectively. The notches at larger τ become less pronounced leading to less effective realization of the TDAS control scheme, i.e., a smaller or no τ interval with negative $\mathrm{Re}(\Lambda)$.

In the case of an unstable periodic orbit, the optimal time delay is equal to the period of the orbit to be stabilized. Note that in the case of an unstable steady state, however, the time delay is not so obviously related to a parameter of the system. We will see later which combinations of the feedback gain K and the time delay τ lead to successful control.

Panel (b) of Fig. 4.3 displays the time evolution of $x(t)$ and its time-delayed counterpart $x(t - \tau)$ in the case of a combination of $K = 0.3$ and $\tau = 1$ that leads to successful control as in panel (d) of Fig. 4.2. The x component of the control force can be calculated from the difference of the two curves and subsequent multiplication by K. Since $x(t)$ tends to zero in the limit of large t (the system reaches the focus

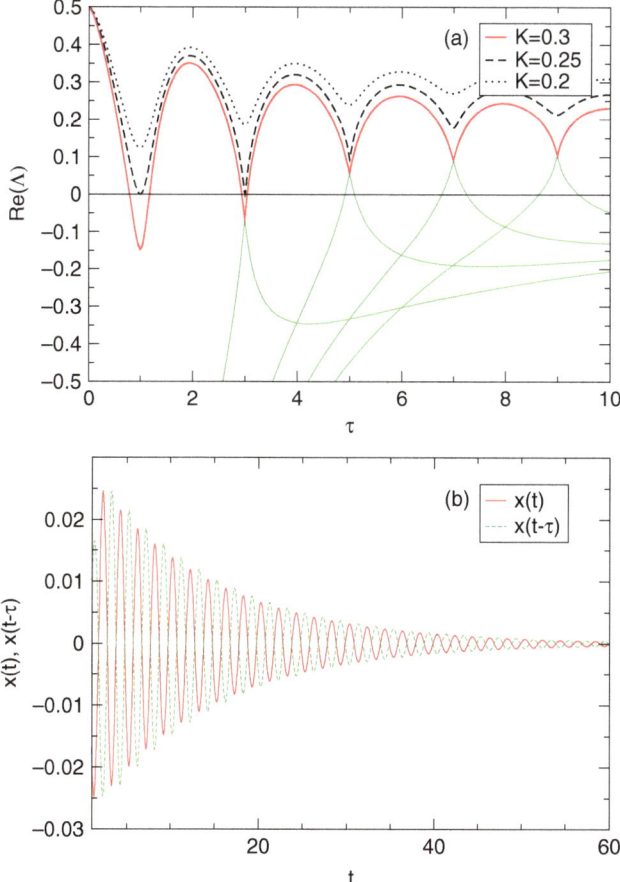

Fig. 4.3 (a) Largest real part of the complex eigenvalues Λ vs. τ for $\lambda = 0.5$ and $\omega = \pi$ for different K. Some lower eigenvalues are also displayed for $K = 0.3$ (*green*). (b) Time series of the x component of the unstable focus: the solid line (*red*) corresponds to $x(t)$ and the dashed line (*green*) to the delayed x component $x(t - \tau)$ with $\tau = 1$. The parameters of the unstable focus and the control scheme are as in panel (d) of Fig. 4.2 [48]

located at the origin), the control force vanishes if the system is stabilized. Thus the control scheme is non-invasive. Note that the current signal (red) and its delayed counterpart (green) are in antiphase.

In the following discussion, it is helpful to consider the real and imaginary part of (4.8) separately in order to gain some analytic information about the domain of control:

$$p + K\left[1 - e^{-p\tau} \cos\left(q\tau\right)\right] = \lambda \tag{4.10}$$

$$q + Ke^{-p\tau} \sin\left(q\tau\right) = \omega$$

with $\Lambda = p + iq$.

The calculation can be done analytically for special points by using, for instance, that $p = 0$ at the threshold of control. Furthermore, we will present an expansion around the minimum value of K that reveals further details of the shape of the domain of control.

At the threshold of control the sign of the real part p of the exponent Λ changes. Therefore, setting p equal to zero in the real and imaginary parts, respectively, of (4.10) yields

$$\lambda = K\left[1 - \cos(q\tau)\right] \tag{4.11}$$

and

$$\omega = q + K\sin(q\tau). \tag{4.12}$$

Since the cosine is bounded between -1 and 1, the following inequality follows from (4.11):

$$\frac{\lambda}{2} \leq K. \tag{4.13}$$

Thus a minimum value of K, $K_{min} = \lambda/2$, for which the control starts, can be inferred. It corresponds to $q\tau = (2n + 1)\pi$ for $n = 0, 1, 2, \ldots$. It should be noted that a similar characteristic equation as (4.8) holds for the Floquet exponents of a unstable periodic orbit, where the lower bound, $K_{min} = \lambda/2$, of the feedback gain has been shown to correspond to the flip threshold of control [43, 112].

In order to express the values of the time delay τ that correspond to the minimum K in terms of the parameters of the uncontrolled system, it is useful to consider even and odd multiples of π for $q\tau$, i.e., $q\tau = 2n\pi$ and $q\tau = (2n + 1)\pi$ for $n = 0, 1, 2, \ldots$. In both cases, the imaginary part of (4.8) leads to $q = \omega$. Hence, in the latter case, the time delay τ for $K_{min} = \lambda/2$ becomes

$$\tau = \frac{\pi}{\omega}(2n + 1). \tag{4.14}$$

The last expression can be rewritten using the uncontrolled eigenperiod T_0

$$\tau = T_0\frac{2n + 1}{2}, \tag{4.15}$$

where T_0 is defined by

$$T_0 = \frac{2\pi}{\omega}. \tag{4.16}$$

This discussion has shown that $K = \lambda/2$ and $\tau = T_0(2n + 1)/2$ with $n = 0, 1, 2, \ldots$ correspond to points of successful control in the (K, τ) plane with minimum feedback gain.

For even multiples, i.e., $q\tau = 2n\pi$ for $n = 0, 1, 2, \ldots$, no control is possible for finite values of K, since

$$\frac{K - \lambda}{K} = \cos{(q\tau)}|_{q\tau=2n\pi} \tag{4.17}$$

$$\Leftrightarrow 1 - \frac{\lambda}{K} = 1, \tag{4.18}$$

which cannot be satisfied for $\lambda \neq 0$ and finite K. Furthermore, (4.12) yields that for time delays which are integer multiples of the eigenperiod, i.e., $\tau = T_0 n = 2\pi n/\omega$ with $n = 0, 1, 2, \ldots$, the control scheme fails for any feedback gain.

Another result that can be derived from (4.8) is a shift of q for increasing K. For this, taking the square of the real and imaginary part of (4.8) and using trigonometrical identities yields

$$q = \omega \mp \sqrt{(2K - \lambda)\lambda}. \tag{4.19}$$

Inserting (4.19) into the real part of (4.8) leads to an explicit expression for the dependence of time delay τ on the feedback gain K at the threshold of stability, i.e., the boundary of the control domain $p = 0$,

$$\frac{K - \lambda}{K} = \cos{(q\tau)} \tag{4.20}$$

$$\Leftrightarrow \quad \tau(K) = \frac{\arccos\left(\frac{K-\lambda}{K}\right)}{\omega \mp \sqrt{(2K - \lambda)\lambda}}. \tag{4.21}$$

In order to visualize the shape of the domain of control we will investigate how small deviations $\epsilon > 0$ from $K_{\min}$, i.e, $K = \lambda/2 + \epsilon$, influence the corresponding values of the time delay τ. For this, let $\eta > 0$ be small and $\tau = \frac{\pi}{\omega}(2n + 1) \pm \eta$ a small deviation from τ at $K_{\min}$. Inserting the expression for K and τ into (4.20) yields after some Taylor's expansions:

$$-1 + \frac{4}{\lambda}\epsilon = -1 + \frac{1}{2}\left[\omega\eta \mp \frac{\pi}{\omega}(2n + 1)\sqrt{2\lambda}\sqrt{\epsilon}\right]^2 \tag{4.22}$$

$$\Leftrightarrow \quad \eta = \left[\pm\frac{2\sqrt{2}}{\omega\sqrt{\lambda}} + \frac{\sqrt{2}\pi}{\omega^2}(2n + 1)\sqrt{\lambda}\right]\sqrt{\epsilon}. \tag{4.23}$$

This equation describes the shape of the domain of control at the threshold of stabilization, i.e., $p = 0$, near the minimum K value at $\tau = T_0(2n + 1)/2$ in the (K, τ) control plane. Small deviations from τ at $K_{\min}$ are influenced by the square root of small deviations from the minimum feedback gain.

Figure 4.4 displays the largest real part of the eigenvalues Λ in dependence on both the feedback gain K and the time delay τ for $\omega = \pi$ and two different values of λ and summarizes the results of this section. The values of Λ are calculated using

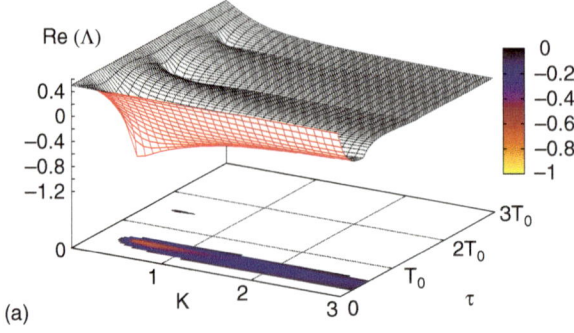

(a)

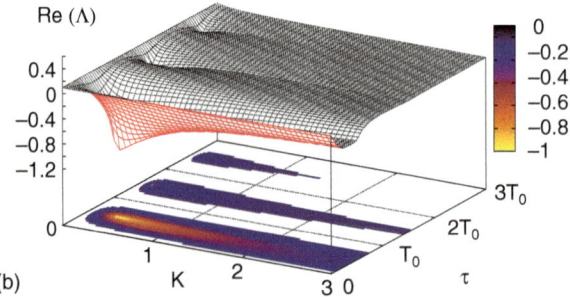

(b)

Fig. 4.4 Domain of control in the (K, τ) plane and largest real part of the complex eigenvalues Λ as a function of K and τ according to (4.9). The two-dimensional projection at the *bottom* shows combinations of τ and K, for which $\text{Re}(\Lambda)$ is negative and thus the control successful [panel (**a**): $\lambda = 0.5$ and $\omega = \pi$; panel (**b**): $\lambda = 0.1$ and $\omega = \pi$] [48]

the analytic solution (4.9) of (4.8). The two-dimensional projections at the bottom of each plot extract combinations of K and τ with negative p, i.e., successful control of the system. In the absence of a control force, i.e., $K = 0$, the real part of Λ starts at λ. Increasing the feedback gain decreases $\text{Re}(\Lambda)$. For $K = K_{\min} = \lambda/2$, the real part of the eigenvalue reaches 0 for certain time delays, i.e., $\tau = T_0(2n + 1)/2$ with $n = 0, 1, 2, \ldots$, and then changes sign. Thus, the system is stabilized. For values of the feedback gain slightly above the minimum value $K_{\min}$, the domain of control shows a square root shape. It can be seen that for time delays of $\tau = T_0 n$, the largest real part of the eigenvalues remains positive for any feedback gain. For a smaller value of λ (Fig. 4.4b), i.e., closer to the instability threshold of the fixed point, the domains of control become larger.

An example of the combination of minimum feedback gain $K_{\min} = \lambda/2$ and corresponding time delay $\tau = T_0(2n + 1)/2$, $n = 0, 1, 2, \ldots$ is shown in panel (c) of Fig. 4.2, where $K = \lambda/2 = 0.25$ and $\tau = T_0/2 = \pi/\omega = 1$. It describes the control threshold case between stable and unstable fixed point.

Socolar et al. introduced an extension of the Pyragas method by taking states into account which are delayed by integer multiples of τ [39]. This method is known as *extended time-delay autosynchronization* or ETDAS. Calculating the difference between two states which are one time unit τ apart yields the following control force, which can be written in three equivalent forms:

$$\mathbf{F}(t) = K \sum_{n=0}^{\infty} R^n \left[\mathbf{x}(t - n\tau) - \mathbf{x}(t - (n+1)\tau) \right] \tag{4.24}$$

$$= K \left[\mathbf{x}(t) - (1 - R) \sum_{n=1}^{\infty} R^{n-1} \mathbf{x}(t - n\tau) \right] \tag{4.25}$$

$$= K \left[\mathbf{x}(t) - \mathbf{x}(t - \tau) \right] + R\mathbf{F}(t - \tau), \tag{4.26}$$

where K and τ denote the (real) feedback gain and the time delay, respectively. $R \in (-1, 1)$ is a memory parameter that takes into account those states that are delayed by more than one time interval τ. Note that $R = 0$ yields the TDAS control scheme introduced by Pyragas [6].

The first form of the control force, (4.24), indicates the non-invasiveness of the ETDAS method because $\mathbf{x}^*(t - \tau) = \mathbf{x}^*(t)$ if the fixed point is stabilized. The third form, (4.26), is suited best for an experimental implementation since it involves states further than τ in the past only recursively.

While the stability of the fixed point in the absence of control is given by the eigenvalues of matrix $\mathbf{A}$, i.e., $\lambda \pm i\omega$, one has to solve the following characteristic equation in the case of an ETDAS control force [50]:

$$\Lambda + K \frac{1 - e^{-\Lambda\tau}}{1 - Re^{-\Lambda\tau}} = \lambda \pm i\omega. \tag{4.27}$$

Due to the presence of the time delay τ, this characteristic equation becomes transcendental and possesses an infinite but countable set of complex solutions Λ. For nonzero memory parameter R, (4.27) must be solved numerically.

Figure 4.5 depicts the dependence of the largest real parts of the eigenvalue Λ upon the time delay τ according to (4.27) for different memory parameters R and fixed feedback gain $K = 0.3$. The dashed, dotted, solid, dash-dotted, and dash-double-dotted curves (red, green, black, blue, and magenta) of Re(Λ) correspond to $R = -0.7, -0.35, 0, 0.35$, and 0.7, respectively. The parameters of the unstable focus are chosen as $\lambda = 0.1$ and $\omega = \pi$. Note that the time delay τ is given in units of the intrinsic period $T_0 = 2\pi/\omega$. When no control is applied to the system, i.e., $\tau = 0$, all curves start at λ which corresponds to the real part of the uncontrolled eigenvalue. For increasing time delay, the real part of Λ decreases and eventually changes sign. Thus, the fixed point becomes stable. Note that there is a minimum of Re(Λ) indicating strongest stability if the time delay τ is equal to half the intrinsic period. For larger values of τ, the real part increases and becomes positive again. Hence, the system loses its stability. Above $\tau = T_0$, the cycle is repeated but the

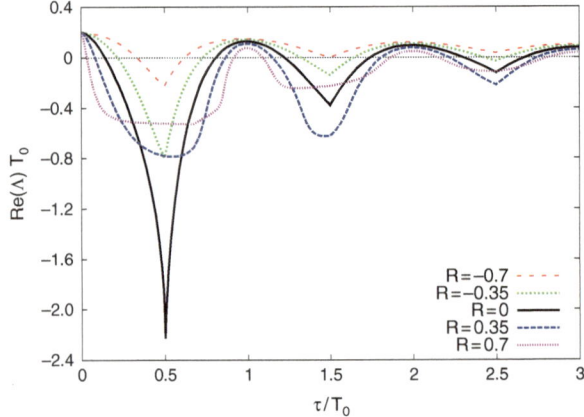

Fig. 4.5 Largest real part of the complex eigenvalues Λ as a function of τ for different values of R. The *dashed, dotted, solid, dash-dotted,* and *dash-double-dotted* curves (*red, green, black, blue,* and *magenta*) correspond to $R = -0.7, -0.35, 0, 0.35$, and 0.7, respectively. The parameters of the unstable focus are chosen as $\lambda = 0.1$ and $\omega = \pi$ which yields an intrinsic period $T_0 = 2\pi/\omega = 2$. The feedback gain K is fixed at $KT_0 = 0.6$ [50]

minimum of $\text{Re}(\Lambda)$ is not so deep. The control method is less effective because the system has already evolved further away from the fixed point. For vanishing memory parameter $R = 0$ (TDAS), the minimum is deepest, however, the control interval, i.e., values of τ with negative real parts of Λ, increases for larger R. Therefore the ETDAS control method is superior in comparison to the Pyragas scheme.

Figure 4.6 shows the domain of control in the plane parametrized by the feedback gain K and time delay τ for different values of R:0, 0.35, 0.7, and -0.35 in panels

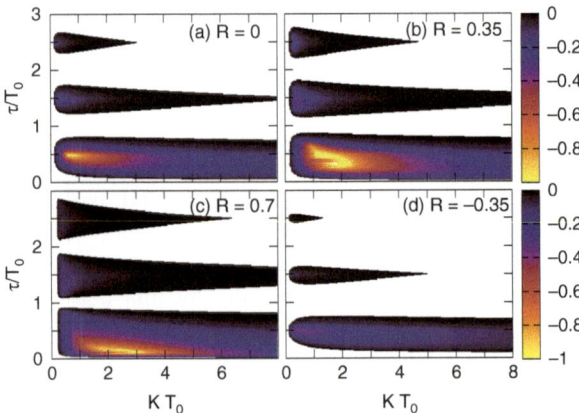

Fig. 4.6 Domain of control in the (K, τ) plane for different values of R:0, 0.35, 0.7, and -0.35 in panels (**a**), (**b**), (**c**), and (**d**), respectively. The grayscale (*color code*) shows only negative values of the largest real part of the complex eigenvalues Λ according to (4.27). The parameters of the system are as in Fig. 4.5 [50]

(a), (b), (c), and (d), respectively. The grayscale (color code) indicates only negative values of the largest real parts of the complex eigenvalue Λ. Therefore, Fig. 4.5 can be understood as a vertical cut through Fig. 4.6 for a fixed value of $KT_0 = 0.6$. Each panel displays several islands of stability which shrink for larger time delays τ. Note that no stabilization is possible if τ is equal to an integer multiple of the intrinsic period T_0. The domains of control become larger if the memory parameter R is closer to 1.

In order to obtain some analytic information of the domain of control, it is helpful to separate the characteristic equation (4.27) into real and imaginary parts. This yields using $\Lambda = p + iq$:

$$K(1 - e^{-p\tau} \cos q\tau) = \lambda - p - Re^{-p\tau}[(\lambda - p)\cos q\tau \pm (\omega - q)\sin q\tau] \quad (4.28)$$

and

$$Ke^{-p\tau} \sin q\tau = \pm(\omega - q) + Re^{-p\tau}[(\lambda - p)\sin q\tau \pm (\omega - q)\cos q\tau]. \quad (4.29)$$

The boundary of the domain of controls is determined by a vanishing real part of Λ, i.e., $p = 0$. With this constraint, (4.28) and (4.29) can be rewritten as

$$K(1 - \cos q\tau) = \lambda - R[\lambda \cos q\tau \pm (\omega - q)\sin q\tau], \quad (4.30)$$
$$K \sin q\tau = \pm(\omega - q) + R[\lambda \sin q\tau \pm (\omega - q)\cos q\tau].$$

At the threshold of control ($p = 0$, $q = \omega$), there is a certain value of the time delay, which will serve as a reference in the following, given by

$$\tau = \frac{(2n + 1)\pi}{\omega} = \left(n + \frac{1}{2}\right) T_0, \quad (4.31)$$

where n is any nonnegative integer. For this special choice of the time delay, the range of possible feedback gains K in the domain of control becomes largest as can be seen in Fig. 4.6. Hence, we will refer to this τ value as optimal time delay in the following. The minimum feedback gain at this τ can be obtained:

$$K_{min}(R) = \frac{\lambda(1 + R)}{2}. \quad (4.32)$$

Extracting an expression for $\sin(q\tau)$ from (4.30) and inserting it into the equation for the imaginary part leads after some algebraic manipulation to a general dependence of K on the imaginary part q of Λ:

$$K(q) = \frac{(1 + R)\left[\lambda^2 + (\omega - q)^2\right]}{2\lambda}. \quad (4.33)$$

Taking into account the multivalued properties of the arcsine function, this yields in turn analytical expressions of the time delay in dependence on q:

$$\tau_1(q) = \frac{\arcsin\left(\dfrac{2\lambda(1-R^2)(\omega-q)}{\lambda^2(1-R^2)^2 + (\omega-q)^2(1+R)^2}\right) + 2n\pi}{q}, \tag{4.34}$$

$$\tau_2(q) = \frac{-\arcsin\left(\dfrac{2\lambda(1-R^2)(\omega-q)}{\lambda^2(1-R^2)^2 + (\omega-q)^2(1+R)^2}\right) + (2n+1)\pi}{q},$$

where n is a nonnegative integer. Together with (4.33), these formulas describe the boundary of the domain of control in Fig. 4.6. Note that two expressions τ_1 and τ_2 are necessary to capture the complete boundary.

For a better understanding of effects due to the memory parameter R, it is instructive to consider the domain of control in the plane parametrized by R and the feedback gain K. The results can be seen in Fig. 4.7, where the black, medium gray, dark gray, and light gray areas (blue, green, red, and yellow) correspond to the domain of control for $\lambda T_0 = 0.2, 1, 5$, and 10, respectively. The other system parameter is chosen as $\omega = \pi$. We keep the time delay constant at $\tau = T_0/2$. Note that the K interval for successful control increases for larger values of R. In fact, while the original Pyragas scheme, i.e., $R = 0$, fails for $\lambda T_0 = 10$, the ETDAS method is still able to stabilize the fixed point. The upper left boundary corresponds to (4.32). The lower right boundary can be described by a parametric representation which can be derived from the characteristic equation (4.27):

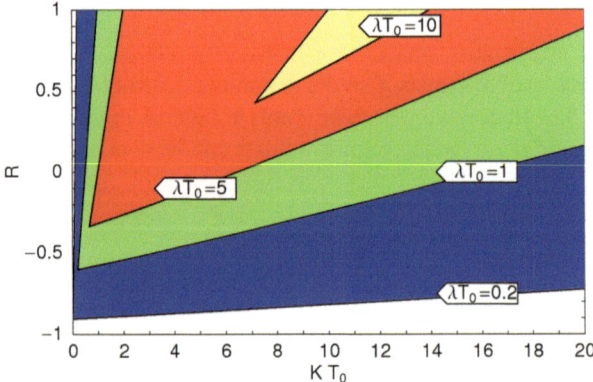

Fig. 4.7 Domain of control in the (K, R) plane for different values of λ. The *black, medium gray, dark gray,* and *light gray* domains (*blue, green, red,* and *yellow*) correspond to $\lambda T_0 = 0.2, 1, 5$, and 10, respectively, as indicated. The time delay is chosen as $\tau = T_0/2$ and $\omega = \pi$ [50]

$$R = \frac{\lambda\tau - \vartheta \tan(\vartheta/2)}{\lambda\tau + \vartheta \tan(\vartheta/2)}, \tag{4.35}$$

$$K\tau = \frac{\vartheta^2 + (\lambda\tau)^2}{\lambda\tau + \vartheta \tan(\vartheta/2)}, \tag{4.36}$$

where we used the abbreviation $\vartheta = (q - \omega)\tau$ for notational convenience. The range of ϑ is given by $\vartheta \in [0, \pi)$. A linear approximation leads to an analytic dependence of R and the feedback gain K given by a function $R(K)$ instead of the parametric equations (4.35) and (4.36). A Taylor expansion around $\vartheta = \pi$ yields

$$K_{\max}(R) = \frac{\lambda^2 + \pi^2}{2\lambda}(R + 1) + 2(R - 1). \tag{4.37}$$

Another representation of the superior control ability of ETDAS is depicted in Fig. 4.8. The domain of control is given in the (K, λ) plan for different values of R. The light gray, dark gray, medium gray, and black areas (yellow, red, green, and blue) refer to $R = -0.35$, 0 (TDAS), 0.35, and 0.7, respectively. The time delay is chosen as $\tau = T_0/2$. One can see that for increasing R, the ETDAS method can stabilize systems in a larger λ range. However, the corresponding K interval for successful control can become small. See, for instance, the black (blue) area ($R = 0.7$) for large λ. A similar behavior was found in the case of stabilization of an unstable periodic orbit by ETDAS [112]. We stress that, as in the case of periodic orbits, the boundaries of the shaded areas can be calculated analytically from the following expression:

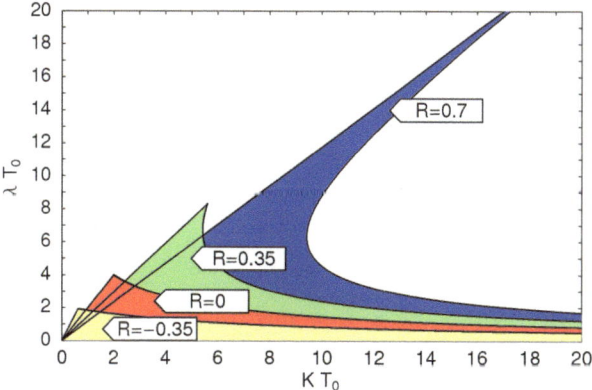

Fig. 4.8 Domain of control in the (K, λ) plane for different memory parameters R. The *light gray*, *dark gray*, *medium gray*, and *black domains* (*yellow*, *red*, *green*, and *blue*) areas correspond to $R = -0.35$, 0 (TDAS), 0.35, and 0.7, respectively. The time delay is fixed at $\tau = T_0/2$ [50]

$$K\tau = \frac{(1-R)\vartheta}{\tan(\vartheta/2)}\left[\left(\frac{1+R}{1-R}\right)^2 + \tan^2(\vartheta/2)\right],\qquad (4.38)$$

$$\lambda\tau = \frac{\vartheta}{\tan(\vartheta/2)}\left(\frac{1+R}{1-R}\right),\qquad (4.39)$$

where we used $\vartheta = (q - \omega)\,\tau$ with $\vartheta \in [0, \pi)$ as in (4.35) and (4.36). The maximum value for λ, which can be stabilized, is given by the special case $\vartheta = 0$:

$$\lambda_{\max}\tau = 2\,\frac{1+R}{1-R}.\qquad (4.40)$$

Extensions to include latency effects associated with the generation and injection of the feedback signal, low-pass and bandpass filtering in the control loop, and non-diagonal control schemes incorporating a feedback phase, have been discussed elsewhere [48, 50].

4.2.2 Asymptotic Properties

It is the purpose of this section to obtain deeper analytical insight into the time-delayed feedback control of steady states for large delay by relating asymptotic properties of the eigenvalue spectrum with the exact solutions and by discussing the shape of the control domain in the space of the control parameters [49].

Three different timescales are of importance in such a control problem: (i) the inverse divergence rate of trajectories around the unstable fixed point $1/\lambda$, (ii) the period of undamped oscillations around the fixed point $T_0 = 2\pi/\omega$, where ω is the oscillation frequency, and (iii) the delay time τ used in the feedback control loop. Here we consider the case $\tau \gg 1/\lambda$ and study again a generic model equation which describes an unstable focus above a Hopf bifurcation and is given by (4.6).

Note that, due to the presence of the delay, (4.8) possesses infinitely many solutions. Nevertheless, the stability of the fixed point is determined by a finite number of critical roots with largest real parts [110]. As a result, the stabilization problem consists in determining these critical eigenvalues and describing their behavior. In particular, successful control is achieved by providing conditions in terms of the control parameters K and τ for which all critical eigenvalues have negative real parts.

Figure 4.9 shows the real parts of the critical eigenvalues Λ as a function of τ for different values of K. The insets show the same eigenvalues as curves in the complex plane parametrized by τ. Note that the eigenvalue originating from the uncontrolled system (red) is the most unstable one for sufficiently small K and does not couple to the eigenvalues generated by the delay (see Fig. 4.9 a,b). The countable set of eigenvalues generated by the delay originates from Re $\Lambda = -\infty$ for $\tau \to 0$ and shows the typical nonmonotonic behavior that leads to stability islands for appropriate τ and K [48]. For larger values of K, the eigenvalue originating from

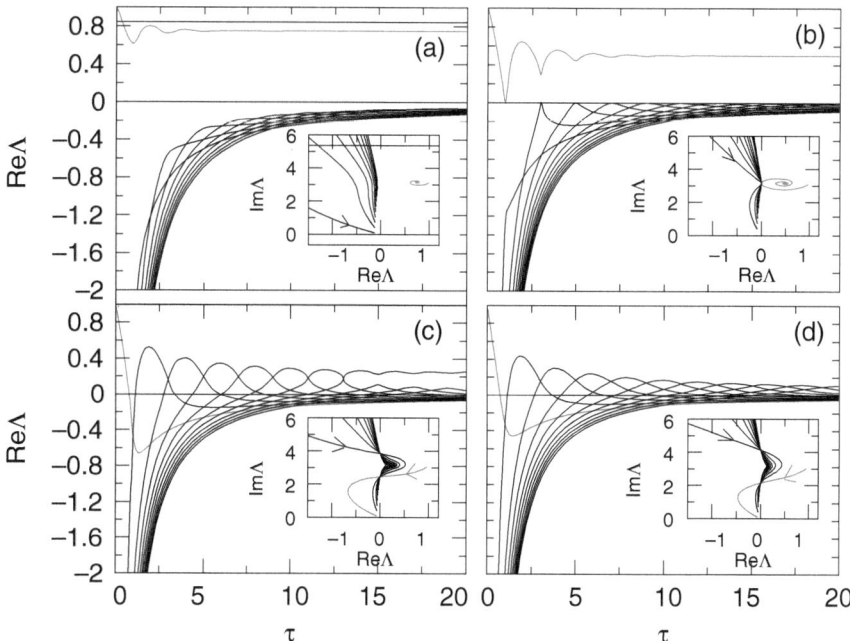

Fig. 4.9 Real parts of the complex eigenvalues Λ as a function of τ calculated from the characteristic equation (4.8) for 10 modes with the largest real parts. (**a**) $K = 0.25$, (**b**) $K = 0.5$, (**c**) $K = 0.75$, and (**d**) $K = 1.0$. Inset: eigenmodes Λ in the complex plane for $\tau \in [0, 20]$. *Red curves*: eigenvalue originating from the uncontrolled system; *black curves*: eigenmodes created by the delay control. Parameters: $\omega = \pi$ and $\lambda = 1$ [49]

the uncontrolled system is no longer separated from those which are generated by the delay (see Fig. 4.9 c,d). Moreover, one can observe a scaling behavior of the real parts of the eigenvalues for large τ in Fig. 4.9(a-c), there is a single eigenvalue retaining a positive real part, whereas all the other real parts tend to zero for large τ. The insets show that the eigenvalues in fact accumulate along the imaginary axis. This observation will be studied in detail in the following.

The scaling behavior of eigenvalues of general linear delay-differential equations for large delay τ has been analyzed in [113]. In particular, it turns out that one can distinguish the following.

(a) *Strongly unstable eigenvalues* Λ_s which have positive real parts that do not tend to zero with increasing τ, i.e., $\Lambda_s \to$ const and Re $\Lambda_s \geq \delta$ for some $\delta > 0$ as $\tau \to \infty$.

(b) *Pseudocontinuous spectrum* of eigenvalues Λ_p with real parts that scale as $1/\tau$, i.e., $\Lambda_p = \frac{1}{\tau}\gamma + i\left(\Omega + \frac{1}{\tau}\varphi\right) + \mathcal{O}\left(\frac{1}{\tau^2}\right)$ with some γ, Ω, and φ. A spectrum with this scaling behavior and positive real part leads to so-called *weak instabilities* (for more details, see [114, 113]).

In order to obtain the strongly unstable eigenvalues, we insert $\Lambda_s = $ const into (4.8) and assume $\tau \to \infty$. Since $\mathrm{Re}\,\Lambda_s > \delta$, the exponential term vanishes and we arrive at the expression for Λ_s:

$$\Lambda_s = \lambda - K \pm i\omega,$$

which holds for $\lambda - K > 0$. Thus we obtain the following statement:

(i) For $K < \lambda$, there exist two eigenvalues of the controlled stationary state, Λ_{s1} and its complex conjugate Λ_{s2}, such that $\Lambda_{s1} \to \lambda - K + i\omega$ as $\tau \to \infty$. The real parts of these eigenvalues are positive and, hence, the stationary state is *strongly unstable* (cf. Fig. 4.9(a-c)).

In order to obtain the asymptotic expression for the remaining pseudo-continuous part of the spectrum, we have to insert the scaling $\Lambda_p = \frac{1}{\tau}\gamma + i\left(\Omega + \frac{1}{\tau}\varphi\right)$ into (4.8). Up to the leading order we obtain the equation

$$i\Omega + K\left(1 - e^{-\gamma}e^{-i\varphi}\right) = \lambda \pm i\omega, \tag{4.41}$$

and the additional condition $\Omega = \Omega^{(m)} = 2\pi m/\tau$, $m = \pm 1, \pm 2, \pm 3, ...,$ (4.41) can be solved with respect to $\gamma(\Omega)$:

$$\gamma(\Omega) = -\frac{1}{2}\ln\left[\left(1 - \frac{\lambda}{K}\right)^2 + \left(\frac{\Omega \pm \omega}{K}\right)^2\right]. \tag{4.42}$$

The fact that $\mathrm{Re}\,\Lambda_p \approx \gamma(\Omega)/\tau$ and $\mathrm{Im}\,\Lambda_p \approx \Omega$ up to the leading order means that the eigenvalues Λ_p accumulate in the complex plane along curves $(\gamma(\Omega), \Omega)$, provided that the real axis is scaled as $\tau\,\mathrm{Re}\,\Lambda$. The actual positions of the eigenvalues on the curves can be obtained by evaluating Ω at points $\Omega^{(m)} = 2\pi m/\tau$. With increasing τ, the eigenvalues cover the curves densely [113]. Hence, we obtain the second statement:

(ii) The fixed point of system (4.6) has a set of eigenvalues which behave asymptotically as $\Lambda_p(\Omega^{(k)}) = \frac{1}{\tau}\gamma(\Omega^{(k)}) + i\left(\Omega^{(k)} + \frac{1}{\tau}\varphi(\Omega^{(k)})\right)$ with $\gamma(\Omega)$ given by (4.42). We have *weak instability* if the maximum of $\gamma(\Omega)$ is positive, i.e.,

$$\gamma_{\max} = \max_{\Omega}\gamma(\Omega) = -\ln\left|1 - \frac{\lambda}{K}\right| > 0,$$

which is the case for $K > \lambda/2$.

Figure 4.10 illustrates the spectrum of the fixed point of system (4.6) for $\tau = 20$. One can clearly distinguish the two types of eigenvalues. For $K < \lambda/2$ (Fig. 4.10a), the fixed point has a pair of strongly unstable eigenvalues, whereas the pseudocontinuous spectrum is stable. Note that the symbols (red) show the spectrum computed numerically from the full eigenvalue equation, whereas the dashed lines are the curves $(\gamma(\Omega), \Omega)$ from the asymptotic approximation where the pseudocontinuous

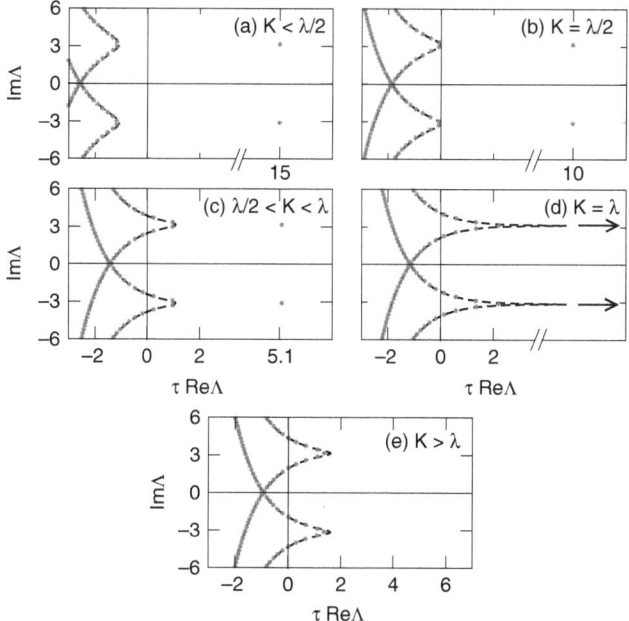

Fig. 4.10 Numerically computed spectrum of eigenvalues for $\tau = 20$ (*asterisks, red*). The *dashed lines* depict the asymptotic pseudocontinuous spectrum. (**a**) Strong instability for $K = 0.25$ ($K < \lambda/2$); (**b**) $K = 0.5 = \lambda/2$, critical case at which the weak instability occurs in addition to the strong one; (**c**) $K = 0.75$ ($\lambda/2 < K < \lambda$), strong and weak instability; (**d**) $K = 1.0 = \lambda$, critical case at which a strong instability disappears via the singularity of the pseudocontinuous spectrum; and (**e**) $K = 1.25$ ($K > \lambda$), weak instability. Parameters: $\omega = \pi$ and $\lambda = 1$ [49]

spectrum accumulates for large τ. At $K = \lambda/2$ (cf. Fig. 4.10b), the pseudocontinuous spectrum touches the imaginary axis resulting in the appearance of a weak instability for $K > \lambda/2$. This leads to the coexistence of strong and weak instabilities for $\lambda/2 < K < \lambda$ (Fig. 4.10c). At $K = \lambda$, the strongly unstable eigenvalues disappear, being absorbed by the pseudocontinuous spectrum, which develops a singularity at this moment, cf. Fig. 4.10(d). Finally, for $K > \lambda$ (Fig. 4.10e), there occurs only a weak instability induced by the pseudocontinuous spectrum.

After inspecting all possibilities given in Fig. 4.10, we conclude that stabilization by the feedback control scheme (4.6) always has an upper limit τ_c such that for $\tau > \tau_c$ it fails. Additionally, we note that for $K < \lambda$ and large delay, the stationary state is strongly unstable with the complex conjugate eigenvalues $\Lambda_{1,2} = \lambda - K \pm i\omega$, and for $K > \lambda$ weakly unstable with a large number of unstable eigenvalues given by (4.41), the real parts of which scale as $1/\tau$.

Next, we show that strongly delayed feedback can stabilize a fixed point in the case when the fixed point is sufficiently close to the Hopf bifurcation. In our case this means that λ is small. In particular, we are going to prove that the delayed feedback control scheme will be successful even for large delay within the range of order $1/\lambda^2$. We will also provide conditions for successful control.

For the fixed point, which is close to the Hopf bifurcation, we assume $K > \lambda$, and hence it has an unstable pseudocontinuous spectrum as shown in Fig. 4.10(e). As λ stays fixed, with increasing τ the curve of the pseudocontinuous spectrum will be densely filled with the eigenvalues ($\Omega^{(m)} = 2\pi m/\tau$). The only possibility for the fixed point to become stable is to assume that λ is also scaled with increasing τ. Particularly, we will show that in order to achieve control we have to scale it as $\lambda = \lambda_0 \varepsilon^2$ with fixed λ_0 (here for convenience we introduce the small parameter $\varepsilon = 1/\tau$).

Figure 4.11 illustrates the part of the curve $\gamma(\Omega)$ which may induce an instability in the system. More precisely, the interval of unstable frequencies is $\Omega_1 < \Omega < \Omega_2$, where Ω_1 and Ω_2 are given by the zeros of $\gamma(\Omega)$:

$$\Omega_{1,2} = \omega \pm K \sqrt{1 - \left(1 - \frac{\lambda}{K}\right)^2}.$$

For small λ we can approximate this as

$$\Omega_{1,2} = \omega \pm \sqrt{2\lambda K}. \tag{4.43}$$

The length of the interval of unstable frequencies is $\Delta\Omega = \Omega_2 - \Omega_1 = 2\sqrt{2\lambda K}$.

We note that the actual position of the eigenvalues on the curve corresponds to the values of $\Omega^{(m)} = 2\pi m\varepsilon$ with any integer m. It is easy to see that the distance between the frequencies of neighboring eigenvalues $\Omega^{(m+1)} - \Omega^{(m)} = 2\pi\varepsilon$ scales as ε. Therefore, the control can be successful if $\lambda = \lambda_0 \varepsilon^2$. In this case the length of the unstable interval is $\Delta\Omega = 2\varepsilon\sqrt{2\lambda_0 K}$ and scales also as ε. The control can be achieved if the length is smaller than the distance between neighboring eigenvalues,

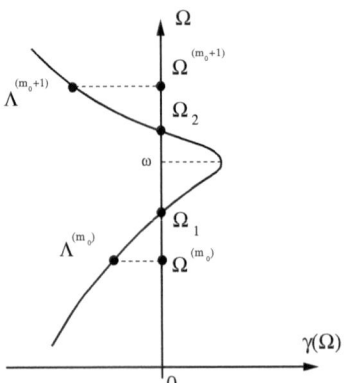

Fig. 4.11 Curve of the pseudocontinuous spectrum. The actual position of the complex eigenvalues $\Lambda = \frac{1}{\tau}\gamma + i\left(\Omega + \mathcal{O}(\frac{1}{\tau})\right)$ on the curve corresponds to $\Omega^{(m)} = 2\pi m\varepsilon$, $m = \pm 1, \pm 2, \pm 3, ...$, and $\varepsilon = 1/\tau$. The fixed point is stable if the imaginary parts of the eigenvalues are outside of the interval $\Omega_1 < \Omega < \Omega_2$. Such a case with $\Omega^{(m_0)} < \Omega_1 < \Omega_2 < \Omega^{(m_0+1)}$ is illustrated, in which the leading eigenvalues $\Lambda^{(m_0)}$ and $\Lambda^{(m_0+1)}$ have negative real parts [49]

i.e., $\Delta\Omega = 2\varepsilon\sqrt{2\lambda_0 K} < 2\pi\varepsilon$, leading to

$$K < \frac{\pi^2}{2\lambda_0}. \qquad (4.44)$$

Equation (4.44) gives a necessary condition for successful control.

The relative phase of the delay plays an additional important role. Depending on this phase, control occurs periodically with τ. In order to quantify this effect, let us introduce $\omega_\tau = 2\pi/\tau$ to be the frequency associated with the delay. Then the ratio of the internal frequency ω and ω_τ is given by $\omega/\omega_\tau = \gamma_\tau$ mod 1. Here $0 < \gamma_\tau < 1$ measures the detuning from the resonance between the internal frequency and the delay-induced one. Using this notation and (4.43), we can rewrite

$$\Omega_{1,2} = m_0\omega_\tau + \gamma_\tau\omega_\tau \pm \varepsilon\sqrt{2\lambda_0 K} = \Omega^{(m_0)} + \varepsilon\left(2\pi\gamma_\tau \pm \sqrt{2\lambda_0 K}\right).$$

Here m_0 is some integer number. The necessary and sufficient condition for the stability is (cf. Fig. 4.11) $\Omega^{(m_0)} < \Omega_1 < \Omega_2 < \Omega^{(m_0+1)}$, which leads to

$$\sqrt{2\lambda_0 K} < 2\pi \min\{\gamma_\tau, 1 - \gamma_\tau\}$$

or

$$K < \frac{2\pi^2}{\lambda_0}(\min\{\gamma_\tau, 1 - \gamma_\tau\})^2 = \frac{2\pi^2}{\lambda_0}\left(\min\left\{\left[\frac{\omega\tau}{2\pi}\right]_f, 1 - \left[\frac{\omega\tau}{2\pi}\right]_f\right\}\right)^2, \qquad (4.45)$$

where $\left[\frac{\omega\tau}{2\pi}\right]_f$ is the fractional part of $\frac{\omega\tau}{2\pi}$. Practically, one has also to satisfy $K > \lambda$, but our scaling assumes the smallness of λ. Figure 4.12 shows the domain of control given by (4.45) for $\lambda = \lambda_0/\tau^2$.

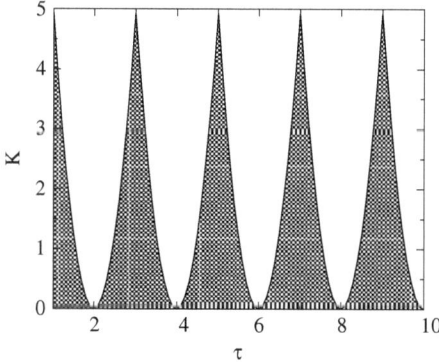

Fig. 4.12 *Shaded region*: domain of control in the (τ, K) plane for the fixed point close to the Hopf bifurcation, given by the asymptotic formula (4.45) for $\lambda = \lambda_0/\tau^2$. Parameters: $\omega = \pi$ and $\lambda_0 = 1$ [49]

In order to return to unscaled parameters, we have to substitute $\lambda_0 = \lambda/\varepsilon^2 = \lambda\tau^2$. Figure 4.13(a) shows the obtained domain of control for fixed small $\lambda = 0.01$. The maximum allowed values of K decrease as $1/\tau^2$. More precisely, we have

$$K_{\max}(\tau) = \frac{\pi^2}{2\lambda\tau^2}. \tag{4.46}$$

The application of the asymptotic analysis allows to reveal many essential features and mechanisms of the stabilization control scheme (4.6) for large delay τ. On the other hand, the obtained approximations are valid as soon as K is much larger than λ. Figure 4.13 shows a comparison of the boundaries of the control domain, which are given by the asymptotic methods and exact analytical formulas derived in the previous section. Very close to the Hopf bifurcation ($\lambda = 0.01$) the agreement is excellent even at small values of τ (Fig. 4.13a), while for larger λ (Fig. 4.13b) the deviations become more visible. In addition, the approximate solution does not give the lower boundary of the control domain for small K which only shows up in Fig. 4.14. The analytical approach also allows us to identify the "peaks" of the control domains, which occur at $\tau_{\max} = (2n + 1)\pi/\omega$, $n = 0, 1, 2, ...$, as double Hopf bifurcation points. The critical time delay, above which control fails, is given by $\tau_c = 2/\lambda$.

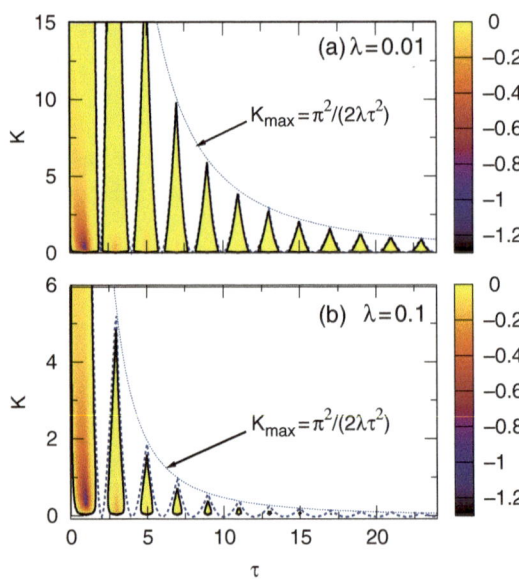

Fig. 4.13 Domain of control in the (τ, K) plane, and largest negative real part of the complex eigenvalues $\Lambda(K, \tau)$ (in *color code*) calculated from the characteristic equation using the Lambert function [(4.9)]. *Dashed lines* (*blue*): asymptotic approximation (4.45) of stability boundary; *dotted lines* (*blue*): approximate maxima (4.46); and *solid lines*: exact stability boundaries. Parameters: (**a**) $\omega = \pi$, $\lambda = 0.01$ and (**b**) $\omega = \pi$, $\lambda = 0.1$ [49]

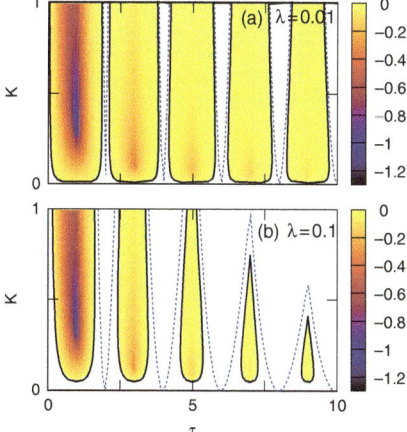

Fig. 4.14 Enlargement of Fig. 4.13: deviation of the asymptotic results (*dashed*) from the exact stability boundary (*solid*) for small K or large λ [49]

An inspection of the islands of stabilization in Figs. 4.13 and 4.14 reveals that the absolute value of the real part of the critical eigenvalue, i.e., the eigenvalue which has the largest real part (but remains negative within those islands), decreases with increasing τ. Hence, the fixed point becomes less stable, and it is expected that the system becomes more sensitive to noise and it will be more difficult to realize stabilization experimentally, if the delay time is chosen several times the system's characteristic time T_0.

4.2.3 Beyond the Odd Number Limitation of Unstable Periodic Orbits

In this section we consider the stabilization of periodic orbits by time-delayed feedback control [52]. Although time-delayed feedback control has been widely used with great success in real-world problems in physics, chemistry, biology, and medicine, e.g., [38, 64, 71–73, 115–122], severe limitations are imposed by the common belief that certain orbits cannot be stabilized for any strength of the control force. In fact, it has been contended that periodic orbits with an odd number of real Floquet multipliers greater than unity cannot be stabilized by the Pyragas method [43, 44, 123–126], even if the simple scheme is extended by multiple delays in form of an infinite series [39]. To circumvent this restriction more complicated control schemes, like an oscillating feedback [127], half-period delays for special, symmetric orbits [128], or the introduction of an additional, unstable degree of freedom [126, 129], have been proposed. Here, we show that the general limitation for orbits with an odd number of real unstable Floquet multipliers greater than unity does not hold: stabilization may be possible for suitable choices of the feedback matrix [52, 53]. Our example consists of an unstable periodic orbit generated by a subcritical Hopf bifurcation. In particular, this refutes the theorem in [44].

Consider the normal form of a subcritical Hopf bifurcation extended by a time-delayed feedback term:

$$\dot{z}(t) = \left[\lambda + i + (1 + i\gamma)|z(t)|^2\right]z(t) + b[z(t - \tau) - z(t)], \qquad (4.47)$$

with $z \in \mathbb{C}$ and real parameters λ and γ. Here the Hopf frequency is normalized to unity. The feedback matrix is represented by multiplication with a complex number $b = b_R + ib_I = b_0 e^{i\beta}$ with real b_R, b_I, β and positive b_0. Note that the nonlinearity $f(\lambda, z(t)) = \left[\lambda + i + (1 + i\gamma)|z(t)|^2\right]z(t)$ commutes with complex rotations. Therefore $\exp(i\vartheta)z(t)$ solves (4.47), for any fixed ϑ, whenever $z(t)$ does. In particular, nonresonant Hopf bifurcations from the trivial solution $z \equiv 0$ at simple imaginary eigenvalues $\eta = i\omega \neq 0$ produce rotating wave solutions $z(t) = z(0)\exp\left(i\frac{2\pi}{T}t\right)$ with period $T = 2\pi/\omega$ even in the nonlinear case and with delay terms. This follows from uniqueness of the emanating Hopf branches.

Transforming Eq. (4.47) to amplitude and phase variables r, θ using $z(t) = r(t)e^{i\theta(t)}$, we obtain at $b = 0$

$$\dot{r}(t) = \left(\lambda + r^2\right)r \qquad (4.48)$$

$$\dot{\theta}(t) = 1 + \gamma r^2. \qquad (4.49)$$

An unstable periodic orbit with $r = \sqrt{-\lambda}$ and period $T = 2\pi/(1 - \gamma\lambda)$ exists for $\lambda < 0$. This is the orbit we will stabilize. We will call it the Pyragas orbit. At $\lambda = 0$ a subcritical Hopf bifurcation occurs. The Pyragas control method chooses the delay time τ as $\tau_P = nT$. This eliminates the feedback term on the orbit, and thus recovers the original T-periodic solution $z(t)$. In this sense the control method is *non-invasive*.

The choice $\tau_P = nT$ defines the local *Pyragas curve* in the (λ, τ) plane for any $n \in \mathbb{N}$

$$\tau_P(\lambda) = \frac{2\pi n}{1 - \gamma\lambda} = 2\pi n(1 + \gamma\lambda + \dots), \qquad (4.50)$$

which emanates from the Hopf bifurcation points $\lambda = 0$, $\tau = 2\pi n$.

Under further nondegeneracy conditions, the Hopf point $\lambda = 0$, $\tau = nT$ ($n \in \mathbb{N}_0$) continues to a Hopf bifurcation curve $\tau_H(\lambda)$ for $\lambda < 0$. We determine this *Hopf curve* next. It is characterized by purely imaginary eigenvalues $\eta = i\omega$ of the transcendental characteristic equation:

$$\eta = \lambda + i + b\left(e^{-\eta\tau} - 1\right), \qquad (4.51)$$

which results from the linearization at the steady state $z = 0$ of the delayed system
(4.47). Separating (4.51) into real and imaginary parts

$$0 = \lambda + b_0[\cos(\beta - \omega\tau) - \cos\beta] \tag{4.52}$$

$$\omega - 1 = b_0[\sin(\beta - \omega\tau) - \sin\beta] \tag{4.53}$$

and using trigonometric identities to eliminate $\omega(\lambda)$ yields an explicit expression for
the multivalued Hopf curve $\tau_H(\lambda)$ for given control amplitude b_0 and phase β:

$$\tau_H = \frac{\pm \arccos\left(\frac{b_0\cos\beta - \lambda}{b_0}\right) + \beta + 2\pi n}{1 - b_0\sin\beta \pm \sqrt{\lambda(2b_0\cos\beta - \lambda) + b_0^2\sin^2\beta}}. \tag{4.54}$$

Note that τ_H is not defined in the case of $\beta = 0$ and $\lambda < 0$. Thus complex b is
a necessary condition for the existence of the Hopf curve in the subcritical regime
$\lambda < 0$. Figure 4.15 displays the family of Hopf curves (4.54), and the Pyragas curve
(4.50) $n = 1$, in the (λ, τ) plane. In Fig. 4.15(b) the domains of instability of the
trivial steady state $z = 0$, bounded by the Hopf curves, are marked by light gray
shading (yellow). The dimensions of the unstable manifold of $z = 0$ are given in
parentheses along the τ-axis in Fig. 4.15(b). By construction, the delay τ becomes
a multiple of the minimum period T of the bifurcating periodic orbits along the
Pyragas curve $\tau = \tau_p(\lambda) = nT$ and the time-delayed feedback term vanishes if the
periodic orbit is stabilized. The inset of Fig. 4.16 displays the Hopf and Pyragas
curves for different values of the feedback b_0. These choices of b_0 are displayed as
full circles in the main figure. For $b_0 > b_0^{crit}$ (a) the Pyragas curve runs partly inside
the Hopf curve. With decreasing magnitude of b_0 the Hopf curves pull back until the

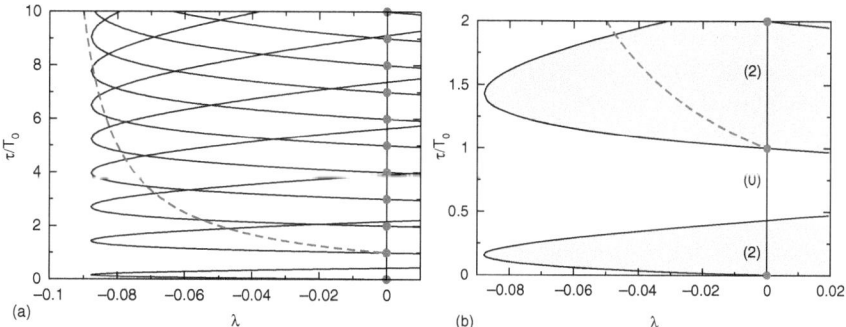

Fig. 4.15 Pyragas (*dashed*) and Hopf (*solid*) curves in the (λ, τ) plane: (**a**) Hopf bifurcation curves
$n = 0, ..., 10$ and (**b**) Hopf bifurcation curves $n = 0, 1$ in an enlarged scale. *Light gray shading
marks* the domains of unstable $z = 0$ and numbers in parentheses denote the dimension of the
unstable manifold of $z = 0$ ($\gamma = -10$, $b_0 = 0.3$, and $\beta = \pi/4$). The time delay is given in units
of the intrinsic timescale T_0 of the trivial fixed point, i.e., $T_0 = 2\pi$ [52]

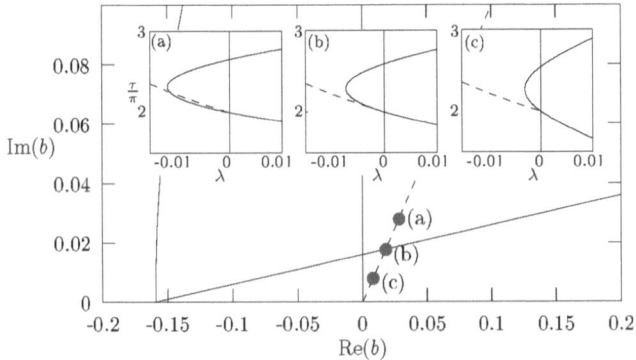

Fig. 4.16 Change of Hopf curves with varying control amplitude b_0. The main figures shows the complex plane of control gain b. The three values marked by *full circles* correspond to the insets (a), (b), and (c), where the Hopf (*solid*) and Pyragas (*dashed*) curves are displayed for $\beta = \frac{\pi}{4}$ and three different choices of b_0: (a) $b_0 = 0.04 > b_0^{\text{crit}}$, (b)$b_0 = 0.025 \approx b_0^{\text{crit}}$, and (c) $b_0 = 0.01 < b_0^{\text{crit}}$ ($\lambda = -0.005$, $\gamma = -10$) [130]

Pyragas curves lie outside (c). At the critical feedback value (b) Pyragas and Hopf curves are tangent at ($\lambda = 0$, $\tau = 2\pi$).

Standard exchange of stability results [131], which hold verbatim for delay equations, then assert that the bifurcating branch of periodic solutions locally inherits linear asymptotic (in)stability from the trivial steady state, i.e., it consists of stable periodic orbits on the Pyragas curve $\tau_P(\lambda)$ inside the shaded domains for small $|\lambda|$. We stress that an unstable trivial steady state is not a sufficient condition for stabilization of the Pyragas orbit. In fact, the stabilized Pyragas orbit can become unstable again if $\lambda < 0$ is further decreased, for instance, in a torus bifurcation. However, there exists an interval for values of λ in our example for which the exchange of stability holds. More precisely, for small $|\lambda|$ unstable periodic orbits possess a single Floquet multiplier $\mu = \exp(\Lambda \tau)$ (with $1 < \mu < \infty$), near unity, which is simple. All other nontrivial Floquet multipliers lie strictly inside the complex unit circle. In particular, the (strong) unstable dimension of these periodic orbits is odd, here 1, and their unstable manifold is two-dimensional. This is shown in Fig. 4.17 panel (a) top, which depicts solutions Λ of the characteristic equation of the periodic solution on the Pyragas curve.

The Floquet exponents of the Pyragas orbit can be calculated explicitly by rewriting (4.47) in polar coordinates $z = r\, e^{i\theta}$

$$\dot{r} = (\lambda + r^2)\, r + b_0[\cos(\beta + \theta(t - \tau) - \theta)\, r(t - \tau) - \cos(\beta)\, r] \quad (4.55)$$

$$\dot{\theta} = 1 + \gamma r^2 + b_0[\sin(\beta + \theta(t - \tau) - \theta)\, \frac{r(t - \tau)}{r} - \sin(\beta)] \quad (4.56)$$

and linearizing around the periodic orbit according to $r(t) = r_0 + \delta r(t)$ and $\theta(t) = \Omega t + \delta\theta(t)$, with $r_0 = \sqrt{-\lambda}$ and $\Omega = 1 - \gamma\lambda$ (see (4.48)). This yields

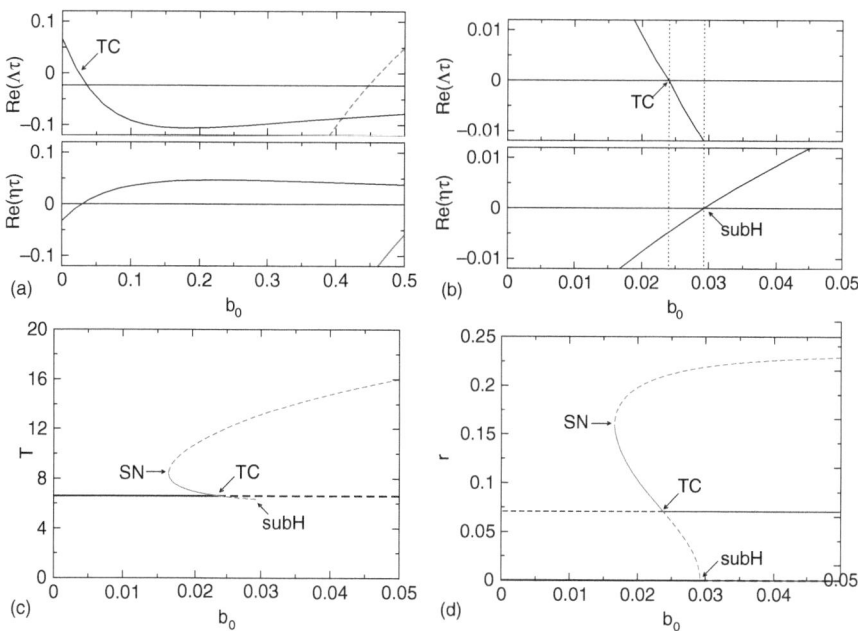

Fig. 4.17 (**a**) *Top*: real part of Floquet exponents Λ of the periodic orbit vs. feedback amplitude b_0. *Bottom*: real part of eigenvalue η of steady state vs. feedback amplitude b_0; (**b**) blowup of (**a**); (**c**) periods; and (**d**) radii of the periodic orbits vs. b_0. The *solid* and *dashed curves* correspond to stable and unstable periodic orbits, respectively. Parameters in all panels: $\lambda = -0.005$, $\gamma = -10$, $\tau = \frac{2\pi}{1-\gamma\lambda}$, and $\beta = \pi/4$ [130]

$$\begin{pmatrix} \delta\dot{r}(t) \\ \delta\dot{\theta}(t) \end{pmatrix} = \begin{bmatrix} -2\lambda - b_0\cos\beta & b_0 r_0 \sin\beta \\ 2\gamma r_0 - b_0 \sin\beta\, r_0^{-1} & -b_0\cos\beta \end{bmatrix} \begin{pmatrix} \delta r(t) \\ \delta\theta(t) \end{pmatrix} \tag{4.57}$$

$$+ \begin{bmatrix} b_0\cos\beta & -b_0 r_0 \sin\beta \\ b_0 \sin\beta\, r_0^{-1} & b_0\cos\beta \end{bmatrix} \begin{pmatrix} \delta r(t-\tau) \\ \delta\theta(t-\tau) \end{pmatrix}. \tag{4.58}$$

With the ansatz

$$\begin{pmatrix} \delta r(t) \\ \delta\theta(t) \end{pmatrix} = u\,\exp(\Lambda t), \tag{4.59}$$

where u is a two-dimensional vector, one obtains the autonomous linear equation

$$\begin{bmatrix} -2\lambda + b_0\cos\beta\,(e^{-\Lambda\tau}-1) - \Lambda & -b_0 r_0 \sin\beta\,(e^{-\Lambda\tau}-1) \\ 2\gamma r_0 + b_0 r_0^{-1}\sin\beta\,(e^{-\Lambda t}-1) & b_0\cos\beta\,(e^{-\Lambda\tau}-1) - \Lambda \end{bmatrix} u = 0. \tag{4.60}$$

The condition of vanishing determinant then gives the transcendental characteristic equation:

$$0 = \left(-2\lambda + b_0 \cos \beta \left(e^{-\Lambda\tau} - 1\right) - \Lambda\right) \left(b_0 \cos \beta \left(e^{-\Lambda\tau} - 1\right) - \Lambda\right) \quad (4.61)$$

$$-b_0 r_0 \sin \beta \left(e^{-\Lambda\tau} - 1\right) \left(2\gamma r_0 + b_0 r_0^{-1} \sin \beta \left(e^{-\Lambda\tau} - 1\right)\right) \quad (4.62)$$

for the Floquet exponents Λ which can be solved numerically.

The largest real part is positive for $b_0 = 0$. Thus the periodic orbit is unstable. As the amplitude of the feedback gain increases, the largest real part of the eigenvalue becomes smaller and eventually changes sign at the point TC (transcritical bifurcation) in Fig. 4.17. Hence the periodic orbit is stabilized. Note that an infinite number of Floquet exponents are created by the control scheme; their real parts tend to $-\infty$ in the limit $b_0 \to 0$, and some of them may cross over to positive real parts for larger b_0 (dashed line in Fig. 4.17(a)), terminating the stability of the periodic orbit.

Panel (a) bottom illustrates the stability of the steady state by displaying the largest real part of the eigenvalues η. The interesting region of the top and bottom panels where the periodic orbit becomes stable and the fixed point loses stability is magnified in panel (b).

Figure 4.18 shows the behavior of the Floquet multipliers $\mu = \exp(\Lambda\tau)$ of the Pyragas orbit in the complex plane with the increasing amplitude of the feedback gain b_0 as a parameter (marked by arrows). There is an isolated real multiplier crossing the unit circle at $\mu = 1$. This is caused by a transcritical bifurcation in which the Pyragas orbit collides with a delay-induced stable periodic orbit. In panels (c) and (d) of Fig. 4.17 the periods and radii of all circular periodic orbits ($r = $ const) are plotted vs. the feedback strength b_0. For small b_0 only the initial (unstable) Pyragas orbit (T and r independent of b_0) and the steady state $r = 0$ (stable) exist. With increasing b_0 a pair of unstable/stable periodic orbits is created in a saddle-node (SN) bifurcation. The stable one of the two orbits (solid) then exchanges stability with the Pyragas orbit in a transcritical bifurcation (TC), and finally ends in a

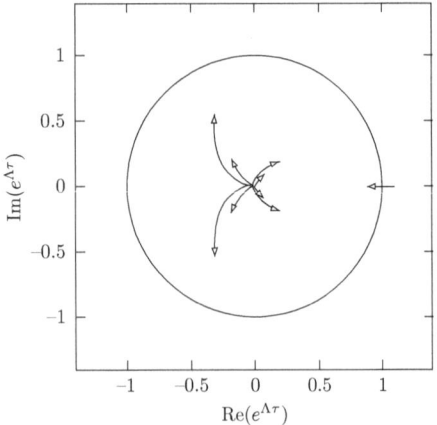

Fig. 4.18 Floquet multipliers $\mu = \exp(\Lambda\tau)$ in the complex plane with the feedback amplitude $b_0 \in [0, 0.3]$. *Arrows* indicate the direction of increasing b_0. Same parameters as in Fig. 4.17 [130]

subcritical Hopf bifurcation (subH), where the steady state $r = 0$ becomes unstable. The Pyragas orbit continues as a stable periodic orbit for larger b_0. Except at TC, the delay-induced orbit has a period $T \neq \tau$ (see Fig. 4.17c). Note that the respective exchanges of stability of the Pyragas orbit (TC) and the steady state (subH) occur at slightly different values of b_0. This is also corroborated by Fig. 4.17(b). The mechanism of stabilization of the Pyragas orbit by a transcritical bifurcation relies upon the possible existence of such delay-induced periodic orbits with $T \neq \tau$, which was overlooked in previous works. Technically, the proof of the odd number limitation theorem in [44] fails because the trivial Floquet multiplier $\mu = 1$ (Goldstone mode of periodic orbit) was neglected there; $F(1)$ in (14) in [44] is thus zero and not less than zero, as assumed. At TC, where a second Floquet multiplier crosses the unit circle, this results in a Floquet multiplier $\mu = 1$ of algebraic multiplicity two.

Next we analyze the conditions under which stabilization of the subcritical periodic orbit is possible. From Fig. 4.15(b) it is evident that the Pyragas curve must lie inside the yellow region, i.e., the Pyragas and Hopf curves emanating from the point $(\lambda, \tau) = (0, 2\pi)$ must locally satisfy the inequality $\tau_H(\lambda) < \tau_P(\lambda)$ for $\lambda < 0$. More generally, let us investigate the eigenvalue crossings of the Hopf eigenvalues $\eta = i\omega$ along the τ-axis of Fig. 4.15. In particular, we derive conditions for the unstable dimensions of the trivial steady state near the Hopf bifurcation point $\lambda = 0$ in our model (4.47). On the τ-axis ($\lambda = 0$), the characteristic equation (4.51) for $\eta = i\omega$ is reduced to

$$\eta = i + b \left(e^{-\eta\tau} - 1 \right), \tag{4.63}$$

and we obtain two series of Hopf points given by

$$0 \leq \tau_n^A = 2\pi, n \tag{4.64}$$

$$0 < \tau_n^B = \frac{2\beta + 2\pi n}{1 - 2b_0 \sin \beta} \quad (n = 0, 1, 2, \dots). \tag{4.65}$$

The corresponding Hopf frequencies are $\omega^A = 1$ and $\omega^B = 1 - 2b_0 \sin \beta$, respectively. Note that series A consists of all Pyragas points, since $\tau_n^A = nT = 2\pi n/\omega^A$. In the series B the integers n have to be chosen such that the delay $\tau_n^B \geq 0$. The case $b_0 \sin \beta = 1/2$, only, corresponds to $\omega^B = 0$ and does not occur for finite delays τ.

We evaluate the crossing directions of the critical Hopf eigenvalues next, along the positive τ-axis and for both series. Abbreviating $\frac{\partial}{\partial \tau}\eta$ by η_τ the crossing direction is given by sign(Re η_τ). Implicit differentiation of (4.63) with respect to τ at $\eta = i\omega$ implies

$$\text{sign(Re } \eta_\tau) = -\text{sign}(\omega)\,\text{sign}(\sin(\omega\tau - \beta)). \tag{4.66}$$

We are interested specifically in the Pyragas–Hopf points of series A (marked by dots in Fig. 4.15), where $\tau = \tau_n^A = 2\pi n$ and $\omega = \omega^A = 1$. Indeed sign(Re η_τ) = sign(sin β) > 0 holds, provided we assume $0 < \beta < \pi$, i.e., $b_I > 0$ for the feedback gain. This condition alone, however, is not sufficient to guarantee stability of the steady state for $\tau < 2n\pi$. We also have to consider the crossing direction sign(Re η_τ) along series B, $\omega^B = 1 - 2b_0 \sin \beta$, $\omega^B \tau_n^B = 2\beta + 2\pi n$, for $0 < \beta < \pi$. Equation (4.66) now implies sign(Re η_τ) = sign(($2b_0 \sin \beta - 1) \sin \beta$) = sign($2b_0 \sin \beta - 1$).

To compensate for the destabilization of $z = 0$ upon each crossing of any point $\tau_n^A = 2\pi n$, we must require stabilization (sign(Re η_τ) < 0) at each point τ_n^B of series B. If $b_0 \geq 1/2$, this requires $0 < \beta < \arcsin(1/(2b_0))$ or $\pi - \arcsin(1/(2b_0)) < \beta < \pi$. The distance between two successive points τ_n^B and τ_{n+1}^B is $2\pi/\omega^B > 2\pi$. Therefore, there is at most one τ_n^B between any two successive Hopf points of series A. Stabilization requires exactly one such τ_n^B, specifically: $\tau_{k-1}^A < \tau_{k-1}^B < \tau_k^A$ for all $k = 1, 2, \ldots, n$. This condition is satisfied if, and only if,

$$0 < \beta < \beta_n^*, \tag{4.67}$$

where $0 < \beta_n^* < \pi$ is the unique solution of the transcendental equation:

$$\frac{1}{\pi}\beta_n^* + 2nb_0 \sin \beta_n^* = 1. \tag{4.68}$$

This holds because the condition $\tau_{k-1}^A < \tau_{k-1}^B < \tau_k^A$ first fails when $\tau_{k-1}^B = \tau_k^A$. Equation (4.67) represents a necessary but not yet sufficient condition that the Pyragas choice $\tau_P = nT$ for the delay time will stabilize the periodic orbit.

To evaluate the remaining condition, $\tau_H < \tau_P$ near $(\lambda, \tau) = (0, 2\pi)$, we expand the exponential in the characteristic equation (4.51) for $\omega\tau \approx 2\pi n$, and obtain the approximate Hopf curve for small $|\lambda|$:

$$\tau_H(\lambda) \approx 2\pi n - \frac{1}{b_I}(2\pi n b_R + 1)\lambda. \tag{4.69}$$

Recalling (4.50), the Pyragas stabilization condition $\tau_H(\lambda) < \tau_P(\lambda)$ is therefore satisfied for $\lambda < 0$ if, and only if,

$$\frac{1}{b_I}\left(b_R + \frac{1}{2\pi n}\right) < -\gamma. \tag{4.70}$$

Equation (4.70) defines a domain in the plane of the complex feedback gain $b = b_R + ib_I = b_0 e^{i\beta}$ bounded from below (for $\gamma < 0 < b_I$) by the straight line

$$b_I = \frac{1}{-\gamma}\left(b_R + \frac{1}{2\pi n}\right). \tag{4.71}$$

Equation (4.68) represents a curve $b_0(\beta)$, i.e.,

$$b_0 = \frac{1}{2n \sin \beta} \left(1 - \frac{\beta}{\pi} \right), \tag{4.72}$$

which forms the upper boundary of a domain given by the inequality (4.67). Thus (4.71) and (4.72) describe the boundaries of the domain of control in the complex plane of the feedback gain b in the limit of small λ. Figure 4.19 depicts this domain of control for $n = 1$, i.e., a time delay $\tau = \frac{2\pi}{1-\gamma\lambda}$. The lower and upper solid curves correspond to (4.71) and (4.72), respectively. The grayscale displays the numerical result of the largest real part, wherever negative, of the Floquet exponent, calculated from linearization of the amplitude and phase equations around the periodic orbit. Outside the shaded areas the periodic orbit is not stabilized. With increasing $|\lambda|$ the domain of stabilization shrinks, as the deviations from the linear approximation (4.69) become larger. For sufficiently large $|\lambda|$ stabilization is no longer possible in agreement with Fig. 4.15(b). Note that for real values of b, i.e., $\beta = 0$, no stabilization occurs at all. Hence, stabilization fails if the feedback matrix B is a multiple of the identity matrix. Figure 4.20 compares the control domain for the same value of $|\lambda|$ for the representation in the planes of complex feedback b (left) and amplitude b_0 and phase β (right).

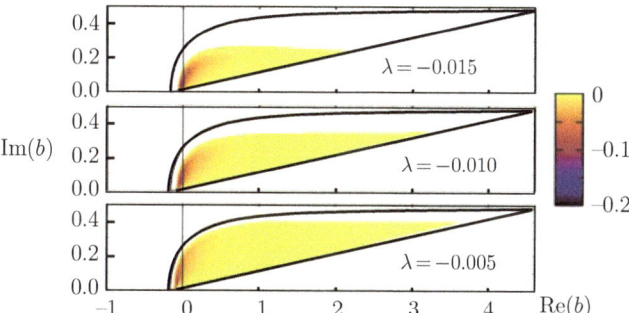

Fig. 4.19 Domain of control in the plane of the complex feedback gain $b = b_0 e^{i\beta}$ for three different values of the bifurcation parameter λ. The *solid curves* indicate the boundary of stability in the limit $\lambda \nearrow 0$, see (4.71) and (4.72). The *shading* shows the magnitude of the largest (negative) real part of the Floquet exponents of the periodic orbit ($\gamma = -10$ and $\tau = \frac{2\pi}{1-\gamma\lambda}$) [52]

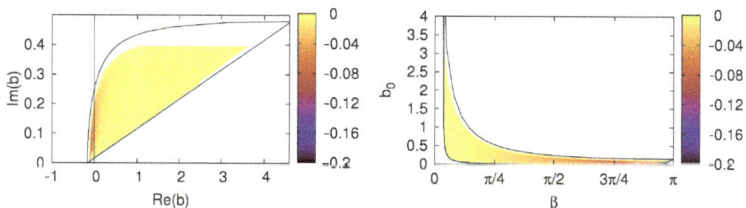

Fig. 4.20 Domain of control in the complex b plane (*left*) and the (β, b_0) plane (*right*) ($\lambda = -0.005$, $\gamma = -10$, and $\tau = \frac{2\pi}{1-\gamma\lambda}$) [130]

4.2.4 Stabilizing Periodic Orbits Near a Fold Bifurcation

Another important example for an unstable periodic orbit which has an odd number of real Floquet multipliers greater than unity is provided by an orbit generated by a fold bifurcation of limit cycles. As a paradigm for fold bifurcation of rotating waves we consider planar systems of the form

$$\dot{z} = g(\lambda, |z|^2)z + ih(\lambda, |z|^2)z. \tag{4.73}$$

Here $z(t)$ is a scalar complex variable, g and h are real valued functions, and λ is a real parameter. Systems of the form (4.73) are S^1 equivariant, i.e., $e^{i\theta}z(t)$ is a solution whenever $z(t)$ is for any fixed $e^{i\theta}$ in the unit circle S^1. In polar coordinates $z = re^{i\varphi}$, this manifests itself by the absence of φ from the right-hand sides of the resulting differential equations:

$$\begin{aligned} \dot{r} &= g(\lambda, r^2)r, \\ \dot{\varphi} &= h(\lambda, r^2). \end{aligned} \tag{4.74}$$

In particular, all periodic solutions of (4.73) are indeed rotating waves, alias harmonic, of the form

$$z(t) = re^{i\omega t}$$

for suitable nonzero real constants r, ω. Specifically, this requires $\dot{r} = 0$ and $\dot{\varphi} = \omega$:

$$\begin{aligned} 0 &= g(\lambda, r^2), \\ \omega &= h(\lambda, r^2). \end{aligned} \tag{4.75}$$

Fold bifurcations of rotating waves are generated by the nonlinearities

$$\begin{aligned} g(\lambda, r^2) &= \left(r^2 - 1\right)^2 - \lambda, \\ h(\lambda, r^2) &= \gamma(r^2 - 1) + \omega_0. \end{aligned} \tag{4.76}$$

Our choice of nonlinearities is generic in the sense that $g(\lambda, r^2)$ is the normal form for a nondegenerate fold bifurcation [132] at $r^2 = 1$ and $\lambda = 0$. See Fig. 4.21 for the resulting bifurcation diagram. We fix coefficients $\gamma, \omega_0 > 0$.

Using (4.75) and (4.76), the amplitude r and frequency ω of the rotating waves then satisfy

$$r^2 = 1 \pm \sqrt{\lambda}, \quad \omega = \omega_0 + \gamma(r^2 - 1) = \omega_0 \pm \gamma\sqrt{\lambda}. \tag{4.77}$$

The signs $\pm$ correspond to different branches in Fig. 4.21, $+$ unstable and $-$ stable.

Our goal is to investigate delay stabilization of the fold system (4.73) by the delayed feedback term

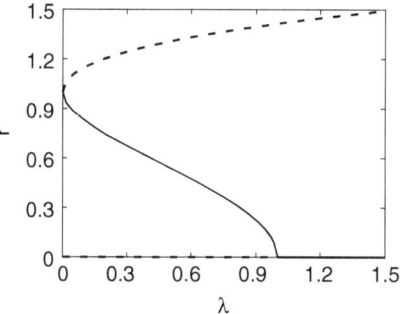

Fig. 4.21 Bifurcation diagram of rotating waves (*solid line*: stable; *dashed line*: unstable) of (4.73) and (4.76) [54]

$$\dot{z} = f(\lambda, |z|^2)z + b_0 e^{i\beta} \left[z(t - \tau) - z(t) \right], \tag{4.78}$$

with real positive control amplitude b_0, delay τ, and real control phase β. Here we have used the abbreviation $f = g + ih$. The Pyragas choice requires the delay τ to be an integer multiple k of the minimum period T of the periodic solution to be stabilized:

$$\tau = kT. \tag{4.79}$$

This choice guarantees that periodic orbits of the original system (4.73) with period T are reproduced exactly and non-invasively by the control system (4.78). The minimum period T of a rotating wave $z = re^{i\omega t}$ is given explicitly by $T = 2\pi/\omega$. Using (4.77), (4.79) becomes

$$\tau = \frac{2\pi k}{\omega_0 \pm \gamma\sqrt{\lambda}}, \tag{4.80}$$

or, equivalently,

$$\lambda = \lambda(\tau) = \left(\frac{2\pi k - \omega_0 \tau}{\gamma \tau} \right)^2. \tag{4.81}$$

In the following we select only the branch of $\lambda(\tau)$ corresponding to the τ value with the $+$ sign, which is associated with the unstable orbit. Condition (4.81) then determines the kth *Pyragas curve* in parameter space (τ, λ) where the delayed feedback is indeed non-invasive. The fold parameter $\lambda = 0$ corresponds to $\tau = 2\pi k/\omega_0$ along the kth Pyragas curve. See Fig. 4.22 for the Pyragas curves in the parameter plane (τ, λ).

For the delay stabilization system (4.78) we now consider τ as the relevant bifurcation parameter. We restrict our study of (4.78) to $\lambda = \lambda(\tau)$ given by the Pyragas

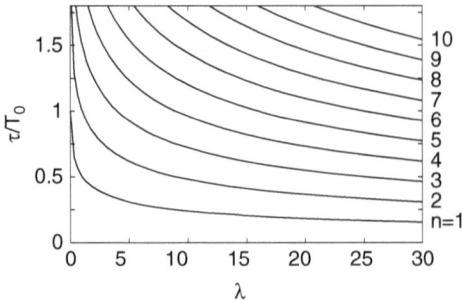

Fig. 4.22 The Pyragas curves $\lambda = \lambda(\tau)$, corresponding to the unstable branch in Fig. 4.21, in the parameter plane (τ, λ); see (4.81). Parameters: $\gamma = \omega_0 = 1$ [54]

curve (4.81), because $\tau = kT$ is the primary condition for non-invasive delayed feedback control.

We begin with the trivial case $b_0 = 0$ of vanishing control. For each $\lambda = \lambda(\tau)$, we encounter two rotating waves given by

$$r^2 = 1 \pm \frac{2\pi k - \omega_0 \tau}{\gamma \tau}, \quad \omega = \omega_0 \pm \left(\frac{2\pi k - \omega_0 \tau}{\tau} \right). \tag{4.82}$$

The two resulting branches form a transcritical bifurcation at $\tau = 2\pi k/\omega_0$. At this stage, the transcriticality looks like an artifact, spuriously caused by our choice of the Pyragas curve $\lambda = \lambda(\tau)$. Note, however, that only one of the two crossing branches features minimum period T such that the Pyragas condition $\tau = kT$ holds. This happens along the branch

$$r^2 = 1 + \frac{2\pi k - \omega_0 \tau}{\gamma \tau}, \quad \omega = 2\pi k/\tau,$$

see Fig. 4.23. We call this branch, which corresponds to '+' in (4.82) the *Pyragas branch*. The other branch has minimum period T with

$$kT = \frac{\pi k}{\omega_0 \tau - \pi k} \tau \neq \tau,$$

except at the crossing point $\omega_0 \tau = 2\pi k$. The minus branch therefore violates the Pyragas condition for non-invasive control, even though it has been generated from the same fold bifurcation.

Our strategy for Pyragas control of the unstable part of the Pyragas branch is now simple. For a nonzero control amplitude b_0, the Pyragas branch persists without change, due to the non-invasive property $\tau = kT$ along the Pyragas curve $\lambda = \lambda(\tau)$. The minus branch, however, will be perturbed slightly for small $b_0 \neq 0$. If the resulting perturbed transcritical bifurcation

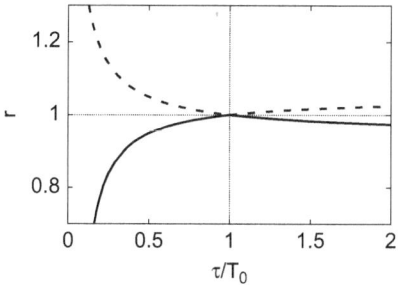

Fig. 4.23 Bifurcation diagram of rotating waves of (4.78) at vanishing control amplitude $b_0 = 0$. Parameters: $T_0 = 2\pi/\omega_0$, $\omega_0 = 1$, and $\gamma = 10$ [54]

$$\tau = \tau_c \qquad (4.83)$$

moves to the left, i.e., below $2\pi k/\omega_0$, then the stability region of the Pyragas branch has invaded the unstable region of the fold bifurcation. Again this refutes the notorious odd number limitation of Pyragas control, see Fiedler et al. [52] and references therein.

Let $\tau = \tau_c$ denote the transcritical bifurcation point on the Pyragas curve $\lambda = \lambda(\tau)$, see (4.81). Let $z(t) = r_c e^{i\omega_c t}$ denote the corresponding rotating wave and abbreviate $\varepsilon \equiv r_c^2 - 1$. Conditions for the transcritical bifurcation in (4.78) can be obtained [54], which yield the following relations between the control amplitude b_c at the bifurcation and ε, τ_c:

$$b_c = -\varepsilon \frac{\omega_0 + \gamma\varepsilon}{k\pi(\gamma\sin\beta + 2\varepsilon\cos\beta)} \qquad (4.84)$$

and

$$b_c = -\frac{2\pi k - \omega_0\tau_c}{\tau_c\left(\frac{1}{2}\gamma^2\tau_c\sin\beta + (2\pi k - \omega_0\tau_c)\cos\beta\right)}. \qquad (4.85)$$

As follows from (4.84) and (4.85), for small ε, alias for τ_c near $2k\pi/\omega_0$, the optimal control angle is $\beta = -\pi/2$ in the limit $\varepsilon \to 0$, and for fixed $k, \omega_0, \gamma, \varepsilon$ this control phase β allows for stabilization with the smallest amplitude $|b_c|$. For $\beta = -\pi/2$ the relations (4.84) and (4.85) simplify to

$$b_c = \frac{\varepsilon}{k\pi}\left(\frac{\omega_0}{\gamma} + \varepsilon\right) \qquad (4.86)$$

and

$$b_c = \frac{2}{(\gamma\tau_c)^2}(2k\pi - \omega_0\tau_c), \qquad (4.87)$$

respectively. For small $b_0 > 0$ we also have the expansions

$$\varepsilon = -\left(k\pi \frac{\gamma}{\omega_0}\sin\beta\right)b_0 + \cdots \tag{4.88}$$

and

$$\tau_c = \frac{2\pi k}{\omega_0} + \left(\frac{1}{2\omega_0}\left(\frac{2k\pi\gamma}{\omega_0}\right)^2 \sin\beta\right)b_0 + \cdots . \tag{4.89}$$

for the location of the transcritical bifurcation. In particular, we see that odd number delay stabilization can be achieved by arbitrary small control amplitudes b_0 near the fold for $\gamma > 0$ and $\sin\beta < 0$. Note that the stability region of the Pyragas curve increases if $\varepsilon = r_c^2 - 1 > 0$, see Fig. 4.21. For vanishing phase angle of the control, $\beta = 0$, in contrast, delay stabilization cannot be achieved by arbitrarily small control amplitudes b_0 near the fold in our system (4.78).

Even far from the fold at $\lambda = 0$ and $\tau = 2k\pi/\omega_0$ the above formulas (4.84), (4.86), and (4.87) hold and indicate a transcritical bifurcation from the (global) Pyragas branch of rotating waves of (4.78) along the Pyragas curve $\lambda = \lambda(\tau)$. This follows by analytic continuation. Delay stabilization, however, may fail long before $\tau = \tau_c$ is reached. In fact, nonzero purely imaginary Floquet exponents may arise, which destabilize the Pyragas branch long before $\tau = \tau_c$ is reached. This interesting point remains open.

A more global picture of the orbits involved in the transcritical bifurcation may be obtained by numerical analysis. Rewriting (4.78) in polar coordinates $z = re^{i\varphi}$ yields

$$\dot{r} = [(r^2 - 1)^2 - \lambda]r + b_0[\cos(\beta + \varphi(t - \tau) - \varphi)\,r(t - \tau) - r\cos\beta] \tag{4.90}$$
$$\dot{\varphi} = \gamma(r^2 - 1) + \omega_0 + b_0[\sin(\beta + \varphi(t - \tau) - \varphi)\,r(t - \tau)/r - \sin\beta]. \tag{4.91}$$

To find all rotating wave solutions we make the ansatz $r = \text{const}$ and $\dot{\varphi} = \omega = \text{const}$ and obtain

$$0 = (r^2 - 1)^2 - \lambda + b_0[\cos(\beta - \omega\tau) - \cos\beta] \tag{4.92}$$
$$\omega = \gamma(r^2 - 1) + \omega_0 + b_0[\sin(\beta - \omega\tau) - \sin\beta]. \tag{4.93}$$

Eliminating r we find a transcendental equation for ω

$$0 = -\gamma^2\lambda + \gamma^2 b_0[\cos(\beta - \omega\tau) - \cos\beta] \tag{4.94}$$
$$+ (\omega - \omega_0 - b_0[\sin(\beta - \omega\tau) - \sin\beta])^2. \tag{4.95}$$

One can now solve this equation numerically for ω and insert the result into

$$r = \left(\frac{\omega - \omega_0}{\gamma} - \frac{b_0}{\gamma} [\sin (\beta - \omega \tau) - \sin \beta] + 1 \right)^{\frac{1}{2}} \qquad (4.96)$$

to obtain the allowed radii (discarding imaginary radii).

The orbit which stabilizes the Pyragas branch in the transcritical bifurcation may be the minus branch or another delay-induced orbit which is born in a fold bifurcation, depending on the parameters. Figure 4.24 displays the different scenarios and the crossover in dependence on the control amplitude b_0. The value of γ is chosen as $\gamma = 9$, 10.5, 10.6, and 13 in panels (a), (b), (c), and (d), respectively. It can be seen that the Pyragas orbit is stabilized by a transcritical bifurcation T_1. As the value of γ increases, a pair of a stable and an unstable orbit generated by a fold bifurcation F_1 approaches the minus branch (see Fig. 4.24a). On this branch, fold bifurcations (F_2 and F_3) occur as shown in Fig. 4.24(b). At $\gamma = 10.6$, the fold points of F_1 and F_2 touch in a transcritical bifurcation T_2 and annihilate (see Fig. 4.24(c, d). Thus, for further increase of γ, one is left with the stable minus branch and the unstable orbit, which was generated at the fold bifurcation F_3. In all panels the radius of the Pyragas orbit is not changed by the control. The radius of the minus branch, however, is altered because the delay time does not match orbit period.

Figure 4.25 shows the region in the (β, b_0) plane where the Pyragas orbit is stable, for a set of parameters. The grayscale (color code) shows only negative values of the largest real part of the Floquet exponents. One can see that the orbit is most stable for feedback phases $\beta \approx -\pi/2$ which agrees with the previous analytic

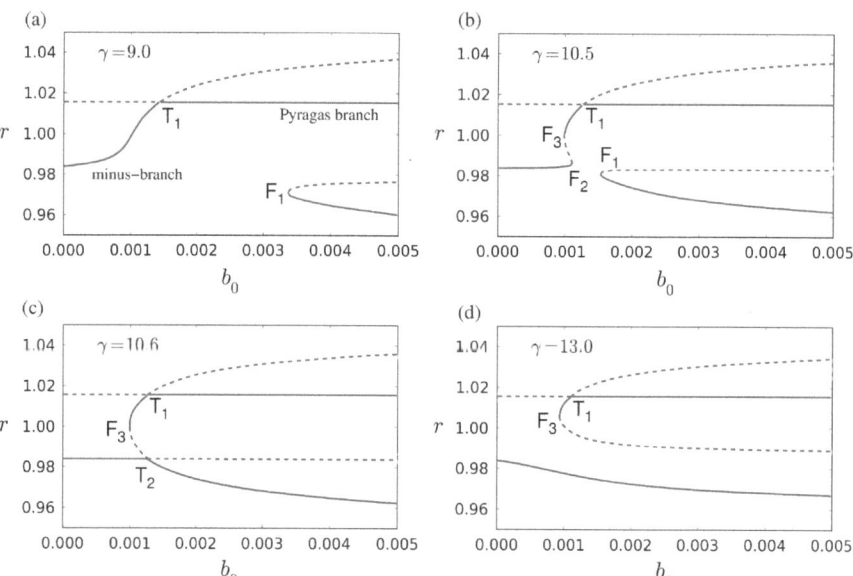

Fig. 4.24 Radii of stable (*solid*) and unstable (*dashed*) rotating wave solutions in dependence on b_0 for different γ. Parameters: $\omega_0 = 1$, $\lambda = 0.001$, and $\beta = -\pi/2$ [54]

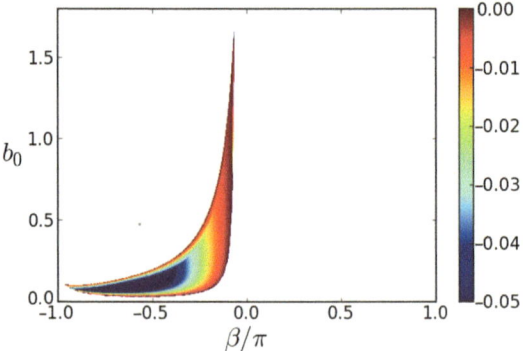

Fig. 4.25 Domain of stability of the Pyragas orbit. The *grayscale* (*color code*) shows only negative values of the largest real part of the Floquet exponents. Parameters: $\omega_0 = 1$, $\lambda = 0.0001$, and $\gamma = 0.1$. [54]

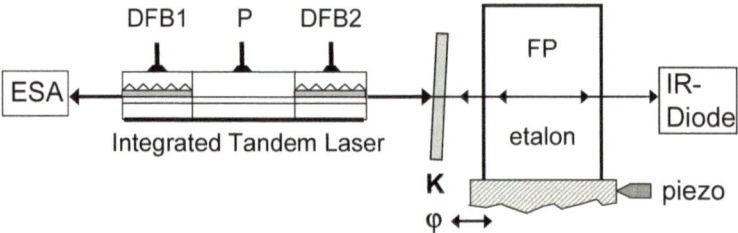

Fig. 4.26 Scheme of an integrated tandem laser with optical feedback from an external Fabry–Perot etalon. Two distributed feedback (DFB) lasers are connected via a passive waveguide section P. Amplitude K and phase φ of the feedback from the FP resonator are controlled by a variable neutral density filter and a piezo positioning, respectively. ESA: electrical spectrum analyzer. IR Diode: power measurement [38]

results for small λ. The picture was obtained by linear stability analysis of (4.90) and (4.91) and numerical solution of the transcendental eigenvalue problem for the Floquet exponents. It clearly shows that the periodic orbit can be stabilized even though it has an odd number of real Floquet multipliers greater than unity.

These results of the simple normal form model can be transferred to a more realistic model of an integrated tandem laser [54], such as the one considered in the next section, see Fig. 4.26, where time-delayed feedback control is realized by a Michelson interferometer.

4.3 Time-Delayed Control of Optical Systems

In this section we will consider semiconductor lasers, where time-delayed feedback control can be readily realized by optical feedback from a mirror or a Fabry–Perot (FP) resonator, and this allows for controlling systems with very fast dynamics still in real-time mode.

4.3.1 Stabilizing Continuous-Wave Laser Emission by Phase-Dependent Coupling

From a practical point of view, it is often desirable to suppress self-sustained oscillations, i.e., intensity pulsations, in order to stabilize continuous-wave (cw) operation of lasers [133, 134]. This amounts to stabilizing an unstable fixed point of the dynamic laser equations.

Here we consider a semiconductor laser device for which control of unstable steady states by time-delayed feedback control has been demonstrated in theory and experiment [38]. Recently, multi-section lasers with their complex dynamical phenomena have opened up new ways in high-speed optical information processing [135, 136]. Their picosecond response times are too short even for a fast electronic realization of time-delayed feedback control. All-optical control is thus the only applicable method so far. The scheme of the setup is shown in Fig. 4.26. An integrated tandem laser [135, 136] is deliberately driven through a Hopf bifurcation into a self-pulsating regime of operation. Suppression of the pulsations and non-invasive stabilization of the steady state is achieved by direct optical feedback from a properly designed external FP etalon. Although proposed one and a half decades ago [39, 137] and despite of some numerical studies [137–139], such non-invasive all-optical control approach has not been implemented experimentally until recently [38]. A novel aspect of our analysis is that it addresses the role of the optical phase as a specific feature of the FP control configuration.

Optical fields emitted by lasers vary generally as $\mathrm{Re}\{E(t)e^{-i\omega_0 t}\}$ where the exponential factor oscillates by orders of magnitude faster than the slow amplitude $E(t)$. The field fed back from the FP resonator has the same shape and, for feedback gain K, its amplitude reads as

$$E_b(t) = Ke^{i\varphi} \sum_{n=0}^{\infty} R^n e^{in\phi} [E(t_n) - e^{i\phi}E(t_{n+1})], \qquad (4.97)$$

with $t_n = t - \tau_l - n\tau$. The delay originates from a single roundtrip between laser and FP resonator, characterized by the latency time τ_l, and n round-trips of time τ within the FP resonator of mirror reflectivity R. Two optical phase shifts $\varphi = \omega_0\tau_l$ and $\phi = \omega_0\tau$ are associated with these delay times. Non-invasive control requires optical target states with $E(t) - e^{i\phi}E(t - \iota)$. Feedback from a FP resonator has been studied previously, see, e.g., [140–143] and references therein. However, those configurations rely on maximum feedback are thus strongly invasive.

For steady states $E(t) = E_0$ non-invasiveness means $e^{i\phi} = 1$, i.e., the FP resonator must be tuned into resonance. While the FP phase is thus fixed, the latency is still arbitrary and makes the feedback phase-sensitive. Conventional time-delayed feedback control corresponds to $\varphi = 0$. However, in the FP geometry, φ is tunable by sub-wavelength changes of the laser-FP separation and thus represents an additional free parameter which all-optical time-delayed feedback control can profit from. In what follows, this is theoretically demonstrated within the simple generic two-variable center-manifold model introduced in Sect. 4.2 [48]. Stabilization of the cw emission has also been shown within a more specific semiconductor laser model of Lang–Kobayashi type, including latency, bandpass filtering in the control loop, and a feedback phase [144].

We consider a nonlinear system closely above a Hopf bifurcation, where it has an unstable fixed point (focus) whose stability is governed by the complex eigenvalues $\lambda \pm i\omega$ (with $\lambda > 0$). For simplicity, we restrict ourself to a single FP roundtrip ($n=1$) and ignore τ_l in the slow amplitude dynamics. An extension to multiple time feedback (ETDAS) is found elsewhere [50]. Linearizing around the fixed point provides a generic equation for the center-manifold coordinates x, y, corresponding to the complex field through $E = E_0 + x + iy$,

$$
\begin{pmatrix} \dot{x} \\ \dot{y} \end{pmatrix} = \begin{pmatrix} \lambda & \omega \\ -\omega & \lambda \end{pmatrix} \begin{pmatrix} x \\ y \end{pmatrix}
$$
$$
- K \begin{pmatrix} \cos\varphi & -\sin\varphi \\ \sin\varphi & \cos\varphi \end{pmatrix} \begin{pmatrix} x(t) - x(t-\tau) \\ y(t) - y(t-\tau) \end{pmatrix}. \tag{4.98}
$$

This equation generalizes the model of (4.6) to phase-sensitive feedback and shows that such feedback creates non-diagonal coupling terms [145]. The characteristic equation for the complex eigenvalues Λ reads as

$$
\Lambda + K e^{\pm i\varphi} \left(1 - e^{-\Lambda\tau} \right) = \lambda \pm i\omega. \tag{4.99}
$$

Note that this characteristic equation can be solved analytically using the Lambert function, which is defined as the inverse function of $g(z) = z e^z$ for complex z.

Figure 4.27 shows the domain of control, i.e., $\text{Re}(\Lambda) < 0$, in dependence on the parameters φ, K, and τ. Unit of time is the intrinsic period $T_0 = 2\pi/\omega$ of the unstable focus and $\lambda T_0 = 0.2$ is chosen in all plots. Panels (a) and (b) represent the (φ, K) plane for fixed values of the time delay $\tau/T_0 = 0.5$ and 0.9, respectively. Note that $\tau = T_0/2$ yields a symmetric domain of control with respect to $\varphi = 0$, which is the case of diagonal coupling [48]. For values other than this optimal time delay, the domain of control is distorted and shrinks. In the situation shown in Fig. 4.27 (b), control can no longer be achieved for $\varphi = 0$, but only for positive phase $\varphi > 0$. Panels (c) and (d) show the domain of control in the (φ, τ) plane for fixed feedback gain $KT_0 = 1$ and 2, respectively. It consists of isolated islands with a horizontal extension that becomes maximum and symmetric with respect to $\varphi = 0$ at delays of $\tau = (n + 1/2)T_0$ ($n = 0, 1, 2, \dots$). No control is possible for integer τ/T_0. For a range of τ values in between, stabilization can be achieved by appropriately chosen φ. When crossing the islands at fixed φ, resonance-type behavior of the damping rate $-\text{Re}(\Lambda)$ occurs. With increasing n, the size of the islands decreases so that they eventually disappear at some critical value determined by the feedback strength K.

These results from the simple generic model have been confirmed by experimental realization of all-optical non-invasive control by means of time-delayed feed-

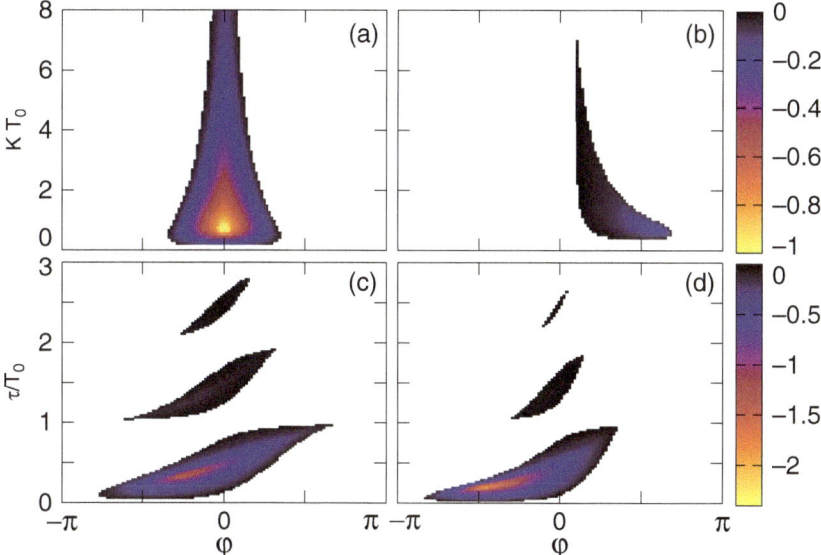

Fig. 4.27 Domain of control in dependence on φ, K, and τ with normalization in units of $T_0 = 2\pi/\omega$. The largest real part of the complex eigenvalues Λ is shown in *colorcode*. (**a**), (**b**): Domain of control in the (φ, K) plane for fixed delay $\tau = T_0/2$ and $0.9T_0$, respectively. (**c**), (**d**): Domain of control in the (φ, τ) plane for fixed feedback gain $K = 1/T_0$ and $2/T_0$, respectively. Fixed parameter: $\lambda = 0.2/T_0$ [38]

back from an external Fabry–Perot cavity [38]. They are also in qualitative agreement with simulations of more realistic laser models of Lang–Kobayashi [144] and traveling-wave type [38, 146].

In conclusion, using phase-dependent feedback, stabilization of the continuous-wave laser output and non-invasive suppression of intensity pulsations has been shown. This study demonstrates the crucial importance of the proper choice of phase of the feedback signal, i.e., of the coupling matrix, which represents a generic feature of all-optical time-delayed feedback control.

4.3.2 Noise Suppression by Time-Delayed Feedback

In this section we investigate the effects of feedback under the influence of noise in a semiconductor laser [95]. A laser with feedback from a conventional mirror can be described by the Lang–Kobayashi equations [147]. Other types of feedback have also been investigated [143, 148]. One particular feedback realizes the delayed feedback control by a Fabry–Perot resonator [38, 139, 144]. A schematic view of this all-optical setup is shown in Fig. 4.28. A fraction of the emitted laser light is coupled into a resonator. The resonator then feeds an interference signal of the actual electric field $E(t)$ and the delayed electric field $E(t - \tau)$ (neglecting multiple reflections) back into the laser.

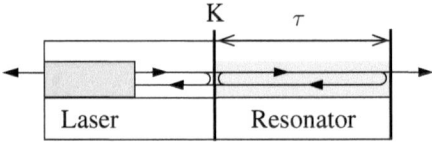

Fig. 4.28 Setup of a laser coupled to a Fabry–Perot resonator realizing the time-delayed feedback control [95]

Scaling (i) time by the photon lifetime $\tau_p \approx 10^{-12}$s, (ii) carrier density (in excess of the threshold carrier density) by the inverse of the differential gain G_N times τ_p, and (iii) electric field by $(\tau_c G_N)^{-1/2}$, where $\tau_c \approx 10^{-9}$s is the carrier lifetime (for details see [149]), one obtains a modified set of non-dimensionalized Lang–Kobayashi equations [139] describing this setup

$$\frac{d}{dt}E = \frac{1}{2}(1 + i\alpha)nE \tag{4.100}$$
$$-e^{i\varphi}K\left[E(t) - e^{i\psi}E(t - \tau)\right] + F_E(t),$$
$$T\frac{d}{dt}n = p - n - (1 + n)|E|^2,$$

where E is the complex electric field amplitude, n is the carrier density in excess of the laser threshold, α is the linewidth enhancement factor, K is the feedback strength, τ is the roundtrip time in the Fabry–Perot resonator, p is the excess pump injection current, $T = \tau_c/\tau_p$ is the timescale parameter, F_E is a noise term describing the spontaneous emission, and φ and ϕ are optical phases.

The phases φ and ψ depend on the sub-wavelength positioning of the mirrors. By precise tuning $\varphi = 2\pi n$ and $\psi = 2\pi m$ one can realize the usual Pyragas feedback control

$$- K\left[E(t) - E(t - \tau)\right]. \tag{4.101}$$

We consider small feedback strength K, so that the laser is not destabilized and no delay-induced bifurcations occur. A sufficient condition [139] is that

$$K < K_c = \frac{1}{\tau\sqrt{1 + \alpha^2}}. \tag{4.102}$$

The noise term F_E in (4.100) arises from spontaneous emission, and we assume the noise to be white and Gaussian

$$\langle F_E \rangle = 0, \qquad \langle F_E(t)\overline{F_E(t')} \rangle = R_{sp}\delta(t - t'), \tag{4.103}$$

with the spontaneous emission rate

$$R_{sp} = \beta(n + n_0), \tag{4.104}$$

where β is the spontaneous emission factor and n_0 is the threshold carrier density. Without noise the laser operates in a steady state (*cw emission*). To find these steady state values, we transform (4.100) into equations for intensity I and phase ϕ by $E = \sqrt{I}\,e^{i\phi}$:

$$\frac{d}{dt}I = nI - 2K\left[I - \sqrt{I}\sqrt{I_\tau}\,\cos(\phi_\tau - \phi)\right] + R_{sp} + F_I(t),$$

$$\frac{d}{dt}\phi = \frac{1}{2}\alpha n + K\frac{\sqrt{I_\tau}}{\sqrt{I}}\,\sin(\phi_\tau - \phi) + F_\phi(t), \qquad (4.105)$$

$$T\frac{d}{dt}n = p - n - (1+n)I,$$

where $I_\tau = I(t - \tau)$, $\phi_\tau = \phi(t - \tau)$, and

$$\langle F_I \rangle = 0, \quad \langle F_\phi \rangle = 0, \qquad (4.106)$$

$$\langle F_I(t)\,F_\phi(t') \rangle = 0, \qquad (4.107)$$

$$\langle F_I(t)\,F_I(t') \rangle = 2R_{sp}\,I\,\delta(t - t') \qquad (4.108)$$

$$\langle F_\phi(t)\,F_\phi(t') \rangle = \frac{R_{sp}}{2I}\,\delta(t - t'). \qquad (4.109)$$

Setting $\frac{d}{dt}I = 0$, $\frac{d}{dt}n = 0$, $\frac{d}{dt}\phi = \text{const}$, and $K = 0$ and replacing the noise terms by their mean values give a set of equations for the mean steady state solutions I_*, n_*, and $\phi = \omega_* t$ without feedback (the solitary laser mode). Our aim is now to analyze the stability (damping rate) of the steady state. A high stability of the steady state, corresponding to a large damping rate, will give rise to small-amplitude noise-induced relaxation oscillations whereas a less stable steady state gives rise to stronger relaxation oscillations. Linearizing (4.105) around the steady state $X(t) = X_* + \delta X(t)$, with $X(t) = (I, \phi, n)$, gives

$$\frac{d}{dt}X(t) = U\,X(t) - V\left[X(t) - X(t - \tau)\right] + F(t), \qquad (4.110)$$

with

$$U = \begin{bmatrix} n_* & 0 & I_* + \beta \\ 0 & 0 & \frac{1}{2}\alpha \\ -\frac{1}{T}(1 + n_*) & 0 & -\frac{1}{T}(1 + I_*) \end{bmatrix}, \qquad (4.111)$$

$$\qquad (4.112)$$

$$V = \text{diag}(K, K, 0), \qquad (4.113)$$

where diag(...) denotes a 3×3 diagonal matrix, and

$$F = (F_I, F_\phi, 0). \qquad (4.114)$$

The Fourier transform of (4.110) gives

$$\widehat{X}(\omega) = \underbrace{[i\omega - U + V(1 - e^{-i\omega\tau})]^{-1}}_{\equiv M} \widehat{F}(\omega).$$ (4.115)

The Fourier-transformed covariance matrix of the noise is

$$\langle \widehat{F}(\omega)\, \widehat{F}(\omega')^\dagger \rangle = \frac{1}{2\pi} \text{diag}(2R_{sp}I_*, \frac{R_{sp}}{2I_*}, 0)\, \delta(\omega - \omega'),$$ (4.116)

with the adjoint †. The matrix-valued power spectral density $S(\omega)$ can then be defined through

$$S(\omega)\, \delta(\omega - \omega') = \langle \widehat{X}(\omega)\, \widehat{X}(\omega)^\dagger \rangle$$ (4.117)

and is thus given by

$$S(\omega) = \frac{1}{2\pi} M\, \text{diag}(2R_{sp}I_*, \frac{R_{sp}}{2I_*}, 0)\, M^\dagger.$$ (4.118)

The diagonal elements of the matrix S are the power spectrum of the intensity $S_{\delta I}$, the phase $S_{\delta\phi}$, and the carrier density $S_{\delta n}$. The frequency power spectrum is related to the phase power spectrum $S_{\delta\phi}(\omega)$ by [150]:

$$S_{\delta\dot{\phi}}(\omega) = \omega^2\, S_{\delta\phi}(\omega).$$ (4.119)

The laser parameters we consider in the following are typical values for a single mode distributed feedback (DFB) laser operating close to threshold [139, 150].

Figures 4.29 and 4.30 display the intensity and the frequency power spectra, respectively, for different values of the delay time τ, obtained analytically from the linearized equations (left) and from simulations of the full nonlinear equations (right). All spectra have a main peak at the relaxation oscillation frequency

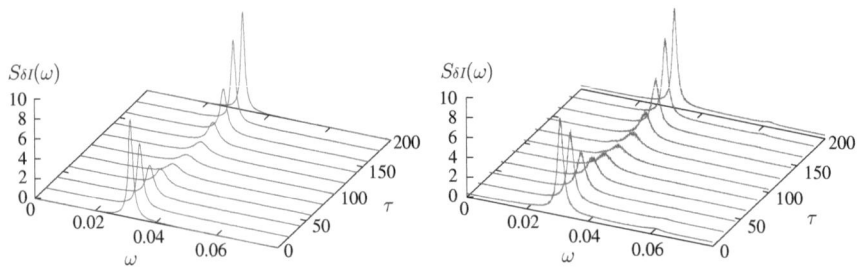

Fig. 4.29 Analytical (*left*) and numerical (*right*) results for the power spectral density $S_{\delta I}(\omega)$ of the intensity for different values of the delay time τ. Parameters: $p = 1$, $T = 1000$, $\alpha = 2$, $\beta = 10^{-5}$, $n_0 = 10$, and $K = 0.002$. (A typical unit of time is the photon lifetime $\tau_p = 10^{-11}$s, corresponding to a frequency of 100 GHz) [95]

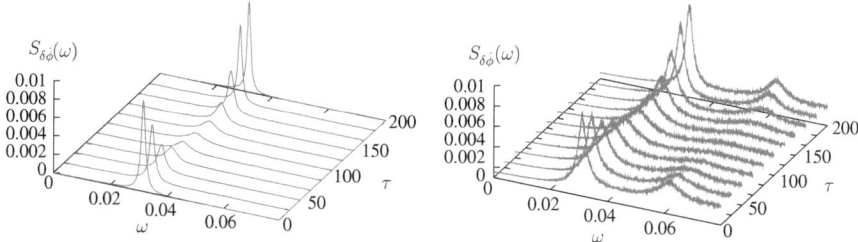

Fig. 4.30 Analytical (*left*) and numerical (*right*) results for the power spectral density $S_{\delta\phi}(\omega)$ of the frequency for different values of the delay time τ. Parameters: $p = 1$, $T = 1000$, $\alpha = 2$, $\beta = 10^{-5}$, $n_0 = 10$, and $K = 0.002$ [95]

$\Omega_{RO} \approx 0.03$. The higher harmonics can also be seen in the spectra obtained from the nonlinear simulations. The main peak decreases with increasing τ and reaches a minimum at

$$\tau_{opt} \approx \frac{T_{RO}}{2} = \frac{2\pi}{2\Omega_{RO}} \approx 100. \tag{4.120}$$

With further increases of τ, the peak height increases again until it reaches approximately its original maximum at $\tau \approx T_{RO}$. A small peak in the power spectra indicates that the relaxation oscillations are strongly damped. This means that the fluctuations around the steady state values I_* and n_* are small. Figure 4.31 displays exemplary time series of the intensity with and without feedback. The time series with feedback shows much less pronounced stochastic fluctuations.

Next, we study the variance of the intensity distribution as a measure for the oscillation amplitude:

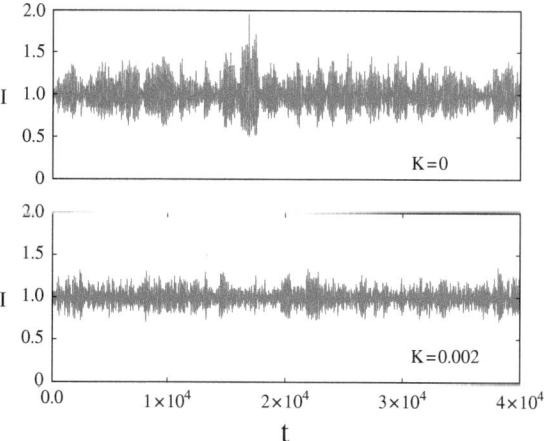

Fig. 4.31 Intensity time series without (*top panel*) and with (*bottom panel*) control. Parameters: $p = 1$, $T = 1000$, $\alpha = 2$, $\beta = 10^{-5}$, $n_0 = 10$, and $\tau = 100 \approx T_0/2$ [95]

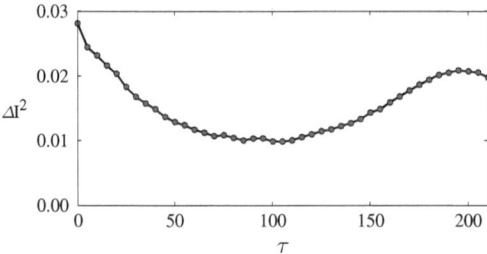

Fig. 4.32 Variance of the intensity I vs. the delay time. Parameters: $p = 1$, $T = 1000$, $\alpha = 2$, $\beta = 10^{-5}$, $n_0 = 10$, and $K = 0.002$ [95]

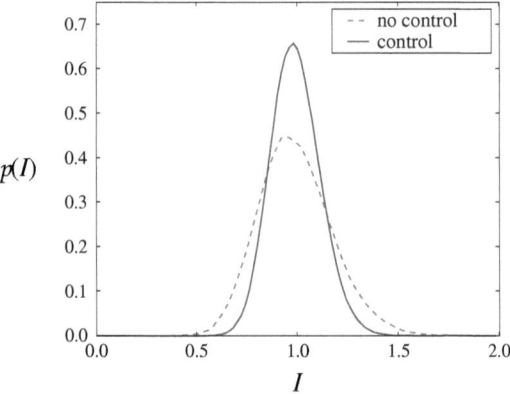

Fig. 4.33 Probability distribution of the intensity I with and without the resonator (simulations). Parameters: $p = 1$, $T = 1000$, $\alpha = 2$, $\beta = 10^{-5}$, $n_0 = 10$, and $K = 0.002$ [95]

$$\Delta I^2 \equiv \left\langle (I - \langle I \rangle)^2 \right\rangle. \tag{4.121}$$

Figure 4.32 displays the variance as a function of the delay time. The variance is minimum at $\tau \approx T_{RO}/2$, thus for this value of τ the intensity is most steady and relaxation oscillations excited by noise have a small amplitude.

Figure 4.33 displays the intensity distribution of the laser without (dashed) and with (solid) optimal control. The time-delayed feedback control leads to a narrower distribution and less fluctuations.

In conclusion, by tuning the cavity roundtrip time of the feedback loop to half the relaxation oscillation period, $\tau_{opt} \approx T_{RO}/2$, noise-induced oscillations in a semiconductor laser can be suppressed to a remarkable degree.

4.4 Time-Delayed Control of Neuronal Dynamics

In this section we study the effect of time-delayed feedback in neural systems [94]. Time delays can occur in the coupling between different neurons due to signal propagation or in a self-feedback loop, e.g., due to neurovascular coupling in the brain. Moreover, time-delayed feedback loops might be deliberately implemented to control neural disturbances, e.g., to suppress undesired synchrony of firing neurons in Parkinson's disease or epilepsy [71, 72]. Here we model the neurons in the framework of the FitzHugh–Nagumo model [151, 152], which is a simple paradigm of excitable dynamics. Time-delayed feedback control of noise-induced oscillations was demonstrated in a single excitable FitzHugh–Nagumo system [26, 27, 89, 91]. The simplest network configuration displaying features of neural interaction consists of two coupled excitable systems. In two coupled FitzHugh–Nagumo systems two situations are studied: (i) stochastic synchronization of instantaneously coupled neurons under the influence of white noise and controlled by local time-delayed feedback [93, 153] and (ii) the emergence of antiphase oscillations in delay-coupled neurons and complex scenarios induced by the additional application of time-delayed self-feedback such as transitions from synchronized in-phase to antiphase oscillations, bursting patterns, or amplitude death [94, 154]. In spatially extended neuronal media time-delayed feedback as well as nonlocal spatial coupling has also been studied, and it has been shown that pulse propagation in a reaction–diffusion system can be suppressed by appropriate choice of the space or timescales of the feedback [22, 23], which suggests failure of feedback as a common mechanism for spreading depolarization waves in migraine aura and stroke. However, in the present section we restrict ourselves to spatially homogeneous coupled FitzHugh–Nagumo systems.

4.4.1 Model of Two Coupled Neurons

In order to grasp the complicated interaction between billions of neurons in large neural networks, those are often lumped into groups of neural populations each of which can be represented as an effective excitable element that is mutually coupled to the other elements [145, 72]. In this sense the simplest model which may reveal features of interacting neurons consists of two coupled neural oscillators. Each of these will be represented by a simplified FitzHugh–Nagumo system, which is often used as a paradigmatic generic model for neurons, or more generally, excitable systems [34].

Neurons are excitable units that can emit spikes or bursts of electrical signals, i.e., the system rests in a stable steady state, but after it is excited beyond a threshold, it emits a pulse. In the following, we consider electrically coupled neurons (Fig. 4.34 a) modeled by the FitzHugh–Nagumo system in the excitable regime:

$$
\begin{aligned}
\epsilon_1 \dot{u}_1 &= u_1 - \frac{u_1^3}{3} - v_1 + C[u_2(t - \tau) - u_1(t)] \\
\dot{v}_1 &= u_1 + a + D_1 \xi_1(t) \\
\epsilon_2 \dot{u}_2 &= u_2 - \frac{u_2^3}{3} - v_2 + C[u_1(t - \tau) - u_2(t)] \\
\dot{v}_2 &= u_2 + a + D_2 \xi_2(t),
\end{aligned}
\tag{4.122}
$$

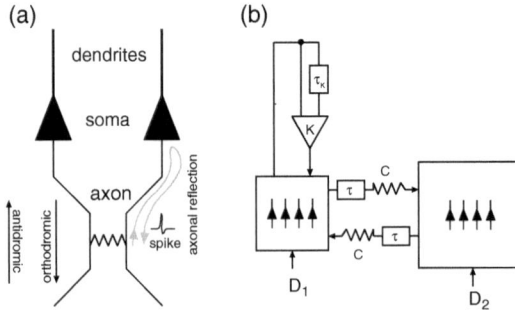

Fig. 4.34 (**a**) Scheme of two axo-axonally coupled neurons (pyramidal cells coupled by an electrical synapse) [155]. (**b**) Two mutually coupled neural populations (delay τ, coupling constant C) with feedback control loop (delay τ_K, coupling constant K) and noise input D_1, D_2 [94]

where the subsystems u_1, v_1 and u_2, v_2 correspond to single neurons (or neuron populations), which are linearly coupled with coupling strength C. The variables u_1 and u_2 are related to the transmembrane voltage and v_1 and v_2 refer to various quantities connected to the electrical conductance of the relevant ion currents. Here a is an excitability parameter whose value defines whether the system is excitable ($a > 1$) or exhibits self-sustained periodic firing ($a < 1$), ϵ_1 and ϵ_2 are the timescale parameters that are usually chosen to be much smaller than unity, corresponding to fast activator variables u_1, u_2 and slow inhibitor variables v_1, v_2.

The synaptic coupling between two neurons is modeled as a diffusive coupling considered for simplicity to be symmetric [156–158]. More general delayed couplings are considered in [159]. The coupling strength C summarizes how information is distributed between neurons. The mutual delay τ in the coupling is motivated by the propagation delay of action potentials between the two neurons u_1 and u_2. Time delays in the coupling must be considered particularly in the case of high-frequency oscillations.

Each neuron is driven by Gaussian white noise $\xi_i(t)$ ($i = 1, 2$) with zero mean and unity variance. The noise intensities are denoted by parameters D_1 and D_2.

Besides the delayed coupling we will also consider delayed self-feedback in the form suggested by Pyragas [6], where the difference $s(t) - s(t - \tau_K)$ of a system variable s (e.g., activator or inhibitor) at time t and at a delayed time $t - \tau_K$, multiplied by some control amplitude K, is coupled back into the same system (Fig. 4.34b). Such feedback loops might arise naturally in neural systems, e.g., due to neurovascular coupling that has a characteristic latency or due to finite propagation speed along cyclic connections within a neuron sub-population or they could be realized by external feedback loops as part of a therapeutical measure, as proposed in [72]. This feedback scheme is simple to implement, quite robust, and has already been applied successfully in a real experiment with time-delayed neurofeedback from real-time magnetoencephalography (MEG) signals to humans via visual stimulation in order to suppress the alpha rhythm, which is observed due to strongly synchronized neural populations in the visual cortex in the brain [160]. One distinct advantage of this

method is its non-invasiveness, i.e., in the ideal deterministic limit the control force
vanishes on the target orbit, which may be a steady state or a periodic oscillation
of period τ. In the case of noisy dynamics the control force, of course, does not
vanish but still remains small, compared to other common control techniques using
external periodic signals, for instance, in deep brain stimulation to suppress neural
synchrony in Parkinson's disease [161].

The phase portrait and the nullclines of a single FitzHugh–Nagumo system without noise and feedback are shown in Fig. 4.35(a). The fixed point A is a stable focus
or node for $a > 1$ (*excitable regime*). If the system is perturbed well beyond point
A' (see inset), it performs a large excursion $A \to B \to C \to D \to A$ in phase space
corresponding to the emission of a spike (Fig. 4.35b). At $a = 1$ the system exhibits
a Hopf bifurcation of a limit cycle, and the fixed point A becomes an unstable focus
for $a < 1$ (*oscillatory regime*).

In the following we choose the excitability parameter $a = 1.05$ in the excitable
regime close to threshold. If noise is present, it will occasionally kick the system
beyond A' resulting in noise-induced oscillations (*spiking*).

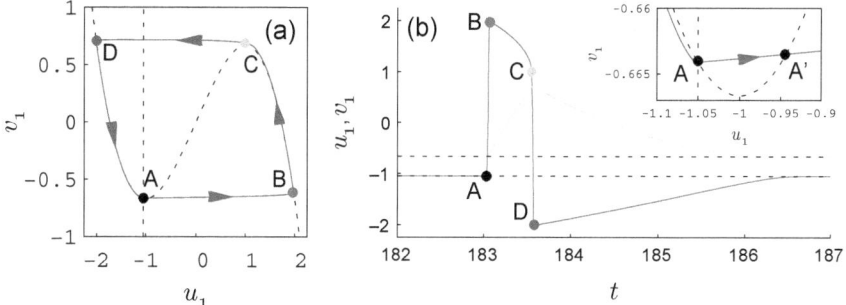

Fig. 4.35 Excitable dynamics of a single FitzHugh–Nagumo system: (**a**) phase portrait (u_1, v_1)
(trajectory: *solid blue* and nullclines: *dashed black*), (**b**) time series of activator $u_1(t)$ (*red*) and
inhibitor $v_1(t)$ (*green*). The colored dots A, B, C, and D mark corresponding points on panels (**a**)
and (**b**). The inset in (**b**) shows a blowup of the phase portrait near A. Parameters: $\epsilon_1 = 0.01$,
$a = 1.05$, and $D_1 = 0$ [94]

4.4.2 Control of Stochastic Synchronization

We shall first consider two coupled FitzHugh–Nagumo systems as in (4.122) albeit
without delay in the coupling ($\tau = 0$). Noise can induce oscillations even though the
fixed point is stable. The noise sources then play the role of stimulating the excitable
subsystems. Even if only one subsystem is driven by noise, it induces oscillations
of the whole system through the coupling. In this subsection, we consider two non-
identical neurons, described by different timescales $\epsilon_1 = 0.005$ and $\epsilon_2 = 0.1$, and
set the noise intensity D_2 in the second subsystem equal to a small value, $D_2 = 0.09$,
in order to model some background noise level. Depending on the coupling strength

C and the noise intensity D_1 in the first subsystem, the two neurons show weak, moderate, or strong stochastic synchronization [93].

If feedback is applied to one of the two interacting subsystems [93, 153], i.e., locally, the dynamical equations are given by:

$$\epsilon_1 \dot{u}_1 = u_1 - \frac{u_1^3}{3} - v_1 + C\,(u_2 - u_1),$$
$$\dot{v}_1 = u_1 + a + K\,[v_1(t - \tau) - v_1(t)] + D_1\,\xi(t), \qquad (4.123)$$
$$\epsilon_2 \dot{u}_2 = u_2 - \frac{u_2^3}{3} - v_2 + C\,(u_1 - u_2),$$
$$\dot{v}_2 = u_2 + a + D_2\,\xi_2(t), \qquad (4.124)$$

where subsystems (4.123) and (4.124) represent two different neurons, and local feedback with strength K and delay time τ is applied to the first subsystem.

There are various measures of the synchronization of coupled systems [162]. For instance, one can consider the average interspike intervals (ISI) of each subsystem, i.e., $\langle T_1 \rangle$ and $\langle T_2 \rangle$, calculated from the u variable of the respective subsystem. Their ratio $\langle T_1 \rangle / \langle T_2 \rangle$ is a measure of frequency synchronization. Other measures for stochastic synchronization are given by the phase synchronization index [93] or the mean phase synchronization intervals [153].

First, we consider subsystems (4.123) and (4.124) with $D_1 = 0.6$ and $C = 0.2$, which corresponds to a moderately synchronized uncontrolled system. We aim to find out if the feedback can make the subsystems more, or less, synchronous, and their global dynamics more or less coherent. In particular, we are interested if perfect 1:1 synchronization can be induced by the local feedback or if the existing synchronization can be destroyed. The ratio of ISIs and the synchronization index $\gamma_{1,1}$ are shown by color code in Fig. 4.36 for a large range of the values of the feedback delay τ and strength K. The lighter areas are associated with the stronger 1:1 synchronization, and the values at $K = 0$ and at $\tau = 0$ characterize the original state of the system without feedback. As seen from Fig. 4.36, the locally applied delayed feedback is able to move the system's state closer to the 1:1 synchronization with suitable feedback parameters. On the other hand, for $\tau \approx 2.5$ (black area), 1:1 synchronization is suppressed.

Next, we consider weakly synchronized subsystems (4.123) and (4.124) that are further from the 1:1 synchronization region under the influence of the controlling feedback. For $D_1 = 0.6$ and $C = 0.1$, the ratio of ISIs and the synchronization index $\gamma_{1,1}$ are shown by color code in Fig. 4.37. Again, the stochastic synchronization can be strongly modulated by changing the delay time, i.e., one can either enhance or suppress synchronization by appropriate choice of the local feedback delay.

Finally, for the system that is very well synchronized from the beginning at $D_1 = 0.15$ and $C = 0.2$ again delayed feedback can either enhance or suppress synchronization (Fig. 4.38). In view of applications, where neural synchronization is often pathological, e.g., in Parkinson's disease or epilepsy, it is interesting to note

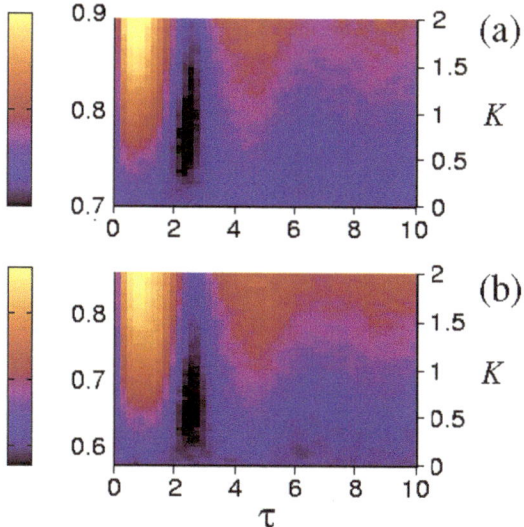

Fig. 4.36 Effect of delayed feedback on frequency and phase synchronizations between the two subsystems at $D_1 = 0.6$ and $C = 0.2$ (moderate synchronization). (**a**) Ratio of average interspike intervals $\langle T_1 \rangle / \langle T_2 \rangle$ from the two systems and (**b**) synchronization index $\gamma_{1,1}$ vs. the control strength K and the time delay τ [93]

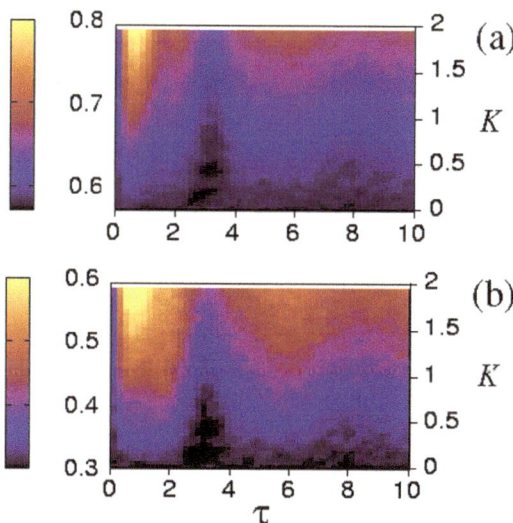

Fig. 4.37 Same as Fig. 4.36 for $D_1 = 0.6$ and $C = 0.1$ (weak synchronization) [93]

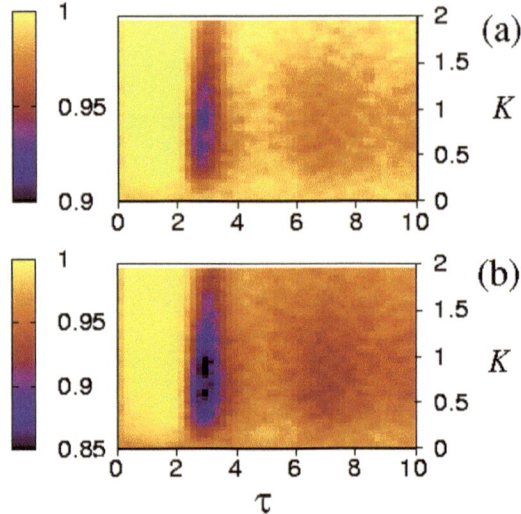

Fig. 4.38 Same as Fig. 4.36 for $D_1 = 0.15$ and $C = 0.2$ (strong synchronization) [93]

that there are cases where a proper choice of the local feedback control parameters leads to desynchronization of the coupled system (dark regions in all three figures).

4.4.3 Dynamics of Delay-Coupled Neurons

In this section we study the influence of a delay in the coupling of two neurons [94, 154], rather than a delayed self-feedback ($K = 0$). We set the noise terms in (4.122) equal to zero, $D_1 = D_2 = 0$, but consider a time delay τ in the coupling. In the deterministic system the delayed coupling plays the role of a stimulus which can induce self-sustained oscillations in the coupled system even if the fixed point is stable. In this sense the delayed coupling has a similar effect as the noise term in the previous section. Here the bifurcation parameters for delay-induced bifurcations are the coupling parameters C and τ.

In the following we shall choose symmetric timescales $\epsilon_1 = \epsilon_2 = \epsilon = 0.01$ and fix $a = 1.05$, where each of the two subsystems has a stable fixed point and exhibits excitability.

The unique fixed point of the system is symmetric and is given by $\mathbf{u}^* \equiv (u_1^*, v_1^*, u_2^*, v_2^*)$, where $u_i^* = -a$ and $v_i^* = a^3/3 - a$. Linearizing (4.122) around the fixed point $\mathbf{u}^*$ by setting $\mathbf{u}(t) = \mathbf{u}^* + \delta\mathbf{u}(t)$, one obtains:

$$\delta\dot{\mathbf{u}} = \frac{1}{\epsilon}\begin{pmatrix} \xi & -1 & 0 & 0 \\ \epsilon & 0 & 0 & 0 \\ 0 & 0 & \xi & -1 \\ 0 & 0 & \epsilon & 0 \end{pmatrix}\delta\mathbf{u}(t) + \frac{1}{\epsilon}\begin{pmatrix} 0 & 0 & C & 0 \\ 0 & 0 & 0 & 0 \\ C & 0 & 0 & 0 \\ 0 & 0 & 0 & 0 \end{pmatrix}\delta\mathbf{u}(t-\tau), \qquad (4.125)$$

where $\xi = 1 - a^2 - C$. The ansatz

$$\delta\mathbf{u}(t) = e^{\lambda t}\mathbf{u}, \tag{4.126}$$

where $\mathbf{u}$ is an eigenvector of the Jacobian matrix, leads to the characteristic equation for the eigenvalues λ:

$$(1 - \xi\lambda + \epsilon\lambda^2)^2 - (\lambda Ce^{-\lambda\tau})^2 = 0, \tag{4.127}$$

which can be factorized giving

$$1 - \xi\lambda + \epsilon\lambda^2 \pm \lambda Ce^{-\lambda\tau} = 0. \tag{4.128}$$

This transcendental equation has infinitely many complex solutions λ. The real parts of all eigenvalues are negative throughout, i.e., the fixed point of the coupled system remains stable for all C. This can be shown analytically for $a > 1$ by demonstrating that no delay-induced Hopf bifurcation can occur. Substituting the ansatz $\lambda = i\omega$ into (4.128) and separating into real and imaginary parts yields for the imaginary part

$$\xi = \pm C\cos(\omega\tau). \tag{4.129}$$

This equation has no solution for $a > 1$ since $|\xi| = a^2 - 1 + C > C$, which proves that a Hopf bifurcation cannot occur.

Delay-induced oscillations in excitable systems are inherently different from noise-induced oscillations. The noise term continuously kicks the subsystems out of their respective rest states and thus induces sustained oscillations. Instantaneous coupling without delay then produces synchronization effects between the individual oscillators [93, 153]. For *delayed* coupling the case is entirely different. Here the impulse of one neuron triggers the other neuron to emit a spike, which in turn, after some delay, triggers the first neuron to emit a spike. Hence self-sustained periodic oscillations can be induced without the presence of noise (Fig. 4.39). It is evident that the oscillations of the two neurons have a phase lag of π. The period of the oscillations is given by $T = 2(\tau + \delta)$ with a small quantity $\delta > 0$.

In order to understand this additional phase shift δ, we shall now consider in detail the different stages of the oscillation as marked in Fig. 4.35. Due to the small value of $\epsilon \ll 1$ there is a distinct timescale separation between the fast activators and the slow inhibitors, and a single FitzHugh–Nagumo system performs a fast horizontal transition $A \rightarrow B$, then travels slowly approximately along the right stable branch of the u_1 nullcline $B \rightarrow C$ (*firing*), then jumps back fast to D, and returns slowly to the *rest state* A approximately along the left stable branch of the u_1 nullcline (*refractory phase*). If a is close to unity, these four points are approximately given by $A = (-a, -a + \frac{a^3}{3})$, $B = (2, -\frac{2}{3})$, $C = (1, \frac{2}{3})$, and $D = (-2, \frac{2}{3})$. A rough estimate for A' is $(a - 2, -a + \frac{a^3}{3})$. The two slow phases $B \rightarrow C$ and $D \rightarrow A$ can be approximated by $v_1 \approx u_1 - \frac{u_1^3}{3}$ and hence $\dot{v}_1 \approx \dot{u}_1(1 - u_1^2) = u_1 + a$ which gives

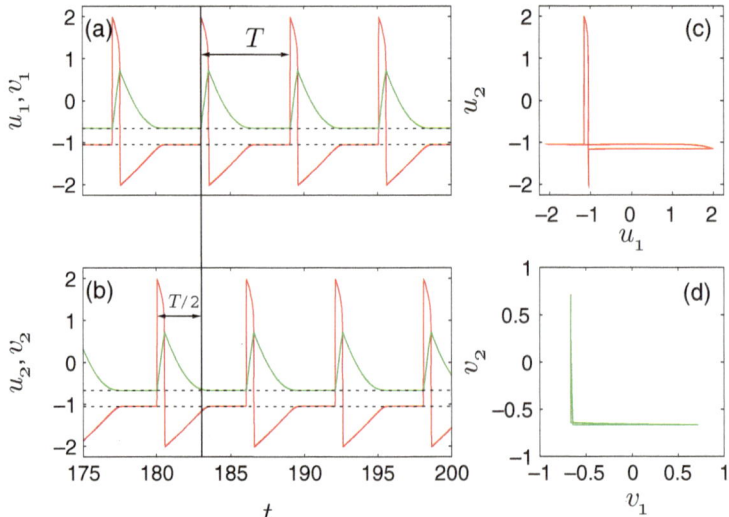

Fig. 4.39 Delay-induced oscillations. (**a**), (**b**): Time series of both subsystems (*red solid lines*: activator u_i, *green solid lines*: inhibitor v_i, and *black dashed lines*: fixed point values of activator and inhibitor). (**c**), (**d**): Phase portraits of activators (**c**) and inhibitors (**d**). Parameters: $a = 1.05$, $\epsilon = 0.01$, $C = 0.5$, and $\tau = 3$ [94]

$$\dot{u}_1 = \frac{u_1 + a}{1 - u_1^2}, \tag{4.130}$$

which can be solved analytically, describing the firing phase (+) and the refractory phase (−):

$$\int_{\pm 2}^{u} dx_1 \frac{1 - u_1^2}{u_1 + a} = (a^2 - 1)\ln\frac{a \pm 2}{a + u} - a(\pm 2 - u) + 2 - \frac{u^2}{2} = t. \tag{4.131}$$

Integrating from B to C gives the firing time

$$T_f = \int_2^1 dx_1 \frac{1 - u_1^2}{u_1 + a} = (a^2 - 1)\ln\frac{a + 2}{a + 1} - a + \frac{3}{2}. \tag{4.132}$$

For $\epsilon = 0.01$, $a = 1.05$ the analytical solution is in good agreement with the numerical solution in Fig. 4.35(b), including the firing time $T_f = 0.482$ (analytical approximation: 0.491).

For a rough estimate, in the following we shall approximate the spike by a rectangular pulse

$$u_1(t) \approx \begin{cases} 2 & \text{if } t < T_f, \\ -a & \text{if } t \geq T_f. \end{cases} \tag{4.133}$$

If the first subsystem is in the rest state, and a spike of the second subsystem arrives at $t = 0$ (after the propagation delay τ), we can approximate the initial dynamic response by linearizing u_1, v_1 around the fixed point (u_1^*, v_1^*) and approximating the feedback by a constant impulse during the firing time T_f. The fast dynamic response along the u_1 direction is then given by

$$\epsilon \delta \dot{u}_1 = \xi \delta u_1 + 2C \qquad (4.134)$$

with $\xi < 0$. This inhomogeneous linear differential equation can be solved with the initial condition $u_1(0) = -a$:

$$u_1(t) = -a + \frac{2C}{|\xi|}(1 - e^{-\frac{|\xi|}{\epsilon}t}). \qquad (4.135)$$

Note that this equation is not valid for large t since (i) the linearization breaks down, and (ii) the pulse duration T_f is exceeded. For small t (4.135) can be expanded as

$$u_1(t) = -a + \frac{2C}{\epsilon}t, \qquad (4.136)$$

which is equivalent to neglecting the upstream flow field $-|\xi|\delta u_1$ in (4.134) near the stable fixed point A compared to the pulling force $2C$ of the remote spike which tries to excite the system toward B. Once the system has crossed the middle branch of the u_1 nullcline at A', the intrinsic flow field accelerates the trajectory fast toward B, initiating the firing state. Therefore there is a turn-on delay δ, given by the time the trajectory takes from A to A', i.e., $u_1(\delta) \approx a - 2$, according to (4.136):

$$\delta = (a - 1)\frac{\epsilon}{C}. \qquad (4.137)$$

Since the finite rise time of the impulse has been neglected in our estimate, the exact solution δ is slightly larger and does not vanish at $a = 1$.

With increasing a the distance $A-A'$ increases and so does δ. The small additional phase shift δ between the spike $u_1(t)$ and the delayed pulse $u_2(t - \tau)$ results in a non-vanishing coupling term at the beginning and at the end of the spike $u_1(t)$. It is the reason (i) that the spike is initiated and (ii) that it is terminated slightly before the turning point of the u_1 nullcline. The latter effect becomes more pronounced if a is increased or τ is decreased (Fig. 4.40). Both lead to a shift of the initial starting point of the spike emission on the left branch of the nullcline toward D, and hence to a longer distance up to the middle branch of the nullcline which has to be overcome by the impulse u_2, hence to a larger turn-on delay δ, and therefore to an earlier termination of the spike u_1. This explains that the firing phase is shortened, and the limit cycle loop is narrowed from both sides with increasing a or decreasing τ, see Fig. 4.40. In the case of $a = 1.05$ and $\tau = 3$ (Fig. 4.40 a), the delay time is large enough for the two subsystems to nearly approach the fixed point A before being perturbed again by the remote signal. If the delay time becomes much smaller, e. g., for $\tau = 0.8$ (Fig. 4.40 b), the excitatory spike of the other subsystem arrives while

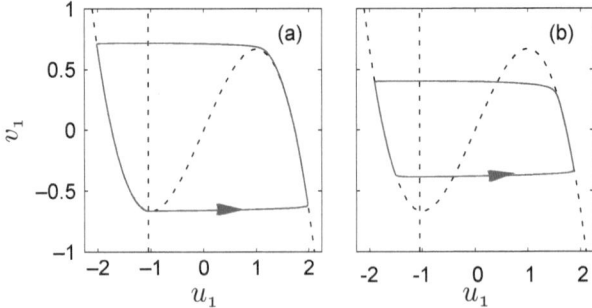

Fig. 4.40 Phase portraits of delay-coupled excitable system (u_1, v_1) for different delay times τ (trajectories: *solid blue* and nullclines: *dashed black*). (**a**) $\tau = 3$ ($\delta = 0.009$) and (**b**) $\tau = 0.8$ ($\delta = 0.015$). Other parameters: $a = 1.05$, $\epsilon = 0.01$, and $C = 0.5$ [94]

the first system is still in the refractory phase, so that it cannot complete the return $D \to A$ to the fixed point. In this case, a in (4.137) has to be substituted by a larger value $\tilde{a}$ with $a < \tilde{a} < 1.7$ in order to get a better estimate of δ. Note that without the phase shift δ the coupling term $C[u_2(t - \tau) - u_1(t)]$ would always vanish in the 2τ periodic state.

Next, we shall investigate conditions upon the coupling parameters C and τ allowing for limit cycle oscillations. On one hand, if τ becomes smaller than some $\tau_{\min}$, the impulse from the excitatory neuron arrives too early to trigger a spike, since the system is still early in its refractory phase. On the other hand, if C becomes too small, the coupling force of the excitatory neuron is too weak to excite the system above its threshold and pull it far enough toward B.

In Fig. 4.41 the regime of oscillations is shown in the parameter plane of the coupling strength C and coupling delay τ. The oscillation period is color coded. The boundary of this colored region is given by the minimum coupling delay $\tau_{\min}$ as a function of C. For large coupling strength, $\tau_{\min}$ is almost independent of C, with decreasing C it sharply increases, and at some small minimum C no oscillations exist at all. At the boundary, the oscillation sets in with finite frequency and amplitude as can be seen in the insets of Fig. 4.41 which show a cut of the parameter plane at $C = 0.8$. The oscillation period increases linearly with τ. The mechanism that generates the oscillation is a saddle-node bifurcation of limit cycles (see inset (b) of Fig. 4.41), creating a pair of a stable and an unstable limit cycle. The unstable limit cycle separates the two attractor basins of the stable limit cycle and the stable fixed point.

4.4.4 Delayed Self-Feedback and Delayed Coupling

In this section we consider the simultaneous action of delayed coupling and delayed self-feedback [94]. Here we choose to apply the self-feedback term symmetrically to both activator equations, but other feedback schemes are also possible:

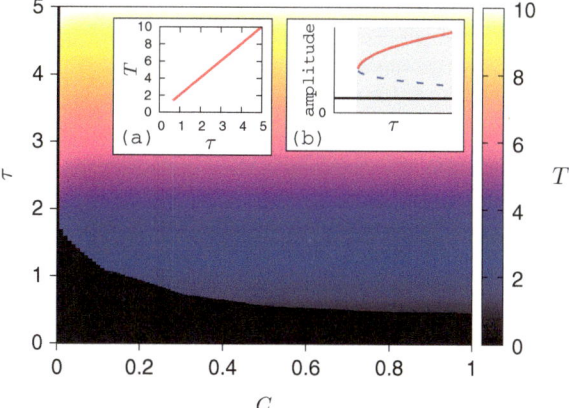

Fig. 4.41 Regime of oscillations in the (τ, C) parameter plane for initial conditions corresponding to single-pulse excitation in one system. The oscillation period T is color coded. The transition between *black* and *color* marks the bifurcation line. Inset (**a**) shows the oscillation period vs. τ in a cut at $C = 0.8$. Inset (**b**): Schematic plot of the saddle-node bifurcation of a stable (*red solid line*) and unstable (*blue dashed*) limit cycle. The maximum oscillation amplitude is plotted vs. the delay time τ and the stable fixed point is plotted as a *solid black line*. The *gray background* marks the bistable region. Parameters: $a = 1.05$, and $\epsilon = 0.01$ [94]

$$\epsilon_1 \dot{u}_1 = u_1 - \frac{u_1^3}{3} - v_1 + C[u_2(t - \tau) - u_1(t)] + K[u_1(t - \tau_K) - u_1(t)]$$
$$\dot{v}_1 = u_1 + a$$
$$\epsilon_2 \dot{u}_2 = u_2 - \frac{u_2^3}{3} - v_2 + C[u_1(t - \tau) - u_2(t)] + K[u_2(t - \tau_K) - u_2(t)]$$
$$\dot{v}_2 = u_2 + a. \tag{4.138}$$

By a linear stability analysis similar to Sect. 4.4.3 it can be shown that the fixed point remains stable for all values of K and τ_K in the case of $a > 1$, as without self-feedback. Redefining $\xi = 1 - a^2 - C - K$, one obtains the factorized characteristic equation

$$1 - \xi\lambda + \epsilon\lambda^2 = \lambda K e^{-\lambda\tau_K} \pm \lambda C e^{-\lambda\tau} \tag{4.139}$$

Substituting the Hopf condition $\lambda = i\omega$ and separating into real and imaginary parts yields for the imaginary part

$$-\xi = K \cos(\omega\tau_K) \pm C \cos(\omega\tau) \tag{4.140}$$

This equation has no solution for $a > 1$ since $|\xi| = a^2 - 1 - |C| - K > C + K$.

The adopted form of control allows for the synchronization of the two cells not only for identical values of τ and τ_K but also generates an intricate pattern of synchronization islands or stripes in the control parameter plane (Fig. 4.42) corresponding to single-spike in-phase and antiphase oscillations with constant

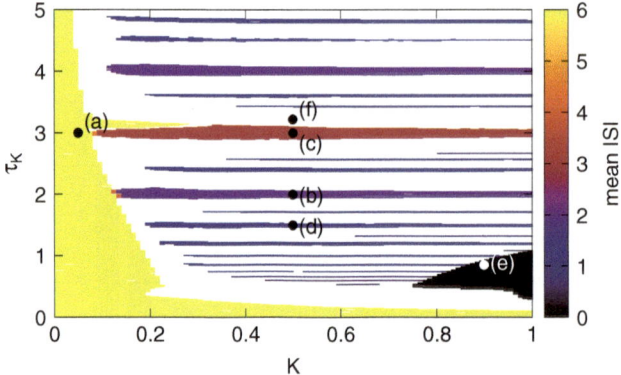

Fig. 4.42 Influence of delayed self-feedback upon coupled oscillations. The mean interspike interval (ISI) is color coded in the control parameter plane of the self-feedback gain K and delay τ_K. *White* areas mark regimes of irregular oscillations where the ISI variance becomes large (> 0.01). Time series corresponding to points (a)–(f) are shown in Fig. 4.43. Other parameters: $a = 1.3$, $\epsilon = 0.01$, $C = 0.5$, and $\tau = 3$ [94]

interspike intervals, see also Fig. 4.43(a–d). Further, for adequately chosen parameter sets of coupling and self-feedback control, we observe effects such as bursting patterns (Fig. 4.43f) and oscillator death (Fig. 4.43e). In addition to these effects, there exists a control parameter regime in which the self-feedback has no effect on the oscillation periods (shaded yellow).

Figure 4.42 shows the control parameter plane for coupling parameters of the uncontrolled system in the oscillatory regime ($C = 0.5$ and $\tau = 3$). We observe three principal regimes: (i) control has no effect on the oscillation period (yellow), although the form of the stable limit cycle is slightly altered (Fig. 4.43a); (ii) islands of in-phase and antiphase synchronization (color coded, see Fig. 4.43 (b–d)); and (iii) oscillator death (black) Fig. 4.43 e).

Figure 4.44 shows the average phase synchronization time as a function of the coupling delay τ and self-feedback delay τ_K for fixed $K = 0.5$. The bright straight rays at rational τ_K/τ indicate long intervals during which both subsystems remain synchronized. A particularly long average synchronization time is found if the two delay times are equal.

In conclusion, we have shown that delayed feedback from other neurons or self-feedback from the same neuron can crucially affect the dynamics of coupled neurons. In the case of noise-induced oscillations in instantaneously coupled neural systems, time-delayed self-feedback can enhance or suppress stochastic synchronization, depending upon the delay time. In the case of delay-coupled neurons without driving noise sources, the propagation delay of the spikes fed back from other neurons can induce periodic antiphase oscillations for sufficiently large coupling strength and delay times. If self-feedback is applied additionally, synchronous zero-lag oscillations can be induced in some ranges of the control parameters, while in other regimes antiphase oscillations or oscillator death as well as more complex bursting patterns can be generated.

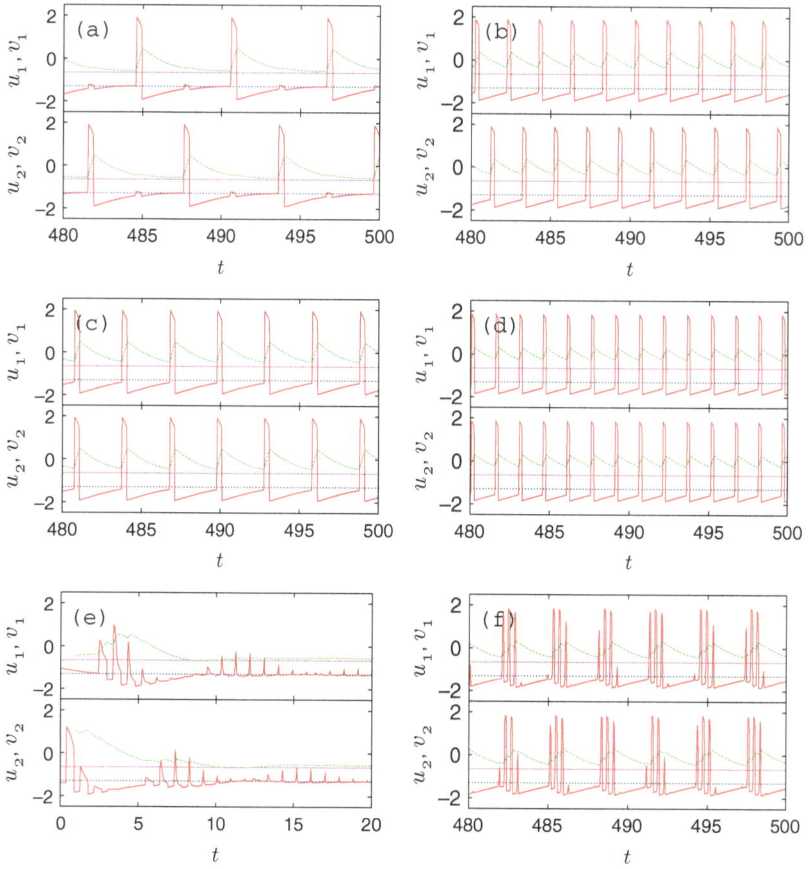

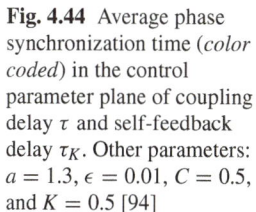

Fig. 4.43 Different modes of oscillation corresponding to different self-feedback parameters K, τ (*red solid lines*: activators $u_i(t)$ and *green solid lines*: inhibitors $v_i(t)$). (**a**), (**b**): Antiphase oscillations for (**a**) $K = 0.05, \tau_K = 3$ (period $T = 6$) and (**b**) $K = 0.5, \tau_K = 2$ ($T = 2$); (**c**), (**d**): in-phase oscillations for (**c**) $K = 0.5, \tau_K = 3$ (period $T = 3$) and (**d**) $K = 0.5, \tau_K = 1.5$ ($T = 1.5$); (**e**): oscillator death for $K = 0.9, \tau_K = 0.9$; and (**f**): bursting pattern for $K = 0.5, \tau_K = 3.2$. Other parameters: $a = 1.3$, $\epsilon = 0.01$, $C = 0.5$, and $\tau = 3$ [94]

Fig. 4.44 Average phase synchronization time (*color coded*) in the control parameter plane of coupling delay τ and self-feedback delay τ_K. Other parameters: $a = 1.3$, $\epsilon = 0.01$, $C = 0.5$, and $K = 0.5$ [94]

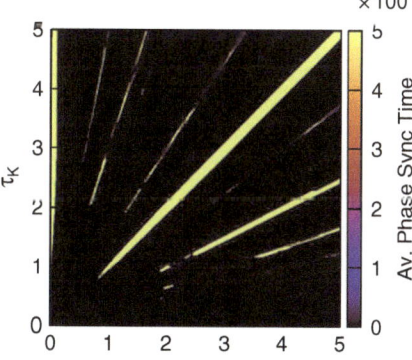

Acknowledgments This work was supported by DFG in the framework of Sfb 555. We are indebted to stimulating collaboration and discussion with A. Amann, A. Balanov, K. Blyuss, S. Brandstetter, T. Dahms, B. Fiedler, D. Gauthier, M. Georgi, B. Hauschildt, F. Henneberger, G. Hiller, N. Janson, W. Just, Y. Kyrychko, A. Panchuk, H. Rittmann-Frank, S. Schiff, S. Schikora, F. Schneider, J. E. S. Socolar, V.Z. Tronciu, M. Wolfrum, H. J. Wünsche, and S. Yanchuk.

References

1. E. Schöll and H. G. Schuster (Eds). *Handbook of Chaos Control*. Wiley-VCH, Weinheim, 2008, second completely revised and enlarged edition.
2. H. Nijmeijer and A. V. D. Schaft. *Nonlinear Dynamical Control Systems*. 3rd ed Springer, New York, 1996.
3. K. Ogata. *Modern Control Engineering*. Prentice-Hall, New York, 1997.
4. A. L. Fradkov, I. V. Miroshnik, and V. O. Nikiforov. *Nonlinear and Adaptive Control of Complex Systems*. Kluwer, Dordrecht, 1999.
5. E. Ott, C. Grebogi, and J. A. Yorke. Controlling chaos. *Phys. Rev. Lett.*, 64, 1196, 1990.
6. K. Pyragas. Continuous control of chaos by self-controlling feedback. *Phys. Lett. A*, 170, 421, 1992.
7. D. J. Gauthier. Resource letter: Controlling chaos. *Am. J. Phys.*, 71, 750, 2003.
8. K. Pyragas. Delayed feedback control of chaos. *Phil. Trans. R. Soc. A*, 364, 2309, 2006.
9. G. Franceschini, S. Bose, and E. Schöll. Control of chaotic spatiotemporal spiking by time-delay autosynchronisation. *Phys. Rev. E*, 60, 5426, 1999.
10. M. Kim, M. Bertram, M. Pollmann, A. von Oertzen, A. S. Mikhailov, H. H. Rotermund, and G. Ertl. Controlling chemical turbulence by global delayed feedback: Pattern formation in catalytic CO oxidation on Pt(110). *Science*, 292, 1357, 2001.
11. O. Beck, A. Amann, E. Schöll, J. E. S. Socolar, and W. Just. Comparison of time-delayed feedback schemes for spatio-temporal control of chaos in a reaction-diffusion system with global coupling. *Phys. Rev. E*, 66, 016213, 2002.
12. N. Baba, A. Amann, E. Schöll, and W. Just. Giant improvement of time-delayed feedback control by spatio-temporal filtering. *Phys. Rev. Lett.*, 89, 074101, 2002.
13. J. Unkelbach, A. Amann, W. Just, and E. Schöll. Time–delay autosynchronization of the spatiotemporal dynamics in resonant tunneling diodes. *Phys. Rev. E*, 68, 026204, 2003.
14. J. Schlesner, A. Amann, N. B. Janson, W. Just, and E. Schöll. Self-stabilization of high frequency oscillations in semiconductor superlattices by time–delay autosynchronization. *Phys. Rev. E*, 68, 066208, 2003.
15. C. Beta, M. Bertram, A. S. Mikhailov, H. H. Rotermund, and G. Ertl. Controlling turbulence in a surface chemical reaction by time-delay autosynchronization. *Phys. Rev. E*, 67, 046224, 2003.
16. C. Beta and A. S. Mikhailov. Controlling spatiotemporal chaos in oscillatory reaction-diffusion systems by time-delay autosynchronization. *Physica D*, 199, 173, 2004.
17. K. A. Montgomery and M. Silber. Feedback control of travelling wave solutions of the complex Ginzburg-Landau equation. *Nonlinearity*, 17, 2225, 2004.
18. E. Schöll, J. Hizanidis, P. Hövel, and G. Stegemann. Pattern formation in semiconductors under the influence of time-delayed feedback control and noise. In: L. Schimansky-Geier, B. Fiedler, J. Kurths, and E. Schöll (eds.), *Analysis and control of complex non-linear processes in physics, chemistry and biology*. World Scientific, Singapore, 2007, pp. 135–183.
19. C. M. Postlethwaite and M. Silber. Stabilizing unstable periodic orbits in the lorenz equations using time-delayed feedback control. *Phys. Rev. E*, 76, 056214, 2007.
20. A. Ahlborn and U. Parlitz. Controlling spatiotemporal chaos using multiple delays. *Phys. Rev. E*, 75, 65202, 2007.

21. A. Ahlborn and U. Parlitz. Control and synchronization of spatiotemporal chaos. *Phys. Rev. E*, 77, 016201, 2008.
22. M. A. Dahlem, F. M. Schneider, and E. Schöll. Failure of feedback as a putative common mechanism of spreading depolarizations in migraine and stroke. *Chaos*, 18, 026110, 2008.
23. F. M. Schneider, E. Schöll, and M. A. Dahlem. Controlling the onset of traveling pulses in excitable media by nonlocal spatial coupling and time delayed feedback. *Chaos*, 19, 015110, 2009.
24. M. Kehrt, P. Hövel, V. Flunkert, M. A. Dahlem, P. Rodin, and E. Schöll. Stabilization of complex spatio-temporal dynamics near a subcritical Hopf bifurcation by time-delayed feedback. *Eur. Phys. J. B*, 68, 557, 2009.
25. Y. N. Kyrychko, K. B. Blyuss, S. J. Hogan, and E. Schöll. Control of spatio-temporal patterns in the Gray-Scott model. *Chaos* 19, 043126, 2009.
26. N. B. Janson, A. G. Balanov, and E. Schöll. Delayed feedback as a means of control of noise-induced motion. *Phys. Rev. Lett.* 93, 010601, 2004.
27. A. G. Balanov, N. B. Janson, and E. Schöll. Control of noise-induced oscillations by delayed feedback. *Physica D*, 199, 1, 2004.
28. J. Pomplun, A. Amann, and E. Schöll. Mean field approximation of time-delayed feedback control of noise-induced oscillations in the Van der Pol system. *Europhys. Lett.*, 71, 366, 2005.
29. N. B. Janson, A. G. Balanov, and E. Schöll. Control of noise-induced dynamics. In: E. Schöll and H. G. Schuster (eds.), *Handbook of Chaos Control*. Wiley-VCH, Weinheim, 2008, chap. 11, pp. 223–274, second completely revised and enlarged edition.
30. G. Hu, T. Ditzinger, C. Z. Ning, and H. Haken. Stochastic resonance without external periodic force. *Phys. Rev. Lett.*, 71, 807, 1993.
31. A. Pikovsky and J. Kurths. Coherence resonance in a noise-driven excitable system. *Phys. Rev. Lett.*, 78, 775, 1997.
32. J. García-Ojalvo and J. M. Sancho. Noise in Spatially Extended Systems. Springer, New York, 1999.
33. C. Masoller. Noise-induced resonance in delayed feedback systems. *Phys. Rev. Lett.*, 88, 034102, 2002.
34. B. Lindner, J. García-Ojalvo, A. Neiman, and L. Schimansky-Geier. Effects of noise in excitable systems. *Phys. Rep.*, 392, 321, 2004.
35. F. Sagués, J. M. Sancho, and J. García-Ojalvo. Spatiotemporal order out of noise. *Rev. Mod. Phys.*, 79, 829, 2007.
36. D. J. Gauthier, D. K. Sukow, H. M. Concannon, and J. E. S. Socolar. Stabilizing unstable periodic orbits in a fast diode resonator using continuous time-delay autosynchronization. *Phys. Rev. E*, 50, 2343, 1994.
37. J. N. Blakely, L. Illing, and D. J. Gauthier. Controlling fast chaos in delay dynamical systems. *Phys. Rev. Lett.*, 92, 193901, 2004.
38. S. Schikora, P. Hövel, H. J. Wünsche, E. Schöll, and F. Henneberger. All-optical noninvasive control of unstable steady states in a semiconductor laser. *Phys. Rev. Lett.*, 97, 213902, 2006.
39. J. E. S. Socolar, D. W. Sukow, and D. J. Gauthier. Stabilizing unstable periodic orbits in fast dynamical systems. *Phys. Rev. E*, 50, 3245, 1994.
40. J. E. S. Socolar and D. J. Gauthier. Analysis and comparison of multiple-delay schemes for controlling unstable fixed points of discrete maps. *Phys. Rev. E*, 57, 6589, 1998.
41. I. Harrington and J. E. S. Socolar. Design and robustness of delayed feedback controllers for discrete systems. *Phys. Rev. E*, 69, 056207, 2004.
42. M. E. Bleich and J. E. S. Socolar. Stability of periodic orbits controlled by time-delay feedback. *Phys. Lett. A*, 210, 87, 1996.
43. W. Just, T. Bernard, M. Ostheimer, E. Reibold, and H. Benner. Mechanism of time-delayed feedback control. *Phys. Rev. Lett.*, 78, 203, 1997.
44. H. Nakajima. On analytical properties of delayed feedback control of chaos. *Phys. Lett. A*, 232, 207, 1997.

45. K. Pyragas. Analytical properties and optimization of time-delayed feedback control. *Phys. Rev. E*, 66, 26207, 2002.
46. W. Just, H. Benner, and E. Schöll. Control of chaos by time–delayed feedback: a survey of theoretical and experimental aspects. In: B. Kramer (ed.), *Advances in Solid State Physics*. Springer, Berlin, 2003, vol. 43, pp. 589–603.
47. P. Hövel and J. E. S. Socolar. Stability domains for time-delay feedback control with latency. *Phys. Rev. E*, 68, 036206, 2003.
48. P. Hövel and E. Schöll. Control of unstable steady states by time-delayed feedback methods. *Phys. Rev. E*, 72, 046203, 2005.
49. S. Yanchuk, M. Wolfrum, P. Hövel, and E. Schöll. Control of unstable steady states by long delay feedback. *Phys. Rev. E*, 74, 026201, 2006.
50. T. Dahms, P. Hövel, and E. Schöll. Control of unstable steady states by extended time-delayed feedback. *Phys. Rev. E*, 76, 056201, 2007.
51. A. Amann, E. Schöll, and W. Just. Some basic remarks on eigenmode expansions of time-delay dynamics. *Physica A*, 373, 191, 2007.
52. B. Fiedler, V. Flunkert, M. Georgi, P. Hövel, and E. Schöll. Refuting the odd number limitation of time-delayed feedback control. *Phys. Rev. Lett.*, 98, 114101, 2007.
53. W. Just, B. Fiedler, V. Flunkert, M. Georgi, P. Hövel, and E. Schöll. Beyond odd number limitation: a bifurcation analysis of time-delayed feedback control. *Phys. Rev. E*, 76, 026210, 2007.
54. B. Fiedler, S. Yanchuk, V. Flunkert, P. Hövel, H. J. Wünsche, and E. Schöll. Delay stabilization of rotating waves near fold bifurcation and application to all-optical control of a semiconductor laser. *Phys. Rev. E*, 77, 066207, 2008.
55. A. G. Balanov, N. B. Janson, and E. Schöll. Delayed feedback control of chaos: Bifurcation analysis. *Phys. Rev. E*, 71, 016222, 2005.
56. J. Hizanidis, R. Aust, and E. Schöll. Delay-induced multistability near a global bifurcation. *Int. J. Bifur. Chaos*, 18, 1759, 2008.
57. K. Pyragas. Control of chaos via extended delay feedback. *Phys. Lett. A*, 206, 323, 1995.
58. A. Ahlborn and U. Parlitz. Stabilizing unstable steady states using multiple delay feedback control. *Phys. Rev. Lett.*, 93, 264101, 2004.
59. A. Ahlborn and U. Parlitz. Controlling dynamical systems using multiple delay feedback control. *Phys. Rev. E*, 72, 016206, 2005.
60. A. Gjurchinovski and V. Urumov. Stabilization of unstable steady states by variable delay feedback control. *Europhys. Lett.*, 84, 40013, 2008.
61. E. Schöll. Delayed feedback control of chaotic spatio-temporal patterns in semiconductor nanostructures. In: E. Schöll and H. G. Schuster (eds), [1], chap. 24, pp. 533–558, second completely revised and enlarged edition.
62. E. Schöll. Pattern formation and time-delayed feedback control at the nano-scale. In: G. Radons, B. Rumpf, and H. G. Schuster (eds.), *Nonlinear Dynamics of Nanosystems*. Wiley-VCH, Weinheim, 2009, pp. 325–367.
63. A. S. Mikhailov and K. Showalter. Control of waves, patterns and turbulence in chemical systems. *Phys. Rep.*, 425, 79, 2006.
64. P. Parmananda, R. Madrigal, M. Rivera, L. Nyikos, I. Z. Kiss, and V. Gáspár. Stabilization of unstable steady states and periodic orbits in an electrochemical system using delayed-feedback control. *Phys. Rev. E*, 59, 5266, 1999.
65. P. Parmananda. Tracking fixed-point dynamics in an electrochemical system using delayed-feedback control. *Phys. Rev. E*, 67, 045202R, 2003.
66. D. Battogtokh and A. S. Mikhailov. Controlling turbulence in the complex Ginzburg-Landau equation. *Physica D*, 90, 84, 1996.
67. J. Schlesner, V. Zykov, H. Engel, and E. Schöll. Stabilization of unstable rigid rotation of spiral waves in excitable media. *Phys. Rev. E*, 74, 046215, 2006.
68. G. J. E. Santos, J. Escalona, and P. Parmananda. Regulating noise-induced spiking using feedback. *Phys. Rev. E*, 73, 042102, 2006.

69. I. Z. Kiss, C. G. Rusin, H. Kori, and J. L. Hudson. Engineering Complex Dynamical Structures: Sequential Patterns and Desynchronization. *Science*, 316, 1886, 2007.
70. Y. Zhai, I. Z. Kiss, and J. L. Hudson. Control of complex dynamics with time-delayed feedback in populations of chemical oscillators: Desynchronization and clustering. *Ind. Eng. Chem. Res.*, 47, 3502, 2008.
71. M. G. Rosenblum and A. Pikovsky. Controlling synchronization in an ensemble of globally coupled oscillators. *Phys. Rev. Lett.*, 92, 114102, 2004.
72. O. V. Popovych, C. Hauptmann, and P. A. Tass. Effective desynchronization by nonlinear delayed feedback. *Phys. Rev. Lett.*, 94, 164102, 2005.
73. K. Hall, D. J. Christini, M. Tremblay, J. J. Collins, L. Glass, and J. Billette. Dynamic control of cardiac alterans. *Phys. Rev. Lett.*, 78, 4518, 1997.
74. F. M. Atay. Distributed delays facilitate amplitude death of coupled oscillators. *Phys. Rev. Lett.*, 91, 094101, 2003.
75. H. Haken. *Brain Dynamics: Synchronization and Activity Patterns in Pulse-Coupled Neural Nets with Delays and Noise.* Springer Verlag GmbH, Berlin, 2006.
76. K. Pyragas, O. V. Popovych, and P. A. Tass. Controlling synchrony in oscillatory networks with a separate stimulation-registration setup. *Europhys. Lett.*, 80, 40002, 2007.
77. M. Gassel, E. Glatt, and F. Kaiser. Time-delayed feedback in a net of neural elements: Transitions from oscillatory to excitable dynamics. *Fluct. Noise Lett.*, 7, L225, 2007.
78. M. Gassel, E. Glatt, and F. Kaiser. Delay-sustained pattern formation in subexcitable media. *Phys. Rev. E*, 77, 066220, 2008.
79. O. D'Huys, R. Vicente, T. Erneux, J. Danckaert, and I. Fischer. Synchronization properties of network motifs: Influence of coupling delay and symmetry. *Chaos*, 18, 037116, 2008.
80. G. C. Sethia, A. Sen, and F. M. Atay. Clustered chimera states in delay-coupled oscillator systems. *Phys. Rev. Lett.*, 100, 144102, 2008.
81. W. Kinzel, J. Kestler, and I. Kanter. *Chaos pass filter: Linear response of synchronized chaotic systems.* 2008. http://arxiv.org/abs/0806.4291
82. M. Zigzag, M. Butkovski, A. Englert, W. Kinzel, and I. Kanter. Zero-lag synchronization of chaotic units with time-delayed couplings. *Europhys. Lett.*, 85, 60005, 2009.
83. E. Schöll and K. Pyragas. Tunable semiconductor oscillator based on self-control of chaos in the dynamic Hall effect. *Europhys. Lett.*, 24, 159, 1993.
84. D. Reznik and E. Schöll. Oscillation modes, transient chaos and its control in a modulation-doped semiconductor double-heterostructure. *Z. Phys. B*, 91, 309, 1993.
85. D. P. Cooper and E. Schöll. Tunable real space transfer oscillator by delayed feedback control of chaos. *Z. f. Naturforsch.*, 50a, 117, 1995.
86. W. Just, S. Popovich, A. Amann, N. Baba, and E. Schöll. Improvement of time–delayed feedback control by periodic modulation: Analytical theory of Floquet mode control scheme. *Phys. Rev. E*, 67, 026222, 2003.
87. E. Schöll, A. G. Balanov, N. B. Janson, and A. Neiman. Controlling stochastic oscillations close to a Hopf bifurcation by time-delayed feedback. *Stoch. Dyn.*, 5, 281, 2005.
88. J. Pomplun, A. G. Balanov, and E. Schöll. Long-term correlations in stochastic systems with extended time-delayed feedback. *Phys. Rev. E*, 75, 040101(R), 2007.
89. T. Prager, H. P. Lerch, L. Schimansky-Geier, and E. Schöll. Increase of coherence in excitable systems by delayed feedback. *J. Phys. A*, 40, 11045, 2007.
90. A. Pototsky and N. B. Janson. Correlation theory of delayed feedback in stochastic systems below Andronov-Hopf bifurcation. *Phys. Rev. E*, 76, 056208, 2007.
91. A. Pototsky and N. B. Janson. Excitable systems with noise and delay, with applications to control: Renewal theory approach. *Phys. Rev. E*, 77, 031113, 2008.
92. A. G. Balanov, V. Beato, N. B. Janson, H. Engel, and E. Schöll. Delayed feedback control of noise-induced patterns in excitable media. *Phys. Rev. E*, 74, 016214, 2006.
93. B. Hauschildt, N. B. Janson, A. G. Balanov, and E. Schöll. Noise-induced cooperative dynamics and its control in coupled neuron models. *Phys. Rev. E*, 74, 051906, 2006.
94. E. Schöll, G. Hiller, P. Hövel, and M. A. Dahlem. Time-delayed feedback in neurosystems. *Phil. Trans. R. Soc. A*, 367, 1079, 2009.

95. V. Flunkert and E. Schöll. Suppressing noise-induced intensity pulsations in semiconductor lasers by means of time-delayed feedback. *Phys. Rev. E*, 76, 066202, 2007.

96. J. Hizanidis, A. G. Balanov, A. Amann, and E. Schöll. Noise-induced oscillations and their control in semiconductor superlattices. *Int. J. Bifur. Chaos*, 16, 1701, 2006.

97. J. Hizanidis, A. G. Balanov, A. Amann, and E. Schöll. Noise-induced front motion: signature of a global bifurcation. *Phys. Rev. Lett.*, 96, 244104, 2006.

98. J. Hizanidis and E. Schöll. Control of noise-induced spatiotemporal patterns in superlattices. *phys. stat. sol. (c)*, 5, 207, 2008.

99. J. Hizanidis and E. Schöll. Control of coherence resonance in semiconductor superlattices. *Phys. Rev. E*, 78, 066205, 2008.

100. G. Stegemann, A. G. Balanov, and E. Schöll. Delayed feedback control of stochastic spatiotemporal dynamics in a resonant tunneling diode. *Phys. Rev. E*, 73, 016203, 2006.

101. E. Schöll, N. Majer, and G. Stegemann. Extended time delayed feedback control of stochastic dynamics in a resonant tunneling diode. *phys. stat. sol. (c)*, 5, 194, 2008.

102. N. Majer and E. Schöll. Resonant control of stochastic spatio-temporal dynamics in a tunnel diode by multiple time delayed feedback. *Phys. Rev. E*, 79, 011109, 2009.

103. K. Pyragas, V. Pyragas, I. Z. Kiss, and J. L. Hudson. Stabilizing and tracking unknown steady states of dynamical systems. *Phys. Rev. Lett.*, 89, 244103, 2002.

104. S. Bielawski, M. Bouazaoui, D. Derozier, and P. Glorieux. Stabilization and characterization of unstable steady states in a laser. *Phys. Rev. A*, 47, 3276, 1993.

105. A. Chang, J. C. Bienfang, G. M. Hall, J. R. Gardner, and D. J. Gauthier. Stabilizing unstable steady states using extended time-delay autosynchronisation. *Chaos*, 8, 782, 1998.

106. K. Pyragas, V. Pyragas, I. Z. Kiss, and J. L. Hudson. Adaptive control of unknown unstable steady states of dynamical systems. *Phys. Rev. E*, 70, 026215, 2004.

107. E. M. Wright. The linear difference-differential equation with constant coefficients. *Proc. R. Soc. Edinburgh, Sect. A: Math. Phys. Sci.*, 62, 387, 1949.

108. E. M. Wright. A non-linear difference-differential equation. *J. Reine Angew. Math.*, 194, 66, 1955.

109. R. Bellmann and K. L. Cooke. *Differential-Difference Equations*. Academic Press, New York, 1963.

110. J. K. Hale. *Functional Differential Equations*. Applied Mathematical Sciences Vol. 3, Springer, New York, 1971.

111. F. M. Asl and A. G. Ulsoy. Analysis of a system of linear delay differential equations. *ASME J. Dyn. Syst., Meas., Control*, 125, 215, 2003.

112. W. Just, E. Reibold, H. Benner, K. Kacperski, P. Fronczak, and J. Holyst. Limits of time-delayed feedback control. *Phys. Lett. A*, 254, 158, 1999.

113. S. Yanchuk and M. Wolfrum. Instabilities of equilibria of delay-differential equations with large delay. In: *Proc. 5th EUROMECH Nonlinear Dynamics Conference ENOC-2005, Eindhoven*, edited by D. H. van Campen, M. D. Lazurko, and W. P. J. M. van den Oever (Eindhoven University of Technology, Eindhoven, Netherlands, 2005), pp. 08–010, eNOC Eindhoven (CD ROM), ISBN 90 386 2667 3.

114. S. Lepri, G. Giacomelli, A. Politi, and F. T. Arecchi. High-dimensional chaos in delayed dynamical-systems. *Physica D*, 70, 235, 1994.

115. K. Pyragas and A. Tamaševičius. Experimental control of chaos by delayed self-controlling feedback. *Phys. Lett. A*, 180, 99, 1993.

116. S. Bielawski, D. Derozier, and P. Glorieux. Controlling unstable periodic orbits by a delayed continuous feedback. *Phys. Rev. E*, 49, R971, 1994.

117. T. Pierre, G. Bonhomme, and A. Atipo. Controlling the chaotic regime of nonlinear ionization waves using time-delay autosynchronisation method. *Phys. Rev. Lett.*, 76, 2290, 1996.

118. D. W. Sukow, M. E. Bleich, D. J. Gauthier, and J. E. S. Socolar. Controlling chaos in a fast diode resonator using time-delay autosynchronisation: Experimental observations and theoretical analysis. *Chaos*, 7, 560, 1997.

119. O. Lüthje, S. Wolff, and G. Pfister. Control of chaotic taylor-couette flow with time-delayed feedback. *Phys. Rev. Lett.*, 86, 1745, 2001.

120. J. M. Krodkiewski and J. S. Faragher. Stabilization of motion of helicopter rotor blades using delayed feedback - modelling, computer simulation and experimental verification. *J. Sound Vib.*, 234, 591, 2000.

121. T. Fukuyama, H. Shirahama, and Y. Kawai. Dynamical control of the chaotic state of the current-driven ion acoustic instability in a laboratory plasma using delayed feedback. *Phys. Plasmas*, 9, 4525, 2002.

122. C. von Loewenich, H. Benner, and W. Just. Experimental relevance of global properties of time-delayed feedback control. *Phys. Rev. Lett.*, 93, 174101, 2004.

123. H. Nakajima and Y. Ueda. Limitation of generalized delayed feedback control. *Physica D*, 111, 143, 1998.

124. I. Harrington and J. E. S. Socolar. Limitation on stabilizing plane waves via time-delay feedback. *Phys. Rev. E*, 64, 056206, 2001.

125. K. Pyragas, V. Pyragas, and H. Benner. Delayed feedback control of dynamical systems at subcritical Hopf bifurcation. *Phys. Rev. E*, 70, 056222, 2004.

126. V. Pyragas and K. Pyragas. Delayed feedback control of the Lorenz system: An analytical treatment at a subcritical Hopf bifurcation. *Phys. Rev. E*, 73, 036215, 2006.

127. H. G. Schuster and M. B. Stemmler. Control of chaos by oscillating feedback. *Phys. Rev. E*, 56, 6410, 1997.

128. H. Nakajima and Y. Ueda. Half-period delayed feedback control for dynamical systems with symmetries. *Phys. Rev. E*, 58, 1757, 1998.

129. K. Pyragas. Control of chaos via an unstable delayed feedback controller. *Phys. Rev. Lett.*, 86, 2265, 2001.

130. B. Fiedler, V. Flunkert, M. Georgi, P. Hövel, and E. Schöll. Beyond the odd number limitation of time-delayed feedback control. In: E. Schöll and H. G. Schuster (eds.), *Handbook of Chaos Control*. Wiley-VCH, Weinheim, 2008, pp. 73–84, second completely revised and enlarged edition.

131. O. Diekmann, S. A. van Gils, S. M. Verduyn Lunel, and H. O. Walther. *Delay Equations*. Springer-Verlag, New York, 1995.

132. Y. A. Kuznetsov. *Elements of Applied Bifurcation Theory*. Springer, New York, 1995.

133. Z. Gills, C. Iwata, R. Roy, I. B. Schwartz, and I. Triandaf. Tracking unstable steady states: Extending the stability regime of a multimode laser system. *Phys. Rev. Lett.*, 69, 3169, 1992.

134. A. Ahlborn and U. Parlitz. Chaos control using notch feedback. *Phys. Rev. Lett.*, 96, 034102, 2006.

135. S. Bauer, O. Brox, J. Kreissl, B. Sartorius, M. Radziunas, J. Sieber, H. J. Wünsche, and F. Henneberger. Nonlinear dynamics of semiconductor lasers with active optical feedback. *Phys. Rev. E*, 69, 016206, 2004.

136. H. J. Wünsche, S. Bauer, J. Kreissl, O. Ushakov, N. Korneyev, F. Henneberger, E. Wille, H. Erzgräber, M. Peil, W. Elsäßer, and I. Fischer. Synchronization of delay-coupled oscillators: A study of semiconductor lasers. *Phys. Rev. Lett.*, 94, 163901, 2005.

137. W. Lu and R. G. Harrison. Controlling chaos using continuous interference feedback: proposal for all optical devices. *Opt. Commu.*, 109, 457, 1994.

138. C. Simmendinger and O. Hess. Controlling delay-induced chaotic behavior of a semiconductor laser with optical feedback. *Phys. Lett. A*, 216, 97, 1996.

139. V. Z. Tronciu, H. J. Wünsche, M. Wolfrum, and M. Radziunas. Semiconductor laser under resonant feedback from a Fabry-Perot: Stability of continuous-wave operation. *Phys. Rev. E*, 73, 046205, 2006.

140. B. Dahmani, L. Hollberg, and R. Drullinger. Frequency stabilization of semiconductor lasers by resonant optical feedback. *Opt. Lett.*, 12, 876, 1987.

141. P. Laurent, A. Clairon, and C. Breant. Frequency noise analysis of optically self-locked diode lasers. *IEEE J. Quantum Electron.*, 25, 1131, 1989.

142. M. Peil, I. Fischer, and W. Elsäßer. Spectral braodband dynamics of seminconductor lasers with resonant short cavities. *Phys. Rev. A*, 73, 23805, 2006.

143. H. Erzgräber, B. Krauskopf, D. Lenstra, A. P. A. Fischer, and G. Vemuri. Frequency versus relaxation oscillations in a semiconductor laser with coherent filtered optical feedback. *Phys. Rev. E*, 73, 055201(R), 2006.

144. T. Dahms, P. Hövel, and E. Schöll. Stabilizing continuous-wave output in semiconductor lasers by time-delayed feedback. *Phys. Rev. E*, 78, 056213, 2008.

145. M. G. Rosenblum and A. Pikovsky. Delayed feedback control of collective synchrony: An approach to suppression of pathological brain rhythms. *Phys. Rev. E*, 70, 041904, 2004.

146. H. J. Wünsche, S. Schikora, and F. Henneberger. Noninvasive control of semiconductor lasers by delayed optical feedback. In: E. Schöll and H. G. Schuster (eds.), *Handbook of Chaos Control*. Wiley-VCH, Weinheim, 2008, second completely revised and enlarged edition.

147. R. Lang and K. Kobayashi. External optical feedback effects on semiconductor injection laser properties. *IEEE J. Quantum Electron*, 16, 347, 1980.

148. G. P. Agrawal and G. R. Gray. Effect of phase-conjugate feedback on the noise characteristics of semiconductor-lasers. *Phys. Rev. A*, 46, 5890, 1992.

149. P. M. Alsing, V. Kovanis, A. Gavrielides, and T. Erneux. Lang and Kobayashi phase equation. *Phys. Rev. A*, 53, 4429, 1996.

150. G. P. Agrawal and N. K. Dutta. *Semiconductor Lasers*. Van Nostrand Reinhold, New York, 1993.

151. R. FitzHugh. Impulses and physiological states in theoretical models of nerve membrane. *Biophys. J.*, 1, 445, 1961.

152. J. Nagumo, S. Arimoto, and S. Yoshizawa. An active pulse transmission line simulating nerve axon. *Proc. IRE*, 50, 2061, 1962.

153. P. Hövel, M. A. Dahlem, and E. Schöll. Control of synchronization in coupled neural systems by time-delayed feedback. *Int. J. Bifur. Chaos*, 2010, in print (arxiv:0809.0819v1).

154. M. A. Dahlem, G. Hiller, A. Panchuk, and E. Schöll. Dynamics of delay-coupled excitable neural systems. *Int. J. Bifur. Chaos*, 19, 745, 2009.

155. D. Schmitz, S. Schuchmann, A. Fisahn, A. Draguhn, E. H. Buhl, E. Petrasch-Parwez, R. Dermietzel, U. Heinemann, and R. D. Traub. Axo-axonal coupling. a novel mechanism for ultrafast neuronal communication. *Neuron*, 31, 831, 2001.

156. D. T. J. Liley and J. J. Wright. Intracortical connectivity of pyramidal and stellate cells: estimates of synaptic densities and coupling symmetry. *Network: Comput. Neural Syst.*, V5, 175, 1994.

157. R. D. Pinto, P. Varona, A. R. Volkovskii, A. Szücs, H. D. I. Abarbanel, and M. I. Rabinovich. Synchronous behavior of two coupled electronic neurons. *Phys. Rev. E*, 62, 2644, 2000.

158. F. F. De-Miguel, M. Vargas-Caballero, and E. García-Pérez. Spread of synaptic potentials through electrical synapses in Retzius neurones of the leech. *J. Exp. Biol.*, 204, 3241, 2001.

159. N. Buric and D. Todorovic. Dynamics of FitzHugh-Nagumo excitable systems with delayed coupling. *Phys. Rev. E*, 67, 066222, 2003.

160. V. Hadamschek. Brain stimulation techniques via nonlinear delayed neurofeedback based on MEG inverse methods. PhD Thesis, TU Berlin, 2006.

161. P. Tass. Effective desynchronization with bipolar double-pulse stimulation. *Phys. Rev. E*, 66, 036226, 2002.

162. M. G. Rosenblum, A. Pikovsky, and J. Kurths. *Synchronization – A universal concept in nonlinear sciences*. Cambridge University Press, Cambridge, 2001.

Chapter 5
Finite Propagation Speeds in Spatially Extended Systems

Axel Hutt

5.1 Introduction

In recent decades finite propagation speeds have been observed experimentally in spatially extended systems. For instance, in neural and biological systems they have been found to evoke novel spatio-temporal phenomena [4, 8–10, 12, 15, 21, 22, 24, 26, 28, 29, 37, 39]. This effect may be understood by the similarity of the delay caused by the finite propagation speed and other intrinsic timescales. For instance, axonal propagation speeds in intracortical neural connections are found in the range of 0.1–1 m/s, while intracortical distances vary in a range of 10^{-4}–10^{-3} m [38]. Hence the intrinsic propagation delay of cortical areas may vary in the range 10^{-4}–10^{-2} s. In comparison, chemical synapses in neural systems respond to incoming activity at similar timescales between 10^{-4} and 10^{-1} s [31] and thus interact with the activity on the timescale of propagation delays [24, 21].

In spatially extended physical systems, the effect of a finite or maximum propagation speed on the system's spatio-temporal dynamics is not much studied. In this context diffusion processes are very interesting. Normal and anomalous diffusive activity [36, 30] is known to spread infinitely fast over space. Although this infinite propagation speed is unphysical, most diffusion processes studied have been modeled successfully by diffusion equations, as diffusion processes in spatial systems evolve at a much lower speed than the maximum propagation speed of the system. In recent years, however, experiments have detected ultra-fast activity pulse propagation in solids and plasma [33, 35, 42]. To describe these effects mathematically, models have been extended successfully by introducing an additional timescale corresponding to the finite propagation speed [7, 33, 42]. The present chapter extends this approach and studies finite propagation speed effects in some detail for general spatially extended systems.

Our work is inspired by achievements on the effect of propagation delays in neural systems. According to the intrinsic spatial structure in biological neural systems,

A. Hutt (✉)
INRIA CR Nancy - Grand Est Équipe CORTEX, CS20101, 54603 Villers-ls-Nancy Cedex, France
e-mail: axel.hutt@loria.fr

F.M. Atay (ed.), *Complex Time-Delay Systems*, Understanding Complex Systems,
DOI 10.1007/978-3-642-02329-3_5, © Springer-Verlag Berlin Heidelberg 2010

most corresponding mathematical studies examine an integral–differential equation which involves spatial nonlocal interactions and finite propagation speed. Interestingly, this model type generalizes several well-known pattern formation systems in one spatial dimension such as standard partial differential equation systems [21, 37]. This can be understood easily by the equation

$$\int_{-\infty}^{\infty} dy K(x - y) S(V(y)) = \sum_{n=0}^{\infty} (-1)^n K_n \frac{\partial^n S(V(x))}{\partial x^n}, \qquad (5.1)$$

with $K_n = \int d\eta K(\eta) \eta^n / n!$ $\forall n \in \mathbb{N}$. The integral on the left-hand side of (5.1) represents the nonlocal spatial interaction and the spatial kernel $K(x - y)$ may be interpreted as the probability density of the spatial interaction between two spatial locations x and y. On the right-hand side of (5.1), the infinite sum of partial spatial derivatives sums up spatial interactions of all orders. For instance, truncating this series after the second order yields locally interacting processes, i.e., diffusion processes [21]. Subsequently, the integral–differential equations studied below generalizes a broad class of pattern forming systems involving finite propagation speeds.

The next section studies spatial systems subject to an external input constant in space and time. Stability criteria are derived and studied with respect to the finite propagation speed. Further, the characteristic temporal scale of the system is introduced and we shall learn how it defines the stability of the system. In addition, the phase speed of traveling waves is derived and compared to the intrinsic propagation speed. Then the subsequent section takes into account external random fluctuations and we discuss the stability of equilibria. The application to a specific model extracts additional effects of the propagation delay on the stationary distribution of the random fluctuations.

5.2 Dynamics in the Absence of Noise

In the following we study spatial systems involving a propagation delay while neglecting external random fluctuations. At first, the study of a specific neural model serves to illustrate the mathematical problem and discusses the analysis steps applied in some detail. Then the subsequent section generalizes these analysis steps by the application to a generic model. The application to a non-standard spatial interaction type elicits the power of this generalization and reveals a non-standard pattern forming mechanism.

5.2.1 A Neural Field Model

Let us start with a well-studied model, which allows for the mathematical description of the spatio-temporal activity evolution in intracortical neural populations [24]. The neural activity $V(x, t)$ obeys the integral–differential equation

$$\left[\frac{\partial^2}{\partial t^2} + 2\frac{\partial}{\partial t} + 1\right] V(x,t) = \int_{-\infty}^{\infty} dx' K(|x - x'|) S\left[V\left(x', t - \frac{|x - x'|}{v}\right)\right] + I(x,t)$$

$$(5.2)$$

$$K(|x - x'|) = \frac{a_e}{2} e^{-|x-x'|} - \frac{a_i r}{2} e^{-|x-x'|r}$$

with the sigmoid function $S[V] = 1/(1 + \exp(-1.8(V - 3.0)))$ [14, 40] and the external input $I(x,t)$. We remark that this equation type is different from integro-differential equations, which exhibits the derivative and the integral in the same variable. In contrast, (5.2) shows a temporal derivative and a spatial integral. The temporal operator on the left-hand side of (5.2) reflects the linear response of chemical synapses and specifies the temporal scale to 1. The integral on the right-hand side takes into account the nonlocal spatial interactions along axonal fibers between neurons, which are defined by the spatial interaction kernel $K(x)$ (Fig. 5.1). The parameter r represents the relation of the spatial range of excitation to the spatial range of inhibition. For $r < 1$ ($r > 1$), the excitation spread is shorter (longer) than the spatial spread of inhibition. Thus r represents the characteristic spatial scale of the system [2]. Further, the interaction is delayed by the finite propagation speed v.

First insights into the spatio-temporal dynamics of the system are gained by the analysis of the linear stability of the system. We assume that the system rests at an equilibrium and shows no spatial structure. In other words, the system is constant in space and time with $V(x,t) = V_0 = $ constant and $I(x,t) = I_0 = $ constant. Now the question arises how the system evolves if small perturbations are applied to the system. It is well known that these deviations may cause various spatio-temporal phenomena [1, 11, 19] if the system parameters fulfill certain conditions. Recall that an equilibrium is called stable if small perturbations applied to this state yields a new system state which evolves in a small neighborhood of the equilibrium for all time. Further, an equilibrium is said to be asymptotically stable if the small perturbations applied to the state are damped out and the perturbed system converges to the equilibrium after infinite time [16]. To study the dynamics of these deviations, we

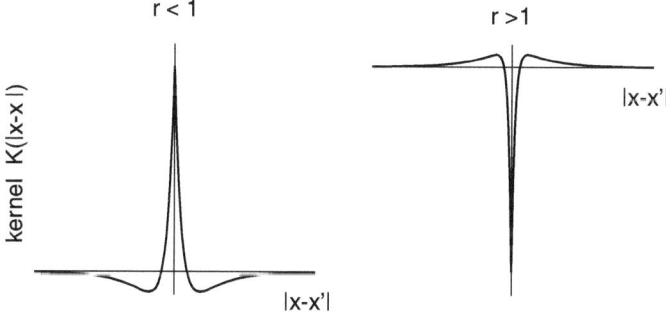

Fig. 5.1 Two cases of spatial interactions with $K(|x-x'|) = a_e \exp[-|x-x'|] - a_i r \exp[|x-x'|r]$. For $r < 1$, the system exhibits local excitation and lateral inhibition, while $r \geq 1$ represents local inhibition and lateral excitation

apply a linear stability analysis to examine the model with respect to the emergence of spatio-temporal pattern.

5.2.1.1 The Stationary Solution

From (5.2), the constant solution V_0 obeys the implicit equation

$$(a_e - a_i)S(V_0) - V_0 + I_0 = 0, \tag{5.3}$$

which can be solved numerically. We point out that a_e and a_i represent the total excitation and total inhibition in the system, respectively. Thus the constant equilibrium state is independent of the choice of the spatial interaction, i.e., independent of r. Figure 5.2 shows a graphical construction of (5.3) for $a_e > a_i$. For $-0.32 < I_0 < 1.32$, there are three solutions, while outside that interval there is a single solution. We observe that the external input I_0 determines the solutions of (5.3) and thus represents the new control parameter of the system.

5.2.1.2 The Stability Threshold

To gain some information on the stability of each of the equilibria, we examine small deviations from the equilibria which may vary in space and time

$$V(x,t) = V_0 + \int_{-\infty}^{\infty} dk\, u(k)e^{ikx+\lambda(k)t}, \tag{5.4}$$

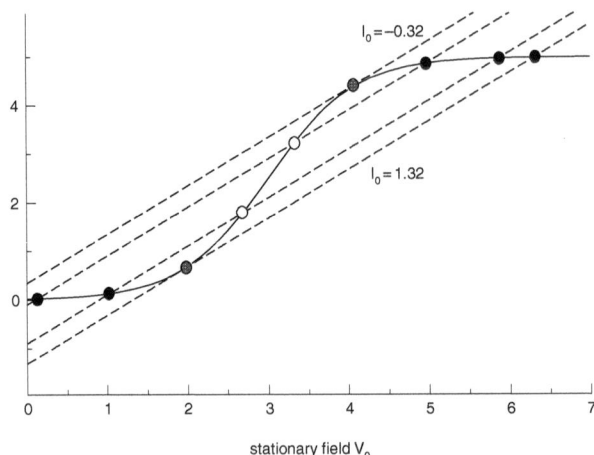

Fig. 5.2 Construction of the constant stationary equilibria of (5.3). *Filled* and *void circles* represent stable and unstable equilibria. At the critical values $I = -0.32$ and $I = 1.32$, there is a saddle node solution (*hatched circle*) synchronous to a stable equilibrium. Parameters are $a_e = 10, a_i = 5$

with constants in time $u(k)$. Here the spatial deviations are formulated by their Fourier transform which may exponentially increase (decrease) in time for positive (negative) real parts of the Lyapunov exponent $\lambda(k)$. This formulation represents a decomposition of the spatial deviations into spatial modes with the corresponding wave number k. Hence the sign of the Lyapunov exponent $\lambda(k)$ defines the stability of spatial modes with wave number k. By inserting (5.4) into (5.2) and utilizing (5.3), we obtain

$$
\begin{aligned}
0 &= \int_{-\infty}^{\infty} dk\, u(k) e^{\lambda(k)t} \times \\
&\quad \left[\left(\lambda^2(k) + 2\lambda(k) + 1 \right) e^{ikx} - S' \int_{-\infty}^{\infty} dx'\, K(x-x') e^{ikx' - \lambda(k)|x-x'|/v} \right] \\
&= \int_{-\infty}^{\infty} dk\, u(k) e^{ikx + \lambda(k)t} \left[\lambda^2(k) + 2\lambda(k) + 1 - S' \int_{-\infty}^{\infty} dz K(z) e^{-\lambda(k)|z|/v - ikz} \right]
\end{aligned}
$$

with the functional derivative $S' = \delta S/\delta V$ computed at $V = V_0$, which can be modified by changing the control parameter I_0. In the neuroscience literature, S' is called the nonlinear gain.

Thus, it turns out that the different spatial modes $u(k)$ decouple, leading to a characteristic polynomial of sixth order for λ [24]

$$
\lambda^2 + 2\lambda + 1 - S' \left[\frac{1 + \lambda/v}{k^2 + (1 + \lambda/v)^2} a_e - \frac{r + \lambda/v}{k^2 + (r + \lambda/v)^2} r a_i \right] = 0. \qquad (5.5)
$$

Focusing on the spatially constant mode with $k = 0$, we observe that the center equilibrium in Fig. 5.2 is unstable while all other equilibria are stable. Figure 5.3 illustrates this finding and shows the equilibria, their corresponding stability, and the nonlinear gain S' with respect to V_0.

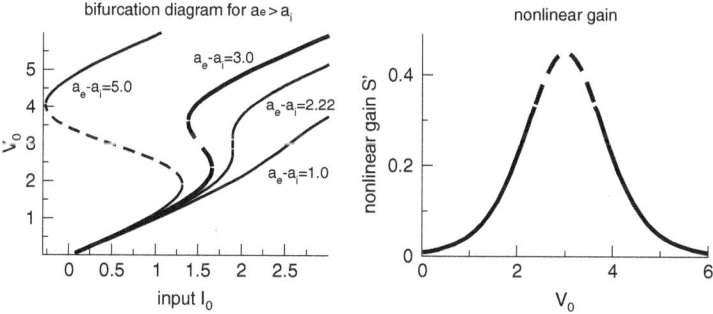

Fig. 5.3 Bifurcation diagram and nonlinear gain for constant bifurcations for $a_e > a_i$. *Left panel*: Stability of the stationary state V_0 with respect to external input I_0. For $a_e - a_i >= 2.22$, stable (*solid line*) and unstable branches (*dashed line*) exist, while for $a_e - a_i \leq 2.22$ there is only a single stable solution. *Right panel*: The nonlinear gain S' with respect to the constant state V_0

5.2.1.3 The Stationary Instability

According to the classification of spatio-temporal instabilities in extended systems (see e.g. [11]), we may distinguish several important cases. The stability threshold is given by the vanishing real part of $\lambda(k)$. For real Lyapunov exponents $\lambda \in \mathbb{R}$, stationary instabilities may occur if $\lambda = 0$, yielding the condition

$$\frac{1}{S_c'} = \left[\frac{a_e}{k_c^2 + 1} - \frac{a_i r^2}{k_c^2 + r^2} \right].$$

For $S' = 0$, the system is stable according to (5.5). Hence, increasing the control parameter S' from low values at which the equilibrium is stable, the equilibrium becomes unstable if $S' \geq S_c'$. This threshold defines the critical wave number k_c. In other words, below the threshold all spatial modes are stable, while the spatial modes with wave number $\pm k_c$ become marginally stable if $S' = S_c'$. In the case of $k_c \neq 0$, this instability is called Turing instability [41, 6]. The resulting state of the system for $t \to \infty$ is stationary, i.e., time independent and space dependent. Moreover, the threshold condition (5.5) implies $r < 1$ if $a_e > a_i$. This means that Turing instabilities may emerge only if the system's total excitation is stronger than its total inhibition while the spatial range of excitation is shorter than the range of inhibition. In other words, Turing instabilities may emerge in case of locally excitatory and laterally inhibitory interaction. Figure 5.1 illustrates this interaction type. Moreover, the critical control parameter S_c' is independent of the propagation speed v.

5.2.1.4 The Non-Stationary Instability

Let us now discuss the oscillatory instability of the equilibrium state, where the system dynamics is expected to be space- and time dependent. Inserting $\lambda = i\omega$, i.e., $\mathrm{Re}\,(\lambda) = 0$, into (5.5) gives the two threshold conditions [24]

$$\omega^6 + b_4\omega^4 + b_2\omega^2 + b_0 = 0, \quad b_5\omega^4 + b_3\omega^2 + b_1 = 0. \tag{5.6}$$

The coefficients $b_i = b_i(k^2, r, S', v, a_e, a_i)$, $i = 1, \ldots, 6$, are algebraic expressions of the wave number and the parameters of (5.5). Now we keep constant a_e, a_i, the values of S' by a specific external stimulus and vary r and the propagation velocity v. Then the threshold condition reads $r = r(v, \omega_c^2, k_c^2)$ in the v–r plane. Here k_c and ω_c denote the wave number and the angular frequency, respectively, which belong to the spatial mode first becoming unstable. Figure 5.4 shows the threshold values $r = r(v)$ for various parameter sets a_e, a_i. We observe that there is a critical propagation speed v for each value of r below which two spatial modes with wave numbers $\pm k_c$ and angular frequencies $\pm \omega_c$ become unstable. Thus, according to (5.4) the

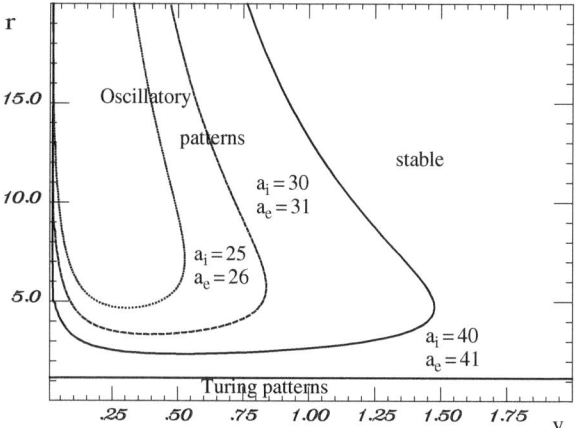

Fig. 5.4 Phase diagram of instabilities. An oscillatory instability may emerge for propagation speeds below a certain threshold. This threshold depends on the parameter r, i.e., the relation of excitatory and inhibitory spatial range. In contrast the Turing instability does not depend on the propagation speed v. Reprinted from [24] by permission

spatio-temporal activity of the neural population is a linear superposition of plane waves

$$V(x,t) = V_0 + u_1 e^{i(k_c x + \omega_c t)} + u_2 e^{i(k_c x - \omega_c t)} + c.c. , \qquad (5.7)$$

where $c.c.$ denotes the complex conjugate and the constants u_i are given by the initial conditions. In the case when $k_c = 0$, such instabilities show global oscillations constant in space and are called Hopf instabilities, while instabilities with $k_c \neq 0$ show a spatial periodicity and are called wave instabilities. The wave number k_c and the angular frequency ω_c represent the spatial and temporal scale of the waves, respectively, which are different from the intrinsic spatial and temporal scales of the system.

In addition, the phase speed of the traveling waves are computed by $v_{ph} = \omega(k_c)/k_c$. Figure 5.5(a) shows the dependence of the phase speed from the propagation speed. It turns out that the wave speed is smaller than the propagation speed in the system, which is expected due to the causality principle in nature. Interestingly v_{ph} grows less for larger propagation speeds v, which indicates a maximum value for $v \to \infty$. Similar behavior has been found in several previous studies [2, 9, 21]. Figure 5.5(b) shows the spatio-temporal activity slightly beyond the stability threshold. We observe the onset of traveling waves that persist for large times, i.e., the traveling wave pattern is stable. At a first glance this stability appears contradictory to the linear instability discussed above. However, Fig. 5.5(b) shows the nonlinear behavior of the full system (5.2), which is different from the linear behavior. Closer examinations of the dynamics reveal that the nonlinear term $S[V]$ in (5.2) confines the activity and thus yields the nonlinear stability. Subsequently the linear stability

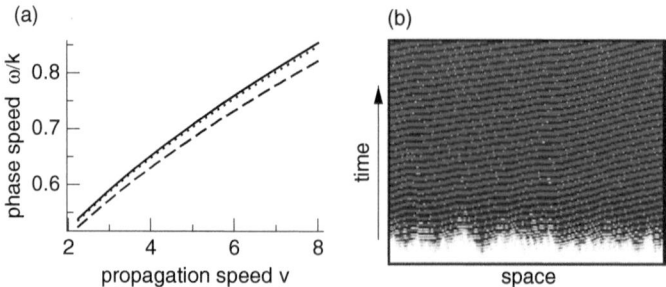

Fig. 5.5 (**a**) The phase speed v_{ph} of traveling waves subject to the propagation speed v. The line types encode the different values $r = 5.0$ (*dashed*), $r = 10.2$ (*dotted*), and $r = 20.2$ (*solid*). (**b**) The space–time plot of traveling waves close to the stability threshold. The parameters are $a_e = 60$, $a_e = 55$, $r = 3$, and $v = 1$

analysis applied above just extracts the conditions for the onset of instabilities but does not give any information on whether the emerging pattern exists for large times. This existence condition can be extracted by a corresponding nonlinear analysis [16, 19].

5.2.1.5 Summary

In the previous paragraphs we have observed that small spatio-temporal deviations from a constant equilibrium of the neural population may be decomposed into spatial Fourier modes yielding an implicit characteristic equation. If the root of the characteristic equation, i.e., the Lyapunov exponent, crosses the imaginary axis at certain wave numbers of spatial modes, the equilibrium becomes unstable. It turns out that the onset of stationary instabilities, such as the Turing instability, is independent of the propagation delay, while the threshold of non-stationary instabilities depend heavily on the propagation delay. In the case of an infinite propagation speed, no non-stationary instability is possible, while propagation speeds below a critical value yield oscillatory instabilities. The following section studies these results in some more detail for a more general model and aims to explain the underlying effects of the propagation delay.

5.2.2 The Generic Model

Let us now examine the more general model equation [2]

$$\hat{L}(\partial/\partial t)V(x,t) = \int_{\Omega} K(x-y)S(V(y,t-|x-y|/v))\,dy + I(x,t) \qquad (5.8)$$

with the spatial domain $\Omega = \mathbb{R}$. Here $\hat{L}(\partial/\partial t)$ denotes the temporal operator with constant coefficients and eigenvalue $L(\lambda)$, i.e., $\hat{L}\exp(\lambda t) = L(\lambda)\exp(\lambda t)$. Further, it is assumed that $L(\lambda)$ is a stable polynomial, i.e., all its roots have negative real parts.

In addition, the intrinsic timescale is set to unity by choosing $L(0) = 1$ without loss of generality [2]. A typical choice is a temporal operator of second order with the eigenvalue

$$L(\lambda) = \eta\lambda^2 + \gamma\lambda + 1, \quad \eta = 0 \text{ or } 1, \ \gamma > 0 . \tag{5.9}$$

The kernel $K : \mathbb{R} \rightarrow \mathbb{R}$ represents the nonlocal spatial interaction. It is continuous, integrable, and symmetric with $K(-z) = K(z)$ for all $z \in \Omega$. The transfer function $S : \mathbb{R} \rightarrow \mathbb{R}$ is assumed to be differentiable and monotone increasing. Further the spatial interaction is delayed by the finite propagation speed v and $I(x,t) \in \mathbb{R}$ denotes the external input.

Similar to the previous section, we aim to gain some insight into the spatio-temporal dynamics by the study of the linear stability of (5.8). For a constant input $I(x,t) = I_0$, a constant equilibrium solution $V(x,t) = V_0$ satisfies

$$V_0 = \int_\Omega K(x - y)S(V_0)\,dy + I_0 = \kappa S(V_0) + I_0 \tag{5.10}$$

with

$$\kappa = \int_\Omega K(z)\,dz = 2\int_0^\infty K(z)\,dz. \tag{5.11}$$

If S is bounded, then (5.10) has a solution V_0 for any $I_0 \in \mathbb{R}$. In addition the uniqueness of V_0 depends on the sign of κ and the shape of S [2]. For instance, if S is positive and increasing on $\mathbb{R}$, such as the sigmoid function in Sect. 5.2.1, and if $\kappa \leq 0$, then the solution V_0 is unique. On the other hand if $\kappa > 0$ then there may be multiple equilibria, cf. Fig. 5.2. Since the external input determines the constant equilibrium solutions, I_0 represents the control parameter. In the following we assume $\kappa > 0$, $S[V] > 0$, and $\delta S[V]/\delta V > 0$.

5.2.2.1 Equilibria and Linearization

Expanding the nonlinear functional about the equilibrium V_0,

$$S[V(x,t)] = S[V_0] + \frac{\delta S[V(x,t)]}{\delta V(x,t)}\Big|_{V(x,t)=V_0} (V(x,t) - V_0)$$

$$+ \frac{1}{2}\frac{\delta^2 S[V(x,t)]}{\delta V(x,t)^2}\Big|_{V(x,t)=V_0} (V(x,t) - V_0)^2 + \cdots ,$$

and considering small deviations $u(x,t) = V(x,t) - V_0$, the linear stability of the equilibrium solution V_0 is determined by the linear variational equation

$$\hat{L}(\partial/\partial t)u(x,t) = S' \int_\Omega K(x - y)u(y, t - |x - y|/v)\,dy \tag{5.12}$$

with $S' = \delta S/\delta V$ computed at V_0. Since V_0 depends on the external input I_0 and S' is unique with respect to choice of V_0, we shall use S' as the control parameter in the following discussion.

Due to the difference kernel $K(x - y)$ the eigenfunctions of the integral operator in (5.12) represent a continuous Fourier basis $\{\tilde{u}(k, t)\}$ with

$$\tilde{u}(k, t) = \frac{1}{\sqrt{2\pi}} \int_\Omega u(x, t) e^{-ikx} dx. \tag{5.13}$$

Then we find the affine delay-differential equation for each spatial mode with wave number k,

$$\hat{L}\tilde{u}(k, t) = 2vS' \int_0^\infty K(v\tau) \cos(kv\tau) \tilde{u}(k, t - \tau) \, d\tau. \tag{5.14}$$

Thus in the linear regime the spatio-temporal dynamics of the system decouples into single modes in Fourier space, while the space-dependent propagation delay transforms to a distribution of constant delays [25].

The subsequent application of the ansatz $\tilde{u}(k, t) \sim \exp(\lambda(k)t)$ to (5.14) yields

$$L(\lambda) = 2S' \int_0^\infty K(z) \cos(kz) \exp(-\lambda z/v) \, dz. \tag{5.15}$$

This relation is the characteristic equation of the linear problem (5.14).

5.2.2.2 Stability Conditions

We have learned in Sect. 5.2.1 that the characteristic equation (5.15) is in general difficult to solve explicitly. The solutions (λ, k) of (5.15) correspond to the perturbations $u(x, t) = e^{\lambda t} e^{ikx}$ about the equilibrium solution, which grow or decay in time depending on whether $\mathrm{Re}\,(\lambda)$ is positive or negative, respectively. Hence in order to determine the stability of V_0, first we give the sufficient conditions for asymptotic stability.

Theorem 5.1 (Asymptotic stability [2]).
  Let $c = S' \int_{-\infty}^\infty |K(z)| \, dz$. If

$$c < \min_\omega |L(i\omega)| \tag{5.16}$$

then the stationary state V_0 is asymptotically stable. In particular, if $L(\lambda) = \lambda + 1$ then the condition $c < 1$ is sufficient for the asymptotic stability of V_0. If $L(\gamma) = \lambda^2 + \gamma\lambda + 1$ with $\gamma > 0$, then the stationary state V_0 is asymptotically stable provided that the condition $\frac{\gamma^2}{2} > 1 - \sqrt{1 - c^2}$ holds in addition to $c < 1$.

The condition (5.16) is valid for all kernels, all temporal operators of up to second order and it is independent of the propagation speed. If it does not hold, then spatio-temporal instabilities may emerge. To be more detailed, let us assume the typical kernel $K(x) = a_e K_e(x) - a_i K_i(x)$, which takes into account excitation and inhibition weighted by a_e and a_i, respectively. The kernels $K_e(x)$ and $K_i(x)$ are normalized to unity, i.e., $\int_\Omega K_{e,i}(x)dx = 1$, and represent the excitatory and inhibitory spatial interaction, respectively. Then Theorem 5.1 states that the equilibria V_0 are asymptotically stable if

$$S' \geq \frac{\gamma\sqrt{1 - \gamma^2/4}}{a_e + a_i} \quad \text{for all } 0 < \gamma < 2,$$

$$S' \geq \frac{1}{a_e + a_i} \quad \text{for all } \gamma \geq 2. \tag{5.17}$$

To gain additional stability conditions, we follow the classification scheme discussed in the previous section. First let us discuss non-stationary instabilities.

Theorem 5.2 (Sufficient condition for non-oscillatory instabilities [21]). *Let the kernel $K(x) = a_e K_e(x) - a_i K(x)$, with $a_e, a_i, K_e(x), K_i(x)$ positive definite. Define the mean spatial interaction ranges of excitation and inhibition, respectively, as*

$$\xi_e = \int_\Omega |z| K_e(z)\, dz \quad and \quad \xi_i = \int_\Omega |z| K_i(z)\, dz,$$

and the corresponding mean propagation delays $\tau_e = \xi_e/v$, $\tau_i = \xi_i/v$. Let $L(\gamma) = \lambda^2 + \gamma\lambda + 1$ with $\gamma > 0$ and

$$S'_c = \frac{\gamma}{a_e \tau_e + a_i \tau_i}. \tag{5.18}$$

Then non-oscillatory instabilities are possible only if $S' \geq S'_c$.

Here we have introduced the mean propagation delays τ_e and τ_i of excitatory and inhibitory spatial interactions, respectively. Since all quantities are positive in (5.18), it is clear that at least one of the propagation delays τ_e or τ_i must be non-vanishing for the occurrence of non-stationary bifurcations. This means that non-stationary instabilities are possible only if the propagation delay is finite. Thus Theorem 5.2 generalizes the specific findings in the previous section. This result is valid for temporal operators up to second order. Figure 5.6 summarizes the previous results and elicits the importance of the propagation delays.

5.2.2.3 The Stability Threshold

Now let us investigate the specific threshold conditions of stationary and non-stationary instabilities. For $S' = 0$, the Lyapunov exponents λ are simply given by the roots of L and we find $\text{Re}(\lambda) < 0$ by the assumption that L is a stable polynomial. In this case the equilibrium point is asymptotically stable. As S' is increased, the stability may be lost if a Lyapunov exponent $\lambda(k)$ for some k crosses

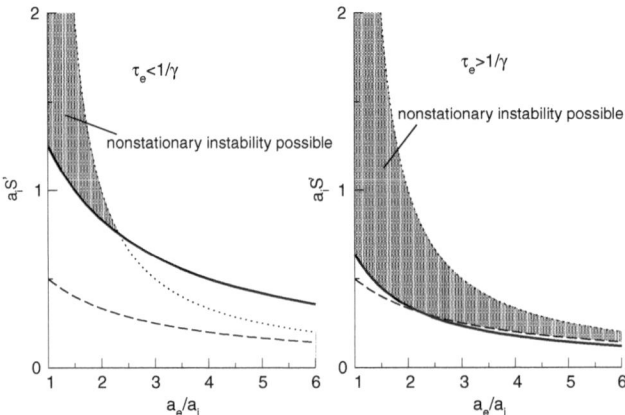

Fig. 5.6 Parameter regimes for non-stationary instabilities for two different excitatory mean propagation delays τ_e. Valid parameters (*hatched area*) are constrained by the threshold of (5.18) plotted as *solid line*, the threshold of the constant instability $S' < 1/(a_e - a_i)$ taken from Sect. 5.2.1 (*dotted line*) and the threshold of the asymptotic stability (5.17) (*dashed line*). Here, it is $\gamma = 2, \tau_i = 0.1$

the imaginary axis, and a dynamically different behavior may result in the original nonlinear equation (5.8). At the stability threshold, $\text{Re}\,(\lambda) = 0$ and thus $\lambda = i\omega$, $\omega \in \mathbb{R}$. Then the characteristic equation (5.15) becomes

$$L(i\omega) = S' \int_{\Omega} K(z) \cos(kz) e^{-i\omega|z|/v} \, dz. \tag{5.19}$$

The possibilities for the resulting behavior when S' is near such a critical value can then be qualitatively classified as follows:

1. For $\omega = 0$, stationary instabilities may emerge, e.g., a Turing instability. Then the stability condition reads

$$1/S' > \int_{\Omega} K(z) \cos(kz) \, dz.$$

   Hence, stationary instabilities are independent of the propagation delay for general kernels and temporal operators up to second order.
2. For $\omega \neq 0$, the non-stationary instability to periodic oscillations from a spatially uniform solution ($k = 0$) or spatially periodic solution ($k \neq 0$) may emerge. The former case is called Hopf instability and the latter is known as wave instability. The corresponding threshold conditions are derived in the subsequent paragraphs.

   In order to study the type of bifurcations that may arise in a given situation, the characteristic equation (5.19) needs to be solved for ω and k. However, explicit solutions are difficult to obtain for general kernel functions. The results of the previous paragraphs imply that for infinite propagation speeds one has a simpler case, where non-stationary instabilities do not exist for temporal operators up to second order. It follows that the role of the propagation delay can be systematically examined by following the changes of the Lyapunov exponent as the value of the transmission speed is decreased from infinity. Hence we consider the change in dynamics as $1/v$ is increased from zero. This yields an approximation scheme that provides valuable insight into the effects of propagation delays in the dynamics of the system [2].

Let us consider the power series estimate

$$\exp\left(-\lambda|z|/v\right) = \sum_{m=0}^{m=N} \frac{(-\lambda|z|/v)^m}{m!} + \mathcal{O}(v^{-(N+1)}).$$

The substitution in the characteristic equation (5.15) at the stability threshold $\lambda = i\omega$ gives a finite series in powers of $1/v$,

$$L(i\omega) = \alpha \int_\Omega K(z)\cos(kz) \left[\sum_{m=0}^{m=N} \frac{(-i\omega|z|)^m}{m!} \left(\frac{1}{v}\right)^m + \mathcal{O}(\varepsilon^{N+1})\right] dz$$

$$= \alpha \sum_{m=0}^{m=N} \frac{(-i\omega)^m}{m!} \left(\frac{1}{v}\right)^m \hat{K}_m(k) + \mathcal{O}(\varepsilon^{N+1}) \tag{5.20}$$

where the terms $\hat{K}_m$ denote the Fourier kernel moments

$$\hat{K}_m(k) = \int_\Omega |z|^m K(z)\cos(kz)\,dz \tag{5.21}$$

and the integrals are assumed to exist. Then the separation of the real and imaginary parts of (5.20) leads to

$$\frac{1}{S'}\operatorname{Re}L(i\omega) = \hat{K}_0(k) - \frac{1}{2}\left(\frac{\omega}{v}\right)\hat{K}_2(k) + \ldots \tag{5.22}$$

$$\frac{1}{S'}\operatorname{Im}L(i\omega) = -\left(\frac{\omega}{v}\right)\hat{K}_1(k) + \frac{1}{6}\left(\frac{\omega}{v}\right)^3 \hat{K}_3(k) - \ldots. \tag{5.23}$$

The number of terms needed for the above series to be useful depends on the value of $1/v$ as well as on the shape of the kernel K. If K is highly concentrated near the origin or if $1/v$ is small, i.e., the spatial interaction is short ranged or the propagation speed is large, then a few terms are sufficient [2]. We assume that at least one of these conditions is satisfied so that a small number of terms suffice to determine the general behavior.

In order to observe the qualitative effects of a finite transmission speed, we thus neglect third and higher order terms in $1/v$ in the series (5.20). Then, for L given by (5.9), (5.22) and (5.23) become

$$\frac{1}{S'}(1 - \eta\omega^2) = \hat{K}(k) - \tfrac{1}{2}\varepsilon^2\omega^2\hat{K}_2(k) \tag{5.24}$$

$$\frac{1}{S'}\gamma\omega = -\varepsilon\omega\hat{K}_1(k) \tag{5.25}$$

where we have substituted the more conventional notation $\hat{K}$ for the Fourier transform $\hat{K}_0$ of the kernel. For stationary instabilities ($\omega = 0$) one obtains from the first equation [2]

$$1/S'_c = \hat{K}(k^*). \tag{5.26}$$

For a non-stationary bifurcation $\omega \neq 0$, (5.25) implies

$$1/S' = -\hat{K}_1(k^*)/v\gamma. \tag{5.27}$$

Since $-\hat{K}_1/v\gamma$ is continuous and $\hat{K}_1(k) \to 0$ as $k \to \pm\infty$, it has a global maximum at some values of k which corresponds to the first mode that loses stability. This loss of stability may happen as v is decreased, $\hat{K}_1$ is decreased, or as S' is increased, see Fig. 5.7 for illustration of the latter case. Thus let

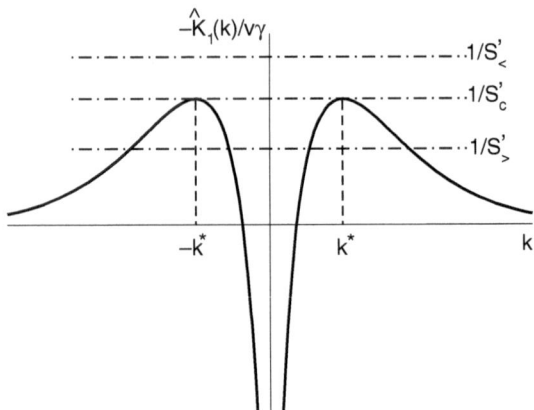

Fig. 5.7 Illustration of the onset of non-stationary instabilities. Below the stability threshold, the control parameters $S' = S'_<$ are smaller than the critical S'_c from (5.27), $1/S'_<$ is larger than $-\hat{K}_1(k^*)/v\gamma$, and the system is stable. For $S' = S'_c$ the condition (5.27) is fulfilled and the crossing points of the horizontal $1/S'_c$ and the curve $-\hat{K}_1(k^*)/v\gamma$ determine the critical wave number $\pm k^*$. In the case of $S' = S'_>$, a band of spatial modes becomes unstable for which $S' > S'_>$, i.e., $1/S' < -\hat{K}_1(k^*)/v\gamma$

$$k^* = \min_k \hat{K}_1(k) = \min_k \int_\Omega |z| K(z) \cos(kz)\, dz, \tag{5.28}$$

and provided that $\hat{K}_1(k^*) < 0$, k^* will be the sought solution of (5.27). Substituting k^* into (5.24) gives the critical angular frequency of the non-stationary instability,

$$\omega^2 = \frac{\alpha \hat{K}(k^*) - 1}{\frac{1}{2}\alpha \hat{K}_2(k^*)/v^2 - \eta}. \tag{5.29}$$

These last analysis steps represent a simple procedure to calculate the pairs (ω, k) satisfying the characteristic equation.

Essentially it remains to determine what type of instability actually occurs. This depends on the mode by which the equilibrium solution loses its stability as the control parameter S' is increased. The procedure described above gives a simple graphical method. Thus, if one plots the curves $\hat{K}(k)$ and $-\hat{K}_1(k)/v\gamma$ in the same graph and adds in mind a horizontal line at $1/S'$ being lowered from infinity ($S' \approx 0$), then the first intersection point specifies the instability type. If the horizontal line touches the graph of $\hat{K}(k)$ first, i.e., the global maximum of $\hat{K}(k)$ is larger than the global maximum of $-\hat{K}_1(k)/v\gamma$, then (5.26) is satisfied and a stationary instability occurs. If, on the other hand, the horizontal line touches $-\hat{K}_1(k)/v\gamma$ first, then (5.27) is satisfied and a non-stationary instability occurs. Furthermore, the value of k at the intersection point being zero or nonzero specifies whether the new solution is spatially constant or not, respectively. It is worthwhile to note that the type of instability that can occur depends only on the extremal values of $\hat{K}(k)$ and $\hat{K}_1$ and not on the exact shapes of their graphs. This observation has the important consequence that the instability type depends only on some general properties of the kernel and not on its precise shape. Figure 5.8 shows different types of spatial kernels and the corresponding functions $\hat{K}(k)$ and $-\hat{K}_1/v\gamma$. We observe that excitatory systems may yield wave instabilities and stationary instabilities with $k = 0$ (Fig. 5.8a), while Hopf instabilities may occur in inhibitory systems (Fig. 5.8b). Moreover, systems with local inhibition and lateral excitation may show stationary bifurcations with $k = 0$ or wave instabilities (Fig. 5.8c) and local excitatory-lateral inhibitory systems may show Turing instabilities, wave instabilities, or Hopf instabilities.

5.2.2.4 Application to a Specific Model

The previous analysis discussed the linear stability of the constant equilibria for general spatial interactions. This allows us to investigate the stability of non-standard spatial interactions. Motivated by experimental findings in the brain of mice [38], the kernel studied represents a family of excitatory interactions and a fast-decaying exponential inhibitory interaction:

$$K(x) = \frac{1}{2\Gamma(p)} |x|^{p-1} e^{-|x|} - \frac{1}{2} r e^{-r|x|}, \tag{5.30}$$

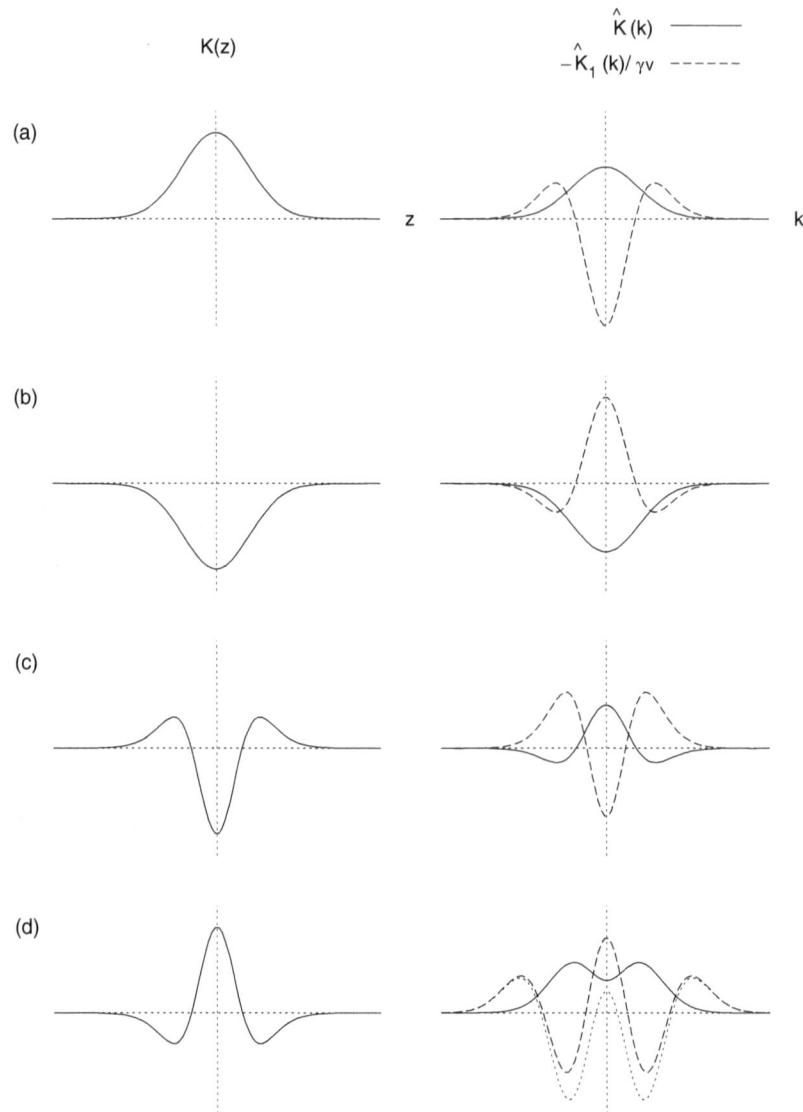

Fig. 5.8 Typical interaction kernels and the corresponding kernel Fourier moments. The first column shows the kernels, with the corresponding Fourier transforms in the second column. In the last subfigure, two distinct possibilities for $-\hat{K}_1/\gamma v$ are shown with *dashed* and *dotted* lines. Reprinted from [2] by permission

with the gamma-function $\Gamma(p)$, the parameter $p > 0$, and the relation of excitatory and inhibitory spatial range r. To be more specific, $p = 1$ leads to an exponentially decaying excitatory kernel, $p > 1$ represents gamma-distributed excitation, and $p < 1$ yields divergent local excitation. In the latter case the additional choice

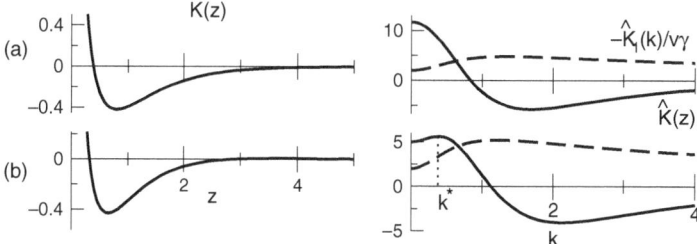

Fig. 5.9 The kernel functions $K(x)$ from (5.30), the scaled negative first kernel Fourier moment $-\hat{K}_1(k)/v\gamma$ (*solid line*) and the Fourier transform of the kernel $\hat{K}(k)$ (*dashed line*) for different parameter sets. The parameters are $a_e = 10, a_i = 8, p = 0.5, v = 0.6, \gamma = 0.2$ and (**a**) $r = 0.8$ and (**b**) $r = 0.7$. The critical wave numbers are (**a**) $k^* = 0.0$ and (**b**) $k^* = 0.4$

$r > 1$ represents local excitation and lateral inhibition. Corresponding to Sect. 5.2.1 and the knowledge from partial differential equation models [19], this interaction type is known to yield Turing instabilities. From a more physical point of view this is reasonable, as the local excitation enhances local activity and the lateral inhibition diminishes lateral activity and thus enhances local structures. In terms of the previous classification procedure, we would expect a global maximum of $\hat{K}(k)$, which exceeds the global maximum of $-\hat{K}_1(k)/v\gamma$. However, Fig. 5.9 shows that there exist parameter sets, which yield a higher global maximum of $-\hat{K}_1(k)/v\gamma$. This is possible due to the low values of $v\gamma$. In other words, if the propagation speed is small enough, oscillatory instabilities emerge. This finding is confirmed numerically for the parameters in Fig. 5.9(b) by integrating the full model (5.8), see Fig. 5.10.

5.2.2.5 Summary

The analysis in this section is useful to gain insights into the effects of the propagation on the dynamical behavior at the stability threshold. At first we have learned that

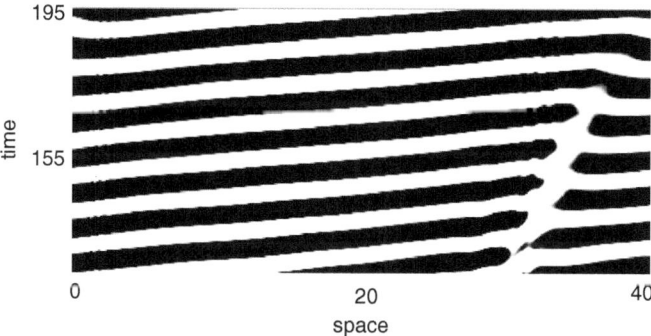

Fig. 5.10 Wave instability for local excitation–lateral inhibition interaction for parameters taken from Fig. 5.9(B)

the propagation delay, which is dependent on space, recasts to a distributed constant delay in Fourier space. Further, we have found the quantity of a mean propagation delay, which defines a sufficient condition for non-oscillatory instabilities. Finally, the Fourier transform and the first Fourier kernel moment are observed to be crucial for stability: the former is instrumental for stationary instabilities while the latter for the non-stationary ones. This finding is valid for fast-decaying kernels or large but finite propagation speeds. Outside of this validity range, the higher moments of the kernel are expected to make contributions to the results, which, however, are not discussed here.

5.3 Dynamics In The Presence of Noise

In this section we aim to investigate how external random fluctuations affect the dynamics of spatial systems involving propagation delay. The final application to a specific model elicits the interaction of random fluctuations and propagation delays.

5.3.1 General Stability Study

The model studied is similar to the one discussed in the previous section and reads

$$\hat{L}(\partial/\partial t)V(x,t) = \int_{\Omega} K(x-y)S(V(y,t-|x-y|/v))\,dy + I(x,t), \qquad (5.31)$$

where $\hat{L}$ represents the temporal operator with eigenvalue $L(\lambda)$, $K(x)$ denotes the spatial interaction kernel, and $S[V]$ is a monotonic increasing functional of the activity variable $V(x,t)$. The system exhibits a finite propagation speed v and is assumed infinite in space, i.e., $\Omega = \mathbb{R}$. In contrast to the previous section, now the external input $I(x,t) = I_0 + \xi(x,t)$ is the sum of a constant input I_0 and random fluctuations $\xi(x,t)$ with

$$\langle \xi(x,t) \rangle = 0, \quad \langle \xi(x,t)\xi(x',t') \rangle = Q\delta(x-x')\delta(t-t'), \qquad (5.32)$$

where $\langle \cdot \rangle$ denotes the ensemble average. The random fluctuations obey a Gaussian distribution with variance Q and are uncorrelated in space and time.

In the previous section, we have shown how to extract stability conditions while considering the propagation delay. Recall that we have determined the constant equilibria and studied the Lyapunov exponent of the linearized problem about the equilibria. At a first glance this approach is not applicable here due to the external random fluctuations. The question arises how we can obtain constant equilibria if a random process disturbs the system for all times. Let us first recall the definition of stability: A system state is said to be stable if the system evolves in a small neighborhood of this state for all time. Hence we argue that the stable system state might be equivalent to the deterministic stable equilibrium state V_0 and the system evolves in a small neighborhood about this stable equilibrium due to the external random fluctuations. Thus the stable equilibrium constant in space and time fulfills

$$V_0 = \kappa S(V_0) + I_0,$$

with $\kappa = 2 \int_0^\infty K(z)\,dz$. Up to this point the stability of V_0 still has to be confirmed. The subsequent paragraphs study the dependence of the system dynamics on the random fluctuations and at last shall answer this question of stability.

Further, presuming the fluctuation variance Q being small may yield small deviations from the equilibria, $u(x,t) = V(x,t) - V_0$ and we obtain the linear delayed stochastic integral–differential equation

$$\hat{L}(\partial/\partial t)u(x,t) = \alpha \int_\Omega K(x-y)u(y, t - |x-y|/v)\,dy + \xi(x,t), \qquad (5.33)$$

with $S' = \delta S/\delta V$ computed at $V = V_0$. In the following sections we shall use S' as the control parameter. Now let us study (5.33) in the Fourier space similar to the deterministic case. With the Fourier transformation

$$\tilde{u}(x,t) = \frac{1}{\sqrt{2\pi}} \int_\Omega \tilde{u}(k,t)e^{ikx}dk,$$

we find the stochastic delay-differential equation for each spatial mode [25],

$$\hat{L}\tilde{u}(k,t) = 2vS' \int_0^\infty K(vt')\cos(kvt')\tilde{u}(k, t - t')\,dt' + \tilde{\xi}(k,t). \qquad (5.34)$$

The new random fluctuations $\tilde{\xi}(k,t)$ of the spatial mode with wave number k obey a Gaussian distribution and are uncorrelated in k-space and in time with

$$\langle \tilde{\xi}(k,t) \rangle = 0, \qquad \langle \tilde{\xi}(k,t)\tilde{\xi}(k',t') \rangle = Q\delta(k-k')\delta(t-t'). \qquad (5.35)$$

Thus, our model corresponds to an infinite set of affine delay-differential equations (see, e.g. [20, 25]) with distributed delays subject to the external force $\tilde{\xi}(k,t)$. Thinking of $\tilde{\xi}(k,t)$ as an external perturbation, linear response theory gives the general solution of (5.34) by

$$\tilde{u}(k,t) = \tilde{u}_h(k,t) + \int_{-\infty}^\infty dt'\, G(k, t-t')s(k,t') . \qquad (5.36)$$

Here $\tilde{u}_h(k,t)$ represents the homogeneous solution of (5.34), i.e., for $\tilde{\xi}(k,t) = 0$, and $G(k, t-t')$ is the Green's function of the spatial mode with wave number k.

Applying standard techniques in linear response theory [27], the Green's function is
given by [23]

$$G(k, t) = \frac{1}{2\pi} \int_{-\infty}^{\infty} d\omega \frac{e^{-i\omega t}}{L(-i\omega) - \bar{K}(k, i\omega)}, \tag{5.37}$$

with

$$\bar{K}(k, i\omega) = 2\nu S' \int_{0}^{\infty} d\tau \, K(\nu\tau) \cos(k\nu\tau) e^{i\omega\tau}. \tag{5.38}$$

It turns out that the denominator in (5.37) is equivalent to the characteristic equation
(5.15). Extending the real domain of ω to the complex plane $\mathcal{P}$ and applying the
residue theorem, we have

$$G(k, t) = \Theta(t) \left[i \sum_{l=1}^{m} \text{Res}_l(e^{-i\Omega_l t}) \right] = \Theta(t) \sum_{l=1}^{m} r_l(k) e^{\lambda_l(k)t}. \tag{5.39}$$

The Heaviside function $\Theta(\cdot)$ guarantees the causality and m denotes the number of
complex roots $\Omega_l(k) \in \mathcal{C}$ of the denominator in (5.37). That is, m is the (in general
infinite) number of complex roots of the characteristic polynomial. Further, Res_l
denotes the residue of the numerator in (5.37) at root $\Omega_l(k)$, $\lambda_l(k) = -i\Omega_l(k)$, and
the constants $r_l \in \mathcal{C}$ are defined by the corresponding residues.

It turns out that the Green's function $G(k, t)$ decays for large times if all variables $\lambda_l(k)$ show negative real parts, while a single root with $\text{Re}(\lambda_l) > 0$ causes
the Green's function to diverge. Since the Green's function defines all dynamical
properties of the system under study [27], the set $\{\lambda_l(k)\}$, $l = 1, \ldots, m$, defines the
stability of the spatial mode k and thus represent the Lyapunov exponents. Subsequently stable spatial modes exhibit roots $\Omega_l(k)$ below the real axis in the complex
plane P, see Fig. 5.11.

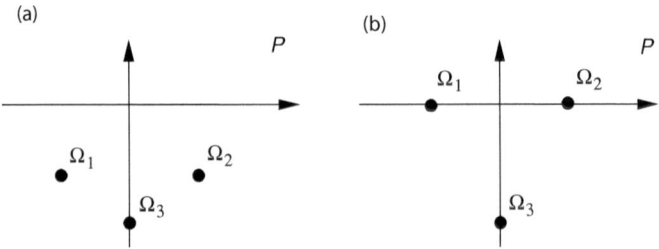

Fig. 5.11 Illustration of the complex plane $\mathcal{P}$ and three roots $\Omega_l(k)$ in the case of stability (**a**) and
at the onset of an oscillatory instability (**b**)

Eventually inserting (5.39) into (5.36) we obtain

$$\tilde{u}(k,t) = \tilde{u}_h(k,t) + \sum_{l=1}^{m} r_l(k) \int_0^t dt' \, e^{\lambda_l(k)(t-t')} \tilde{s}(k,t')$$

and

$$u(x,t) = \tilde{u}_h(x,t)$$
$$+ \frac{1}{\sqrt{2\pi}} \sum_{l=1}^{m} \int_0^t \int_{-\infty}^{\infty} e^{\lambda_l(k)(t-t')} \tilde{s}(k,t') r_l(k) e^{ikx} dk \, dt', \qquad (5.40)$$

assuming the initial time at $t = 0$. If all roots are located in the lower complex plane, i.e., $\text{Im}\,(\Omega_l(k)) = \text{Re}\,(\lambda_l(k)) < 0$, (5.40) has stable solutions for random fluctuations described by a Lévy process [18]. This means that the stability of the equilibria V_0 in the presence of external additive fluctuations for a stochastic path is given by the Lyapunov exponents obtained from the deterministic case.

5.3.2 Application to a Specific Model

The previous paragraphs showed that the stability conditions of the system are independent of the additive fluctuations. Now the question arises how the external random fluctuations affect the dynamics of spatial systems that involve finite propagation speeds. To study these effects, let us assume excitation and inhibition with the specific kernels

$$K_e(x) = \frac{1}{2\sqrt{D}} e^{-|x|/\sqrt{D}} \quad \text{and} \quad K_i(x) = \frac{1}{2}\delta(x - |R|),$$

respectively, and the kernel function is given by $K(x) = a_e K_e(x) - a_i K_i(x)$ with weights a_e, a_i. Here $\sqrt{D}$ and R denote the excitatory and inhibitory interaction range, respectively. Then the linear evolution equation of the spatial mode k (5.34) reads [25]

$$\frac{\partial}{\partial t} \tilde{u}(k,t) = S' \int_0^{\infty} \left[\frac{a_e}{\sqrt{D}} e^{-|x|/\sqrt{D}} - a_i \delta(x - |R|) \right] \cos{(kx)} \tilde{u}(k, t - x/v) \, dx$$

$$+ \xi(k,t)$$

$$\approx -a(k)\tilde{u}(k,t) - b(k)\tilde{u}(k, t - \tau) + \xi(k,t) \qquad (5.41)$$

with the constant delay $\tau = R/v$ and the parameters

$$a(k) = 1 - S' a_e(1 - Dk^2), \quad b(k) = S' a_i \cos{(kR)}$$

Here the finite propagation speed v becomes a constant delay $\tau > 0$. In addition, (5.41) presumes a very short excitation range $\sqrt{D}$ with $\sqrt{D}\omega \ll v$ and $\sqrt{D} \ll k$ at a typical frequency ω and a typical wave number k. This approximation reflects excitatory spatial diffusion [25].

Equation (5.41) is a linear stochastic delay-differential equation for the Fourier amplitudes $\tilde{u}(k, t)$. In recent years several studies have examined the stochastic dynamics of such systems [3, 5, 13, 17, 32, 34]. Küchler and Mensch [32] proved that the deterministic stability of (5.41) guarantees a stationary probability distribution of $\tilde{u}(k, t)$ (Proposition 2.8 in [32]). Moreover, the stationary distribution is Gaussian with zero mean and variance σ^2. This important result allows us to study the properties of $\tilde{u}(k, t)$. After extracting the stability conditions, which coincide with the deterministic conditions, we shall examine the variance and its dependence on the delay time, i.e., the propagation speed.

5.3.2.1 Stationary Instability

The deterministic characteristic equation reads $\lambda = -a - b\exp(-\lambda\tau)$. First let us discuss (5.41) in the context of stationary instabilities. Then the stability threshold is given by $a = -b$ or

$$\frac{1}{S'_c} = a_e(1 - Dk^2) - a_i \cos(kR), \tag{5.42}$$

see Fig. 5.12. The graphical illustration is similar to the previous section: For $S' < S'_c$, the thought horizontal line $1/S'$ exceeds the global maximum of the graph of $a_e(1 - Dk^2) - a_i \cos(kR)$ and all spatial modes are stable with respect to stationary bifurcations. If $1/S' = 1/S'_c$, the crossing points of the horizontal $1/S'$ and the graph of $a_e(1 - Dk^2) - a_i \cos(kR)$ define the critical wave numbers $\pm k^*$, i.e., a Turing instability emerges with wave number k^*.

In addition, the stability condition (5.42) is independent of the delay in accordance to the general results obtained in the previous section. According to [32], the variance of the resulting stationary distribution of $\tilde{u}(k, t)$ is

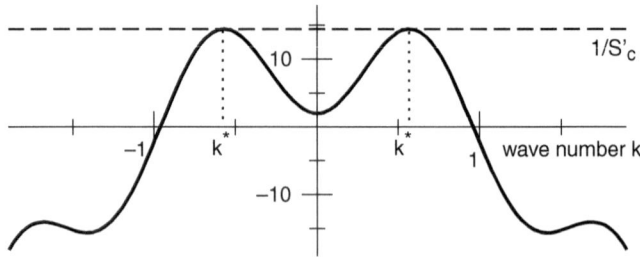

Fig. 5.12 Graphical illustration of the threshold of stationary instabilities (5.42) with critical wave number k^*. The *solid line* denotes the right-hand side of (5.42). Parameters are $a_e = 10, a_i = 8, D = 1, R = 5$

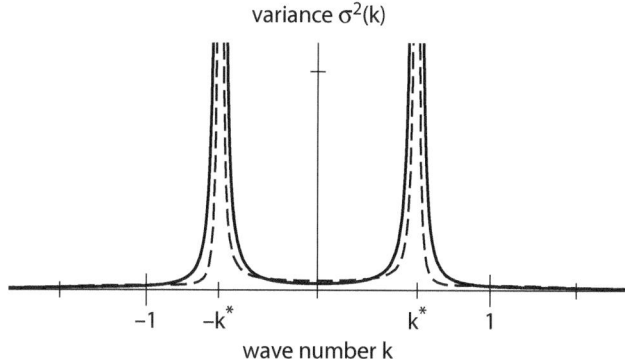

variance $\sigma^2(k)$

wave number k

Fig. 5.13 The variance of the Turing instability constructed in Fig. 5.12 with critical wave number k^* and $\tau = 0.5$ (*solid line*), $\tau = 1.5$ (*dashed line*). Other parameters are taken from Fig. 5.12

$$\sigma^2 = \frac{1 + \tau S' a_i \cos{(kR)}}{4S' a_i \cos{(kR)} + 1 - S' a_e (1 - Dk^2)}, \tag{5.43}$$

which depends on the delay time τ. Hence the delay, i.e., the finite propagation speed, does not affect the threshold condition of stationary instabilities but the stationary distribution of the system. Figure 5.13 shows the variance (5.43) with respect to the wave number k and it diverges at the critical wave numbers $\pm k^*$. This divergence of variance is well known from the study of phase transitions, and the system shows the so-called critical fluctuations.

5.3.2.2 Non-Stationary Instability

Now let us focus on oscillatory instabilities. To this end, we find the stability threshold for the spatial mode with wave number k,

$$\tau_c(k) = \frac{1}{\Omega(k)} \arctan\left(-\frac{a(k)}{b(k)}\right), \quad \Omega(k) = \sqrt{b^2(k) - a^2(k)}, \tag{5.44}$$

where $\Omega(k)$ denotes the angular frequency of the oscillations and the τ_c is the corresponding critical delay. Then the fluctuations $\tilde{u}(k, t)$ are stationary if the conditions

$$0 < \tau < \tau_c \quad \text{and} \quad a_i \cos{(kR)} > 1/\alpha - a_e(1 - Dk^2) > -a_i \cos{(kR)}$$

hold [32]. Figure 5.14 illustrates the stationary regime for a band of wave numbers. At the borders of that band the critical delay τ_c diverges. This happens as the corresponding wave numbers represent the critical wave numbers of the stationary instability (cf. (5.42)) which is independent of the delay. Moreover, Fig. 5.14 reveals a minimum critical delay $\tau_c(0)$ beyond which there is a band of unstable spatial modes. In other words, for delays $\tau > \tau_c(0)$ the system is unstable.

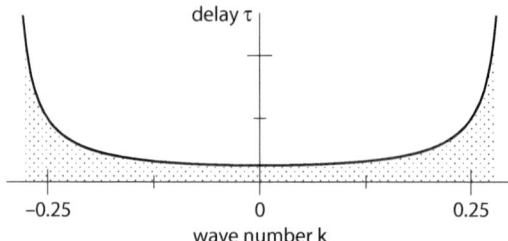

Fig. 5.14 The stationarity regime for various delays in a specific band of wave numbers. The *bold line* denotes the threshold function $\tau_c(k)$ and the *shaded area* below that line denotes the regime of stationarity. The parameter values are $a_e = 10, a_i = 8, \alpha = 0.1, D = 1.0, R = 5$

Finally let us study the fluctuations. They are Gaussian with zero mean and have the variance [25]

$$\sigma^2(k) = \frac{\Omega(k) + b(k)\sin(\Omega\tau)}{2\Omega(k)(a(k) + b(k)\cos(\Omega(k)\tau))}.$$

Figure 5.15 shows σ^2 with respect to the wave numbers for the delay τ just below the critical delay τ_c. The variance of the spatial modes with $k \neq 0$ is finite, while the constant spatial mode exhibits the diverging variance reflecting the critical fluctuations of the marginal stable mode.

5.3.2.3 Summary

In this section we studied spatial systems subject to external random fluctuations by applying linear response theory. It turned out that the Lyapunov exponents determined from the deterministic model define the stability of the stochastic system near the equilibrium. The subsequent study of specific spatial interactions extracted the conditions for stationary and non-stationary instabilities. Near the onset of stationary instabilities, the stability threshold is independent of the propagation delay, while the variance of the corresponding stationary probability depends on the propagation delay. In the case of oscillatory instabilities, both the stability threshold and the variance depend on the propagation delay.

Fig. 5.15 The variance of the stationary distribution just below the onset of the Hopf instability with $\tau \approx \tau_c(0)$. Further parameters are taken from Fig. 5.14

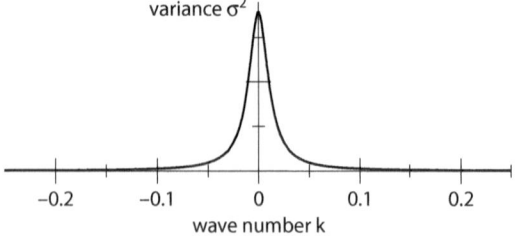

References

1. S. Amari. Dynamics of pattern formation in lateral-inhibition type neural fields. *Biol. Cybernet.*, 27:77–87, 1977.
2. F. M. Atay and A. Hutt. Stability and bifurcations in neural fields with finite propagation speed and general connectivity. *SIAM J. Appl. Math.*, 65(2):644–666, 2005.
3. C. T. H. Baker and E. Buckwar. Exponential stability in p-th mean of solutions, and of convergent euler-type solutions, of stochastic delay differential equations. *J. Comput. Appl. Math.*, 182(2):4004–427, 2005.
4. P. C. Bressloff and S. Coombes. Physics of the extended neuron. *Int. J. Mod. Phys. B*, 11(20):2343–2392, 1997.
5. A. A. Budini and M. O. Caceres. Functional characterization of linear delay langevin equations. *Phys. Rev. E*, 70:046104, 2004.
6. V. Castets, E. Dulos, J. Boissonade, and P. De Kepper. Experimental-evidence of a sustained standing turing-type non-equilibrium chemical-pattern. *Phys. Rev. Lett.*, 64:2953–2956, 1990.
7. C. Cattaneo. A form of heat conduction equation which eliminates the paradox of instantaneous propagation. *Compt. Rend.*, 247:431–433, 1958.
8. S. Coombes. Waves and bumps in neural field theories. *Biol. Cybernet.*, 93:91–108, 2005.
9. S. Coombes, G. J. Lord, and M. R. Owen. Waves and bumps in neuronal networks with axo-dendritic synaptic interactions. *Physica D*, 178:219–241, 2003.
10. S. M. Crook, G. B. Ermentrout, M. C. Vanier, and J. M. Bower. The role of axonal delays in the synchronization of networks of coupled cortical oscillators. *J. Comput. Neurosci.*, 4:161–172, 1997.
11. M. C. Cross and P. C. Hohenberg. Pattern formation outside of equilibrium. *Rev. Mod. Phys.*, 65(3):851–1114, 1993.
12. G. B. Ermentrout, J. McLeod, and J. Bryce. Existence and uniqueness of travelling waves for a neural network. *Proc. Roy. Soc. A*, 123(3):461–478, 1993.
13. T. D. Frank and P. J. Beek. Stationary solutions of linear stochastic delay differential equations: Applications to biological systems. *Phys. Rev. E*, 64:021917, 2001.
14. W. J. Freeman. *Mass Action in the Nervous System*. Academic Press, New York, 1975.
15. W. J. Freeman. Characteristics of the synchronization of brain activity imposed by finite conduction velocities of axons. *Int. J. Bif. Chaos*, 10(10):2307–2322, 2000.
16. J. Guckenheimer and P. Holmes. *Nonlinear Oscillations, Dynamical Systems, and Bifurcations of Vector Fields*, volume 42 of *Applied Mathematical Sciences*. Springer Verlag, New York, 1983.
17. S. Guillouzic, I. L'Heureux, and A. Longtin. Small delay approximation of stochastic delay differential equation. *Phys. Rev. E*, 59(4):3970, 1999.
18. A. A. Gushchin and U. Kuechler. On stationary solutions of delay differential equations driven by a lévy process. *Stochastic Proc. Appl.*, 88:195–211, 2000.
19. H. Haken. *Advanced Synergetics*. Springer, Berlin, 1983.
20. J. K. Hale and S. M. V. Lunel. *Introduction to Functional Differential Equations*. Springer, Berlin, 1993.
21. A. Hutt and F. M. Atay. Analysis of nonlocal neural fields for both general and gamma-distributed connectivities. *Physica D*, 203:30–54, 2005.
22. A. Hutt and F. M. Atay. Effects of distributed transmission speeds on propagating activity in neural populations. *Phys. Rev. E*, 73:021906, 2006.
23. A. Hutt and F. M. Atay. Spontaneous and evoked activity in extended neural populations with gamma-distributed spatial interactions and transmission delay. *Chaos, Solitons and Fractals*, 32:547–560, 2007.
24. A. Hutt, M. Bestehorn, and T. Wennekers. Pattern formation in intracortical neuronal fields. *Network: Comput. Neural Syst.*, 14:351–368, 2003.
25. A. Hutt and T. D. Frank. Critical fluctuations and 1/f -activity of neural fields involving transmission delays. *Acta Phys. Pol. A*, 108(6):1021, 2005.

26. M. A. P. Idiart and L. F. Abbott. Propagation of excitation in neural network models. *Network: Comp. Neural Sys.*, 4:285–294, 1993.
27. J. D. Jackson. *Classical Electrodynamics*. Wiley, New York, 3 edition, 1998.
28. V. K. Jirsa. Connectivity and dynamics of neural information processing. *Neuroinformatics*, 2(2):183–204, 2004.
29. K. H. W. J. ten Tusscher and A. V. Panfilov. Alternans and spiral breakup in a human ventricular tissue model. *Am. J. Physiol. Heart Circ. Physiol.*, 291:H1088–H1100, 2006.
30. J. Klafter and I. M. Sokolov. Anomalous diffusion spreads its wings. *Physics World*, 18(8):29–32, 2005.
31. C. Koch. *Biophysics of Computation*. Oxford University Press, Oxford, 1999.
32. U. Küchler and B. Mensch. Langevin stochastic differential equation extended by a time-delayed term. *Stoch. Stoch. Rep.*, 40:23–42, 1992.
33. E. Lazzaro and H. Wilhelmsson. Fast heat pulse propagation in hot plasmas. *Physics of plasmas*, 5(4):2830–2835, 1998.
34. M. C. Mackey and I. G. Nechaeva. Solution moment stability in stochastic differential delay equations. *Phys. Rev. E*, 52:3366–3376, 1995.
35. Th. Martin and R. Landauer. Time delay of evanescent electromagnetic waves and the analogy to particle tunneling. *Phys. Rev. A*, 45:2611–2617, 1992.
36. R. Metzler and J. Klafter. The random walk's guide to anomalous diffusion: a fractional dynamics approach. *Phys. Rep.*, 339:1, 2000.
37. J. D. Murray. *Mathematical Biology*. Springer, Berlin, 1989.
38. P. L. Nunez. *Neocortical Dynamics and Human EEG Rhythms*. Oxford University Press, New York- Oxford, 1995.
39. D. J. Pinto and G. B Ermentrout. Spatially structured activity in synaptically coupled neuronal networks: I. travelling fronts and pulses. *SIAM J. Applied Math.*, 62(1):206–225, 2001.
40. P. A. Robinson, C. J. Rennie, and J. J. Wright. Propagation and stability of waves of electrical activity in the cerebral cortex. *Phys. Rev. E*, 56(1):826–840, 1997.
41. A. M. Turing. The chemical basis of morphogenesis. *Philos. Trans. R. Soc. London*, 327B:37–72, 1952.
42. D. Y. Tzou and J. K. Chen. Thermal lagging in random media. *J. Thermophys. Heat Transfer*, 12:567–574, 1998.

Chapter 6
Stochastic Delay-Differential Equations

André Longtin

6.1 Introduction

This chapter concerns the effect of noise on linear and nonlinear delay-differential equations. Currently there exists no formalism to exactly compute the effects of noise in nonlinear systems with delays. The standard Fokker–Planck approach is not justified because it is meant for Markovian systems. Delay-differential systems are non-Markovian, although various approximations to them might be Markovian. For example, if the delay is small in comparison with all other timescales of the system, one can approximate the SDDE by a system of ordinary stochastic differential equations [12] (see Sect. 4). A notable exception is the class of linear stochastic differential equations with additive noise (see Sect. 3 below). In this case, while there is no Fokker–Planck formalism, the statistics of the system are Gaussian and one needs only compute the first two moments of the probability density of the variable.

Delay-differential equations (DDEs) occur in a wide variety of natural and man-made systems. They often arise from an approximation to a partial differential equation that describes, e.g., diffusion of some reacting substance or a traveling wave in some medium. For simplicity we confine our discussion to DDEs in one variable, referred heretofore as scalar DDEs. This nomenclature is somewhat misleading because delay-differential equations are infinite-dimensional dynamical systems: an infinite number of initial conditions—a function on the initial delay interval—is needed to uniquely specify their time evolution. The general class of dynamical systems that we focus on in this chapter is

$$dx(t) = f\left(x(t), x(t - \tau)\right) \, dt + \sigma g\left(x(t)\right) \, dW(t) \tag{6.1}$$

where $x \in R$, $W(t)$ is the standard Wiener process (i.e., with zero mean and variance equal to t) and σ is the noise strength.

A. Longtin (✉)
Department of Physics, University of Ottawa, Canada
e-mail: alongtin@uottawa.ca

F.M. Atay (ed.), *Complex Time-Delay Systems*, Understanding Complex Systems,
DOI 10.1007/978-3-642-02329-3_6, © Springer-Verlag Berlin Heidelberg 2010

One key question in this area is the following: Does there exist a Fokker–Planck type of description that can be solved for the time-dependent probability density $\rho(x, t)$? The problem of obtaining a complete description of the time evolution of the probability density for even a scalar delay-differential equation (i.e., a delay-differential equation in one variable) is currently beyond reach. We review a number of approximations to this density that are available for nonlinear systems. These usually involve converting the DDE to a deterministic system that approximates it—and that reduces its dimension by making it finite. In the following we will thus refer to these approximations as reduction methods.

Not surprisingly, since DDEs occur in a variety of applications, the influence of noise on such DDEs is increasingly a focus of investigation. This is true especially in the world of laser physics, where delays arise from finite propagation times inside optical cavities and around optical circuits external to the laser, as well as in optical fiber networks. There have also been very recent studies of the enhancement by noise of the oscillations in a network of delayed-coupled oscillators [20] and in a bistable discrete-time stochastic map [35], of stochastic resonance in a non-Markovian system [38], and of control of noise-induced motion in relaxation oscillators using delayed feedback [17].

In the biological world, delays arise from finite maturation or division time of various cellular species, such as blood cell lines, or the synthesis of various molecular species, as in the immunological system or genetic control systems [15]. In neuroscience, delays arise from the propagation time of nerve impulses down axons and across synapses. In such systems one is often faced with the problem of distributions of delays [1]. Noise has particularly permeated the fields of research of genetics and neuroscience over the last decades, efforts that point to the potential useful synergy between these fields and the development of a common language for the experimental and modeling approaches [41].

The influence of noise in neural systems will be summarized in the last part of our review. We first begin by stating the fundamental problem with the analysis of SDDEs, then describe the linear case, followed by the pros and cons of small delay Taylor expansions of the SDDE. This chapter then discusses other recent approximations that may be useful in different contexts and which may point the reader into new directions of investigation.

6.2 The Fundamental Issue

Recent mathematical treatments of SDDEs have revealed interesting properties on, e.g., the existence of smooth probability densities [2] and exponential stability [29]. Here we focus on the issue of formulating a Fokker–Planck-type equation from an SDDE. Let $x \in (a, b)$. Define $x_\tau \equiv x(t - \tau)$ and $x \equiv x(t)$. Let the bivariate density $P(x_o, t_o; x_\tau, t_\tau | \phi) \, dx_o dx_\tau$ describe the probability that $x(t_o) \in (x_o, x_o + dx_o)$ and $x(t_\tau) \in (x_\tau, x_\tau + dx_\tau)$, conditional on the initial function $x(t) = \phi(t)$ for all $t \in (-\tau, 0)$. We can then show that [12]

$$\frac{\partial}{\partial t} p(x_o, t|\phi) = -\frac{\partial}{\partial x_o} \left(p(x_o, t|\phi) \int_a^b dx_\tau \, f(x_o, x_\tau) p(x_\tau, t - \tau |x_o, t; \phi) \right) \quad (6.2)$$

$$+ \frac{\sigma^2}{2} \frac{\partial^2}{\partial x_o^2} \left(p(x_o, t|\phi) g^2(x_o) \right), \quad (6.3)$$

$$(6.4)$$

where

$$p(x_o, t_o|\phi) \equiv \int_a^b dx_\tau \, p(x_o, t_o; x_\tau, t_\tau |\phi) \quad (6.5)$$

and

$$p(x_\tau, t_\tau |x_o, t_o; \phi) \equiv \frac{p(x_o, t_o; x_\tau, t_\tau)|\phi)}{p(x_o, t_o|\phi)}. \quad (6.6)$$

This PDE looks very much like the usual Fokker–Planck equation associated with the Ito Langevin equation

$$\frac{dx}{dt} = f(x) + \sigma g(x) \xi(t) \quad (6.7)$$

for a Markovian system, where $\xi(t) = dW/dt$ is Gaussian white noise:

$$\frac{\partial p(x, t)}{\partial t} = -\frac{\partial f(x) p(x, t)}{\partial x} + \frac{\sigma^2}{2} \frac{\partial^2 g^2(x) p(x, t)}{\partial x^2} \quad (6.8)$$

where we have dropped the conditioning on the initial state of the system for clarity. There is, however, one important difference: the drift term involves an integral over the conditional probability density

$$p(x_\tau, t - \tau |x_o, t; \phi). \quad (6.9)$$

The problem is that this conditional density must be known in order to solve for the density of interest, $p(x_o, t|\phi)$, so one is faced with a circular problem. Nevertheless this form is useful as a starting point for approximation schemes or when used in conjunction with them. Let us define a "conditional average drift" or CAD as

$$\bar{f}(x_o, t_o|\phi) \equiv \int_a^b dx_\tau \, f(x_o, x_\tau) p(x_\tau, t_o - \tau |x_o, t_o; \phi), \quad (6.10)$$

which is also the average of $(d/dt)x(t)$ at time t_o given that $x(t_o) = x_o$. Using this CAD, (6.4) becomes

$$\frac{\partial}{\partial t}p(x_o, t|\phi) = -\frac{\partial}{\partial x_o}\left\{\bar{f}(x_o, t|\phi)p(x_o, t|\phi)\right\} \tag{6.11}$$

$$+\frac{\sigma^2}{2}\frac{\partial^2}{\partial x_o^2}\left\{g^2(x_o)p(x_o, t|\phi)\right\} \tag{6.12}$$

which is the usual Fokker–Planck equation corresponding to the SDE

$$dx(t) = \bar{f}(x(t), t|\phi)dt + \sigma g(x(t))dW(t). \tag{6.13}$$

As the system approaches a steady state for $t \to \infty$, the functions $\bar{f}(x_o, t|\phi)$ and $p(x_o, t|\phi)$ approach their steady-state equivalents $\bar{f}^s(x_o|\phi)$ and $p(x_o, t|\phi)$, respectively. For reflecting boundary conditions, these two functions are related by the potential solution

$$p^s(x_o|\phi) = \frac{N}{g^2(x_o)}\exp\left(\frac{2}{\sigma^2}\int_c^{x_o}dx'\frac{\bar{f}^s(x'|\phi)}{g^2(x')}\right) \tag{6.14}$$

where $c \in (a, b)$ and N is the normalization constant over (a, b). We will return to these results when we investigate the delayed Langevin equation, perhaps better referred to as the stochastic Wright equation, in the section on the small delay expansion.

6.3 Linear SDDEs

It is possible to obtain a description of the stochastic delayed dynamics in terms of a stationary probability density for the delayed Langevin problem [8, 12, 19]:

$$dx(t) = \alpha x(t - \tau)dt + dW(t). \tag{6.15}$$

Since the zero-mean Gaussian process acts additively on a linear ordinary differential equation, $x(t)$ will also be a zero-mean Gaussian process. This means that one needs only to calculate the second moment. The solution can be obtained in different ways. Here is one way involving the computation of the autocorrelation function. Since $x(t)$ and $\eta(t)$ are jointly weak-sense-stationary (WSS), their auto- and cross-correlation functions will not depend on absolute time. Hence,

$$\frac{d}{dt}R_{x\eta}(t) = -\alpha R_{x\eta}(t - \tau) + R_{\eta\eta}(t) \tag{6.16}$$

and

$$\frac{d}{dt}R_{xx}(t) = \alpha R_{xx}(t + \tau) - R_{x\eta}(t), \tag{6.17}$$

where $t \equiv t_1 - t_2$, i.e., it is the difference of the times at which the two-point correlation functions are evaluated. The power spectral density $S(\omega)$ of the process is given by the Fourier transform of $R(t)$ via the Wiener–Khintchine theorem. One readily obtains

$$S_{xx}(\omega) = \frac{S_{\eta\eta}(\omega)}{\alpha^2 + \omega^2 - 2\alpha\omega\sin(\omega\tau)} . \tag{6.18}$$

Hence the variance of the process is

$$\sigma_{xx}^2 = R_{xx}(0) = \frac{1}{2\pi} \int_{\infty}^{\infty} d\omega \frac{S_{\eta\eta}(\omega)}{\alpha^2 + \omega^2 - 2\alpha\omega\sin(\omega\tau)} . \tag{6.19}$$

This integral is numerically equivalent to the closed-form expression obtained in [19]

$$\sigma_x^2 = \frac{\sigma^2}{2\alpha}(1 + \alpha\tau) . \tag{6.20}$$

Nechaeva and Mackey also considered linear SDDEs [28]. They looked at additive and multiplicative noise and studied the stability and boundedness of these processes by analytically computing the time evolution of the first two moments of the state variable.

6.4 Small Delay Expansion

An expansion of $f(x(t), x(t-\tau))$ in powers of τ using a Taylor expansion around $x(t)$ is valid to quadratic order in τ. In performing such expansions, one must take care to gather all terms at each order of $\sqrt{dt}$ and dt [12]. First, we expand in a Taylor series

$$f(x, x_\tau) \approx f(x, x) + (x_\tau - x)\frac{\partial}{\partial x_\tau}f(x, x) . \tag{6.21}$$

We further expand

$$x_\tau \approx x(t) - \tau\frac{dx(t)}{dt} . \tag{6.22}$$

Substituting the delayed Langevin equation (6.1) for dx/dt in this last expression, and reinserting the expansion (6.21) (which comes from (6.1)), one obtains

$$dx = f_a(x)dt + \sigma g_a(x)dW , \tag{6.23}$$

where the approximate drift and diffusion terms of the resulting approximating non-delayed Langevin equation are

$$f_a(x_o) \equiv f(x_o, x_o) \left(1 - \tau \frac{\partial}{\partial x_\tau} f(x_o, x_o) \right), \tag{6.24}$$

$$g_a(x_o) \equiv g(x_o) \left(1 - \tau \frac{\partial}{\partial x_\tau} f(x_o, x_o) \right) \tag{6.25}$$

and

$$\frac{\partial}{\partial x_\tau} f(x_o, x_o) \equiv \left. \frac{\partial}{\partial x_\tau} f(x_o, x_\tau) \right|_{x_\tau = x_o} . \tag{6.26}$$

This approximate Langevin SDE is readily seen to yield the usual Langevin equation when the delay vanishes. This approximate SDE has been shown to provide accurate values of the variance of the stochastic process. The accuracy deteriorates when the delay is large or when the deterministic DDE associated with the SDDE has complex eigenvalues. This is not surprising, since these Taylor expansions can only yield a first-order ODE, which cannot have complex eigenvalues—and thus cannot exhibit oscillations.

The importance of carefully keeping terms at each order can be illustrated in the following example. Consider the deterministic DDE

$$\frac{dx}{dt} = -\alpha x(t - \tau) \tag{6.27}$$

where α is a positive coefficient. This DDE was studied by Wright [45]. We wish to gain an intuitive understanding of the dependence on the delay of the variance of the Gaussian process that results from adding Gaussian white noise to the right-hand side of Eq. 6.27. Its (deterministic) characteristic equation is

$$\lambda = -\alpha \exp(-\lambda \tau) . \tag{6.28}$$

It has only one root $\lambda = -\alpha$ for $\tau = 0$. As τ increases from zero, this root r_1 becomes more negative. Further, an infinite number of roots come into existence, whose real parts decrease off to $-\infty$. One of these roots, r_2, is real, with $r_2 < r_1$; all other roots are born as complex conjugate pairs. At $\tau = 1/(\alpha e)$, r_1 and r_2 merge. As the delay increases further, this merged root acquires an imaginary part, i.e., they become a complex conjugate pair. The real part of these roots subsequently increases with delay. At $\tau = \pi/2\alpha$, this pair crosses the imaginary axis, and the linear system diverges in an oscillatory manner.

Now imagine what happens when Gaussian white noise also drives this process. Intuitively, the variance will be proportional to how positive the dominant or rightmost root of the system is. In particular, as τ begins increasing from zero, the rightmost root actually moves to the left. This suggests that the system becomes more stable, and thus that the variance decreases—this is similar to what we expect from the Ornstein–Uhlenbeck process $dx/dt = -\alpha x(t) + \xi(t)$ as α increases. In fact, if one non-dimensionalizes (6.27), one obtains $dX/dt = -\alpha \tau X(t-1)$, in which one readily sees that increasing the delay is akin to increasing α, i.e., increasing the restoring spring constant of a (delayed) quadratic potential.

Thus, if we base our intuition on the behavior of the rightmost root, we would expect the variance to first decrease, then increase when the two real roots are close together, merge, and then see their real part increase with delay after having become a complex conjugate pair. However, what a careful analysis shows is that the first part of this story is wrong, and that in fact the variance of this Gaussian process increases monotonically and eventually diverges. This can be readily seen by expanding the integrand in (6.19) and taking the first terms, yielding

$$\sigma_x^2 = \frac{\sigma^2}{2\alpha}(1 + \alpha\tau). \tag{6.29}$$

The result indicates that the variance is indeed a monotonically increasing function of the delay. The paradox can be resolved by looking at the expression for the approximate diffusion (6.25) in the small delay expansion. One sees that the drift actually contributes to the diffusion by an amount proportional to the delay. Thus, even though the delay does not explicitly appear in the diffusion term in the original SDDE, it enhances the influence of the noise on this system.

We complete our illustration with the calculation of the approximate SDE following a small delay expansion and the associated conditional average drift (CAD) and probability densities. The approximate SDE reads

$$dx = -\alpha(1 + \alpha\tau)x\,dt + (1 + \alpha\tau)\sigma\,dW, \tag{6.30}$$

and the corresponding approximate steady-state probability density is

$$p_a^s(x_0|\phi_0) = N_a \exp\left[\frac{-\alpha x_0^2}{\sigma^2(1 + \alpha\tau)}\right] \tag{6.31}$$

from which one can extract the same approximate variance as in (6.29). It is clear from the approximate SDE that, as the delay increases, the eigenvalue $-\alpha(1 + \alpha\tau)$ becomes more negative but the noise intensity increases—and this latter effect dominates the behavior of the variance. One may also consider the CAD $\overline{f}^s(x_0|\phi) = -\alpha(1 - \alpha\tau)x_0$, with the corresponding SDE

$$dx = -\alpha(1 - \alpha\tau)x\,dt + \sigma\,dW. \tag{6.32}$$

In contrast to the SDE obtained from the small delay expansion, the eigenvalue (the coefficient of the drift term) of this SDE actually increases with the delay in line with the behavior of the variance (note that the diffusion is now constant in this

SDE). The CAD thus accounts for the behavior of both the approximate drift and diffusion of the small delay expansion.

We close this section with an important caveat regarding the type of calculus to apply to SDDEs. Both Ito and Stratonovich interpretations are valid, but as usual, one must take care in making sure that the interpretations match for the analytics and numerics—and to apply proper conversions when they do not [4, 12].

6.5 Reduction Techniques

6.5.1 Reducing the Dimensionality

Here we briefly discuss approximations which, when used systematically, may lead to insights into a Fokker–Planck type analysis for SDDEs. First we consider ways to approximate the DDE with a finite set of ODEs. For example, one can use the so-called chain trick to convert a delay-differential equation into a finite number of coupled ordinary differential equations. Consider the integro-differential system

$$\frac{dx}{dt} = f(x(t), z(t)) \tag{6.33}$$

$$z(t) = \int_{-\infty}^{t} K(t-s)x(s)\,ds, \tag{6.34}$$

where $K(t)$ is a memory kernel. For certain families of kernels, this system can be rewritten as a set of ODEs using a recurrence relationship. For example, let

$$K_a^m(t) \equiv \frac{a^{m+1}}{m!} t^m \exp^{-at}. \tag{6.35}$$

For $m = 0$, this kernel is simply the usual exponentially decaying memory kernel. For higher m values, this kernel becomes more and more localized near a mean delay of $\bar{\tau} \equiv (m+1)/a$. In the limit $m \to \infty$, the memory kernel converges to a Dirac delta function $K_a^\infty(t) = \delta(t - \bar{\tau})$. For a given value of m, one can convert (6.34) into a finite set of $(m+2)$ ordinary differential equations:

$$\frac{dy_o}{dt} = f(y_o, y_{m+1}) \tag{6.36}$$

$$\frac{dy_i}{dt} = a(y_{i-1} - y_i) \quad i = 1, 2, ..., m+1. \tag{6.37}$$

$$\tag{6.38}$$

In the limit $m \to \infty$, the IDE becomes equivalent to a DDE, and likewise to an infinite set of ODEs. This analysis can then in principle be carried out for the IDE (6.34) with additive Gaussian white noise

$$\frac{dx}{dt} = f(x(t), z(t)) + \xi(t), \tag{6.39}$$

$$z(t) = \int_{-\infty}^{t} K(t - s)x(s)ds. \tag{6.40}$$

For example, for the stochastic version of the Wright equation (6.27)

$$\frac{dx}{dt} = -\alpha x(t - \tau) + \xi(t), \tag{6.41}$$

the chain trick with $m = 1$ yields the following three-dimensional system of ODEs:

$$\frac{dy_o}{dt} = -\alpha y_2(t) + \xi(t) \tag{6.42}$$

$$\frac{dy_1}{dt} = -\frac{2}{\tau} (y_1(t) - y_o(t)) \tag{6.43}$$

$$\frac{dy_2}{dt} = -\frac{2}{\tau} (y_2(t) - y_1(t)) . \tag{6.44}$$

This is a simple three-dimensional system of linear SDEs with additive white noise. The full time-dependent solution of the Fokker–Planck equation associated with this system can in principle be obtained [16]. However, in practice this involves solving an increasingly complex matrix equation, and further, since we are ultimately interested in the density of the original variable $x(t)$, i.e., $\rho(x, t)$, we must integrate over all the degrees of freedom except y_o to obtain the desired marginal density (Longtin, unpublished results). And the method breaks down if the original SDDE is not linear. Nevertheless, the approach may yield interesting insights into a proper formulation of a Fokker–Planck-type equation for SDDEs; one can see in particular how the known analytical expression for the variance seen above in (6.20) develops from approximate stochastic integro-differential formulations of the noisy delayed problem.

Another reduction method that has been used in the literature is to invert a time evolution operator. To our knowledge there is no rigorous justification for the following operation, so one should really proceed at one's own (potentially high!) risk. Rewrite

$$\frac{dx}{dt} = h(x(t - \tau)) + \xi(t). \tag{6.45}$$

This can be rewritten in terms of a time-shift operator

$$\frac{dx}{dt} = \exp\left[-\tau \frac{d}{dt}\right] h(x(t)) + \xi(t). \tag{6.46}$$

Inverting the operator yields

$$\exp\left[\tau\frac{d}{dt}\right]\frac{dx}{dt} = h\left(x(t)\right) + \tilde{\xi}(t) \tag{6.47}$$

where $\tilde{\xi}(t) \equiv \xi(t + \tau)$, i.e., the forward shifted white noise will yield a dynamical system with the same statistical properties, since it is an external input. This procedure would already be more justified if a small parameter can be found such that terms in the expansion decay quickly (the expansion referring here to the Taylor expansion of the exponential operator). In this case as in the linear chain trick system above, one obtains a system of ODEs of higher order if one wants a better approximation.

Ohira and Milton [33] have also investigated an original approximation to SDDEs via a random walk process in which the step sizes are function of the current position of the walker, i.e., a random walk in a state-dependent potential. Another approach consists of looking at the large delay limit, in which the SDDE can be approximated by a map [26]. Analytical techniques exist for studying the effect of noise on maps; further there is, in this limit, a correspondence between the effects of additive and multiplicative noise, i.e., we can deduce the one by knowing the effect of the other. This is in contrast to what is expected in ODEs, where additive and multiplicative noise can have qualitatively different effects on dynamics and bifurcations.

Frank and his colleagues [8–10] have recently done a large amount of work on SDDEs using a variety of techniques, including some that revolve around nonlinear Fokker–Planck equations.

There have been recent developments in the theoretical analysis of noise in delay systems near a bifurcation point, where a clear separation of timescales exists. One such study has proposed a perturbation theoretical analysis where the system is described by a differential equation on an $O(1)$ timescale and another on a slow $O(\epsilon)$ timescale, where $\epsilon \ll 1$. The resulting coupled Langevin equation approximation produces good quantitative agreement with the numerical solutions, at least in the vicinity of the bifurcation point [21].

Not surprisingly, SDDEs have also appeared in the context of coupled oscillators. Yeung and Strogatz [46] have studied how the Kuramoto transition is altered by the presence of noise and delayed coupling in a collection of coupled phase oscillators. Their heuristic approach is based on an analysis of a Fokker–Planck equation, which allows a good approximation to the equilibrium density for the phase distribution.

Finally there is a class of stochastic dynamical systems that are closely related to SDDEs, and which are framed in terms of Langevin equations with a memory kernel (see [42, 47] and references therein). This kernel often accounts for a memory in the friction of the system, i.e., the dissipation is proportional to an integral of the velocity over some past history of the system. Mathematically these systems are stochastic integro-differential equations, and in the limit where the memory is completely localized at a discrete time in the past, they converge to SDDEs. Consequently, one expects a fruitful interaction between the investigations on these two classes of systems.

6.5.2 Crossing Time Problems

In crossing time problems, the goal is to characterize the properties of the evolution time of a dynamical system toward a specific boundary. This time is a random variable if the system is a stochastic dynamical system. One usually focuses on the mean time to cross a threshold, i.e., the so-called mean first passage time. There have been some attempts to characterize the effect of memory on crossing time (see [13] and references therein). The small delay formalism described above provides means for calculating transition rates between two stable states in the small delay limit. The analysis [13] relies on simplifying the system using a two-state filter, i.e., on discarding the detailed knowledge of the precise state in favor of a binary left–right description of the state. For example, consider the evolution in the delayed quartic (bistable) potential

$$\frac{dx}{dt} = x(t - \tau) - x^3(t - \tau) + \sigma\xi(t). \tag{6.48}$$

This is a quartic potential in the over-damped limit, i.e., $x - x^3$ is the negative of the gradient of a potential $V(x) = -x^2/2 + x^4/4$. One can obtain a density function using the small delay expansion, and further write down a rate equation for the population inside each well using phenomenological rate constants derived from that approximate density.

There has also been a study of the effect of noise on the switching probability between the two wells of a symmetric double-well system with linear delayed feedback [36]:

$$\frac{dx}{dt} = x(t) - x^3(t) + \alpha x(t - \tau) + \xi(t), \tag{6.49}$$

where α is a parameter that controls the strength of the delayed feedback. The deterministic part of this equation finds applications in a variety of areas such as the modeling of the El Nino phenomenon (see [39] and references therein) and of the dynamics of genetic networks [15]. The idea is again to simplify the dynamics by considering a two-state filter of the system, i.e., by coarse-graining the phase space into two boxes: left and right of the origin. Since the feedback can only take on one of two values, it tilts the quartic (bistable) potential one way or the other. One can then write a master equation for the transitions between the two states, which includes terms that depend on where the system was a delay earlier—in other words, one has a system of two delayed kinetic equations governing the probability to the right and left of the origin.

This approach can then include a simple Kramers' type of activation over the barrier; given that there are two possible tilts, there are two possible barriers. One can then describe the steady-state probability densities of the particle being in each well. The authors further extended their analysis to periodic forcing, and in particular to subthreshold forcing where multiple stochastic resonance effects are seen. Their two-state analysis successfully approximates the behavior of the full system in a regime where the system is bistable. However, there are other regimes nearby in the parameter space of the deterministic delayed bistable system, all of which are organized around a Taken–Bogdanov bifurcation known to occur in the deterministic dynamics of (6.49) [39]. Thus one expects that more elaborate analyses will be required to account for noise-induced switching when the system is not strongly bistable.

6.6 Stochastic Delayed Neurodynamics

6.6.1 Neural Noise and Delays

A neuron is characterized by a current balance equation and a set of gating equations for the various ionic species that flow across its membrane. The goal of this description is to properly account for the transmembrane potential, and how it varies in time with or without external currents to the neuron. These inputs come from other neurons via synaptic connections where current flows across specialized ionic channels. Single neurons or collections of neurons ("neural systems") are part of the more general class of excitable systems. Strictly speaking, excitability refers to the nonlinear property of a system that responds mildly to small input signals, but exhibits a large excursion in phase space followed by a brief return back to the vicinity of some fixed point when the input exceeds some threshold. These systems further exhibit autonomous oscillations, in the form of periodic sequences of such brief pulses, when the input signal is maintained above threshold—strictly speaking, they are no longer "excitable," but are nevertheless referred to as such because the individual spikes are large excursions that return to the vicinity of the fixed point.

There are many sources of noise in neurons. The dominant one is synaptic input from a large number of other neurons. The mathematical treatment of the effect of noise on the evolution of the transmembrane potential, and consequently on the firing activity of the neuron, has a long and distinguished history [14, 43]. Novel phenomena that arise from the interaction of noise and nonlinearity in excitable systems has also received much attention in recent times. An excellent recent review can be found in [24]. These efforts have clarified how noise can induce firings or modify otherwise periodic firing. Much of this work focuses on first passage time problems from some resting voltage, and the role of noise in shaping spatio-temporal patterns of firing activity.

Because neurons are highly nonlinear systems, as a consequence (especially) of their threshold, and because there are finite delays in activity propagation between neurons, they are somewhat canonical biological SDDEs. But these descriptions are not restricted to spiking neurons; they also apply to so-called rate descriptions, where the quantity of interest is the rate of firing of neurons in various subpopulations, averaged over a short time interval. This description is particularly relevant when considering large-scale neural systems, such as those involved in reflexes and motor control [6].

6.6.2 Neural Control

Longtin, Milton, and colleagues in the early 1990s studied how simple bifurcations could occur in a neural control system. This work was carried out in the context of the emerging fields of nonlinear dynamics and chaos in biology and medicine, one that scrutinized every variable biological time series for signs of low-dimensional chaotic behavior. The human pupil light reflex is mainly designed to control the light flux on the retina by modifying the area of the iris muscle. This reflex was a tempting puzzle in this context because of the ongoing fluctuations it exhibited, even in constant lighting conditions. A satisfactory explanation for these fluctuations came from experiments that allowed the feedback loop to be opened and then closed electronically using infrared videopupillometry. The gain of this control system was varied systematically and non-invasively, and the fluctuations recorded and analyzed. What they found were fluctuations that were consistent with sweeping across a Hopf bifurcation in the presence of delayed feedback and of neural noise inside the brain [27].

Other investigators applied these ideas in the context of visual control of limb movements (such as a finger) [44]. The motivation is to learn how the dynamics may change in various pathologies. In this context, there are two delays: one for the proprioceptive feedback (which informs the brain on the position of a limb) and one for the visual processing feedback. Ohira and Milton have looked very closely at the problem of posture control with random perturbations [33–35]. A more recent study of "pencil balancing on the finger" by Cabrera an Milton [5] has revealed that the experimental data can be well modeled by a DDE with parametric noise; further, the system has to operate in the vicinity of a bifurcation, which surprisingly allows control on many timescales, including on times shorter than the delay.

6.6.3 Neural Population Dynamics

There is currently no general theory for the dynamics of populations of noisy neurons with delayed interactions. There are, however, a number of techniques that have been developed to approximate and characterize some of their properties. Rather than simulate a large population of firing neurons coupled to one another, Nykamp and Tranchina [32] have proposed a set of self-consistent differential equations that describe the evolution of the probability of firing in the network. While there is no theoretical solution to these equations, their numerical solution provides a close match to full network simulations—at a fraction of the computational cost. Propagation delays can easily be added to this formalism.

Knight and co-workers [18] have developed an approximate theory for the behavior of uncoupled stochastic neurons, and feedback can be included under certain conditions. Delays are particularly well accommodated in the so-called spike response model formalism by Gerstner and colleagues [11]. This formalism focuses on the voltage waveforms due to spikes, synaptic input, and external input and allows for the calculation of single cell or network activity. In the presence of noise, this formalism does not explicitly calculate mean first passage times to threshold, but rather mimics noise-induced or noise-perturbed firing by the use of a phenomenological escape rate function. The latter is a probability of firing that is parametrized by the level of input to the neuron and a "temperature" which sets the steepness of this activation function (a higher temperature, i.e., a noisier neuron, has a shallower activation function). The approximation leads to a satisfactory description of many properties of neural networks with noise and delays.

Brunel and Hakim [3] have developed an analytic theory for oscillations in a neural network of stochastic neurons with delays (see also [30]). This theory is based on a second-order mean field theory, where one writes an equation for a typical neuron that is driven by the mean level of current from the network, and the "network firing rate dependent" noise associated with this mean level:

$$\frac{dV}{dt} = -V(t) + \mu + \sigma\sqrt{\nu(t-\tau)}\xi(t), \qquad (6.50)$$

where V is the membrane voltage and μ is the sum of a bias current and a mean current from the network. Here $\nu(t-\tau)$ is the firing rate of the network at a time τ in the past. In other words, connections between neurons occur with a fixed delay τ, and when their firing rate increases, the mean input to the cells they are coupled to increases, as does the noise (mean and variance being correlated for stochastic point processes). This is a special brand of SDDE, since the delayed variable is not the usual state variable, but rather a quantity that depends on the probability flux of the stochastic process through a threshold. The connectivity matrix is sparse, i.e., each neuron connects on average to only a small fraction of the neurons in the network. The firing rate is calculated as the probability flux of the population through the (fixed) threshold. This analysis revealed the existence of a sharp transition between stationary and oscillatory (weakly synchronized) global activity as the strength of the inhibitory feedback between the cells becomes large enough (for an infinite network), and how global oscillations with a finite coherence time can be found both below and above this critical threshold in a finite network.

Lindner et al. [25] have successfully applied linear response theory to networks of noisy delayed-coupled neurons. It enables computations of single cell properties, such as spike train power spectra which are directly available experimentally. The motivation is to understand network oscillations in response to spatio-temporal noise. The idea is to consider the basic firing activity of an isolated leaky integrate-and-fire neuron driven by a bias current μ and zero-mean Gaussian white noise $\xi(t)$:

$$\frac{dV}{dt} = -V + \mu + \xi(t). \tag{6.51}$$

As above, this system evolves from a "reset" voltage; whenever a fixed firing threshold is reached, a spike is "fired" and the voltage put back to its reset value. The mean firing rate, i.e., mean first passage time to threshold, is known for this process. Likewise, the spectral properties of this spike train are known [23]. In the spirit of linear response theory, one can write the Fourier transform of the resulting spike train as

$$\tilde{X}(\omega) = \tilde{X}_o(\omega) + A(\omega)Y(\omega) \tag{6.52}$$

where $A(\omega)$ is the frequency response (i.e., susceptibility) of the noisy neuron, and Y is the Fourier transform of the input to the neuron, which comes from the external input *as well as* the internal input from delayed feedback. One also uses a self-consistency relationship for the mean rate in the presence of feedback [25, 18].

There has also been renewed interest in SDDEs in the context of the control of tremor using deep brain stimulation in Parkinson's patients [37]. The idea is to record from the abnormally firing neurons and feed back the activity to those neurons, with the goal of desynchronizing their activity. Theoretical analyses predict parameter values for the delay and feedback gain which lead to this destabilization.

6.6.4 Simplified Stochastic Spiking Model with Delay

Morse and Longtin [31] have recently proposed a simple model that may offer some insights into theoretical approaches to stochastic neural networks with delays. The neuron itself is a simple threshold crossing device (TCD), which emits a pulse whenever a set threshold is crossed in a positive-going direction. The firing statistics of such TCDs are well known and depend on various moments of the power spectral density of the noise that drives the system. The goal of that study was to understand the interplay of excitability, noise-induced firing, delayed feedback, and harmonic driving in the form of sinusoidal forcing.

In the absence of harmonic forcing, the system exhibits reverberating activity due to the positive feedback. In other words, any spikes that occur are fed back to the system after a (fixed) delay and cause more spikes to occur—this leads to oscillations with a period on the timescale of the delay. These oscillations can actually be initiated by positive fluctuations of the noise, and later extinguished by negative fluctuations. If feedback spikes are smaller than the threshold, the reverberating oscillation is strongest for a moderate amount of noise—a phenomenon known as coherence resonance.

One can then apply a sinusoidal drive that is subthreshold. In open loop (i.e., without feedback), this system exhibits stochastic resonance, i.e., the sinusoid is optimally expressed in the output spike train by a moderate amount of noise. With positive feedback, the system also exhibits stochastic resonance. Furthermore, if the delay is close to the value of the forcing period, the resonance is much stronger. In other words, a simple neuron (or neural population) fed back onto itself with a delay can resonate on its own as a stochastic oscillator, and it will react ("resonate") very strongly to a forcing signal that has a period close to that of this intrinsic stochastic oscillation. This resonant behavior was analyzed using a DDE built from the known mean firing rate of a TCD and from shot noise theory that approximates the effect of such input by its mean [31]. This analysis pointed to novel regimes where the mean behavior of the system could behave chaotically, with stochastic fluctuations on top of that. This leaves open a array of interesting avenues to explore.

6.7 Conclusion

We have given a review of recent theoretical work on the stochastic dynamics of delay-differential equations. We have provided motivations for their study from different areas of application, with special focus on neural systems where delays are known to play major computational roles.

It is clear from the foregoing discussion that there is still much need for more formal results on the effect of noise on DDEs, and in particular, on the dynamics of networks of excitable systems. One ongoing issue with neural dynamics is the necessity to eventually add more biophysical realism. For example, neural dynamics are complicated by the fact that the delay may in fact be state dependent. A strong input to a population of nerves may activate larger axons on which activity propagates faster, and thus the propagation delay is shorter. If the amplitude of the input is a function of the state of the system, as in the case of feedback systems, then one has all the ingredients for a SDDE with state-dependent delays, about which almost nothing is known, formally or numerically.

Also, in neural systems, the variance of the neural activity is going to be a function of its mean—this is true, for example, for a Poisson (shot noise) process [3]. There are also many challenges ahead in incorporating the effect of correlations between input noises [25], their interaction with intrinsic neuron dynamics which impose their own correlations on the spike trains, the effect of plasticity and stochastic release probability in neurons, and the discrete nature of synaptic input [40]. Delays will lead in certain cases to multistability of neural activity along reverberatory loops [7], and there is a great need to understand how noise influences firing patterns under these conditions.

Finally one can think of situations where the delay time itself is a stochastic quantity. This is the case in neuroscience when activity propagates along feedback pathways with variable amounts of refractoriness depending on the recent history of firing. There are other areas where delays have gone unnoticed in neuroscience: back-propagation of a spike back to the neuron's dendrites and second messenger/genetic signaling. A recent review of how noise affects the dynamics of neural and genetic networks can be found in [41]. And in complex delayed systems where chaos can occur, one has the possibility of approximating, at least for modeling purposes, the feedback terms that generate the chaotic instability as noise sources. This would be an extension of the ideas developed two decades ago for the study of the Ikeda DDE [22].

We hope that, by bringing under one roof this diverse set of results on SDDEs in different areas, this chapter will provide the impetus to pursue original and daring new directions that will lead to a fuller comprehension of the effect of noise on systems with memory.

Acknowledgments This work was supported by NSERC Canada and CIHR Canada.

References

1. F. M. Atay, and A. Hutt. Neural fields with distributed transmission speeds and long-range feedback delays. *SIAM J. Appl. Dynam. Syst.*, **5**: 670–698, 2006.
2. D. R. Bell, and S.-E. A. Mohammed. Smooth densities for degenerate stochastic delay equations with hereditary drift. *Ann. Prob.*, **23**: 1875–1894, 1995.
3. N. Brunel, and V. Hakim. Fast global oscillations in networks of integrate-and-fire neurons with low firing rates. *Neural Comp.*, **11**: 1621–1671, 1999.
4. E. Buckwar, A. Pikovsky, and M. Scheutzow. Stochastic dynamics with delay and memory. Stochastics and Dynamics, 5(2) Special Issue, June 2005.
5. J. L. Cabrera, and J. G. Milton. On-off intermittency in a human balancing task. *Phys. Rev. Lett.*, 89: 157802, 2002.
6. Y. Chen, M. Ding, and J. A. S. Kelso. Long memory processes (1/fa type) in human coordination. *Phys. Rev. Lett.*, **79**: 4501–4504, 1997.
7. J. Foss, A. Longtin, B. Mensour, and J. G. Milton. Multistability and delayed recurrent loops. *Phys. Rev. Lett.*, **76**: 708–711, 1996.
8. T. D. Frank, and P. J. Beek. Stationary solutions of linear stochastic delay differential equations: Application to biological systems. *Phys. Rev. E.*, **64**: 021917, 2001.
9. T. D. Frank, P. J. Beek, and R. Friedrich. Fokker-Planck perspective on stochastic delay systems: Exact solutions and data analysis of biological systems. *Phys. Rev. E*, **68**: 021912, 2003.
10. T. D. Frank. Delay Fokker-Planck equations, perturbation theory, and data analysis for nonlinear stochastic systems with time delays. *Phys. Rev. E*, **71**: 031106, 2005.
11. W. Gerstner, and W. Kistler. *Spiking Neuron Models*. Cambridge, Cambridge University Press 1999.
12. S. Guillouzic, I. L'Heureux, and A. Longtin. Small delay approximation of stochastic delay differential equations. *Phys. Rev. E*, 59: 3970–3982, 1999.
13. S. Guillouzic, I. L'Heureux, and A. Longtin. Rate processes in a delayed, stochastically driven, and overdamped system. *Phys. Rev. E*, 61: 4906–4914, 2000.
14. A. Holden. *Models of the Stochastic Activity of Neurons. Lecture Notes in Biomathematics*. Springer, Berlin 1976.
15. D. Huber, and L. S. Tsimring. Dynamics of an ensemble of noisy bistable elements with global time delayed coupling. *Phys. Rev. Lett.*, **91**: 260601, 2005.

16. H. C. Haken. *Advanced Synergetics*. Springer, Berlin 1983.
17. Janson, N.B. Balanov, A.G. Schöll E.: Delayed feedback as a means of control of noise-induced motion. *Phys. Rev. Lett.*, **93**: 010601, 2004.
18. B. W. Knight, A. Omurtag, and L. Sirovich. The approach of a neuron population firing rate to a new equilibrium: An exact theoretical result. *Neural Comput.*, **12**: 1045–1055, 2000.
19. U. Küchler, and B. Mensch. Langevins stochastic differential equation extended by a time-delayed term. *Stoch. Stoch. Rep.*, **40**: 23–42, 1992.
20. S. Kim, S. H. Park, and H.-B. Pyo. Stochastic resonance in coupled oscillator systems with time delay. *Phys. Rev. Lett.*, **82**: 1620–1623, 1999.
21. M. Klosek, and R. Kuske. Multi-scale analysis for stochastic differential delay equations, *SIAM Multiscale Model. Simul.*, **3**: 706–729, 2005.
22. M. LeBerre, E. Ressayre, A. Tallet, and Y. Pomeau. Dynamic system driven by a retarded force acting as colored noise. *Phys. Rev. A.*, **41**: 6635–6646, 1990.
23. B. Lindner, L. Schimansky-Geier, and A. Longtin. Maximizing spike train coherence and incoherence in the leaky integrate-and-fire model. *Phys. Rev. E*, **66**: 031916, 2002.
24. B. Lindner, J. Garcia-Ojalvo, A. Neiman, and L. Schimansky-Geier. Effects of noise in excitable systems. *Phys. Rep.*, **392**: 321–424, 2004.
25. B. Lindner, B. Doiron, and A. Longtin. Theory of oscillatory firing induced by spatially correlated noise and delayed inhibitory feedback. *Phys. Rev. E.*, **72**: 061919, 2005.
26. A. Longtin. Noise-induced transitions at a Hopf bifurcation in a first order delay-differential equation. *Phys. Rev. A*, **44**: 4801–4813, 1991.
27. A. Longtin, J. G. Milton, J. E. Bos, and M. C. Mackey. Noise-induced transitions in the human pupil light reflex, *Phys. Rev. A*, **41**: 6992–7005, 1990.
28. M. C. Mackey, and I. G. Nechaeva. Solution moment stability in stochastic differential delay equations. *Phys. Rev. E*, **52**: 3366–3376, 1995.
29. X. Mao. Razumikhin-type theorems on exponential stability of neutral stochastic functional differential equations. *SIAM J. Math. Anal.*, **28**: 389–401, 1997.
30. M. Mattia, and P. Del Giudice. Population dynamics of interacting spiking neurons. *Phys. Rev. E*, **66**: 051917, 2002.
31. R. Morse, and A. Longtin. Coherence and stochastic resonance in threshold crossing detectors with delayed feedback. *Phys. Lett. A*, **359**: 640–646, 2006.
32. D. Nykamp, and D. Tranchina. A population density approach that facilitates large-scale modeling of neural networks: Analysis and an application to orientation tuning. *J. Comput. Neurosci.*, **8**: 19–50, 2000.
33. T. Ohira, and J. G. Milton. Delayed random walks. *Phys. Rev. E.*, **52**: 3277–3280, 1995.
34. T. Ohira. Oscillatory correlation of delayed random walks. *Phys. Rev. E*, **55**: 1255–1258, 1997.
35. T. Ohira, and Y. Sato. Resonance with noise and delay. *Phys. Rev. Lett.*, **82**: 2811–2815, 1999.
36. A. Pikovsky, and L. S. Tsimring. Noise-induced dynamics in bistable systems with delay. *Phys. Rev. Lett.*, **87**: 250602, 2001.
37. O.V. Popovych, C. Hauptmann, and P.A. Tass. Effective desynchronization by nonlinear delayed feedback. *Phys. Rev. Lett.*, **94**: 164102, 2005.
38. T. Prager, and L. Schimansky-Geier. Stochastic resonance in a non-markovian discrete state model for excitable systems. *Phys. Rev. Lett.*, **91**: 230601, 2003.
39. B. Redmond, V. LeBlanc, and A. Longtin. Bifurcation analysis of a class of first-order nonlinear delay-differential equations with reflectional symmetry. *Physica D*, **166**: 131–146, 2002.
40. M. Richardson, and W. Gerstner. Synaptic shot noise and conductance fluctuations affect the membrane voltage with equal significance. *Neural Comput.*, **17**: 923–947, 2005.
41. P. Swain, and A. Longtin. Noise in neural and genetic networks. *Chaos*, **16**: 026101, 2006.
42. S. Trimper, and K. Zabrocki. Memory driven pattern formation. *Phys. Lett. A*, **331**: 423–431, 2004.
43. H. C. Tuckwell. *Stochastic Processes in the Neurosciences*. CBMS-NSF Regional Conference Series in Applied Mathematics Vol. 56. Society for Industrial and Applied Mathematics, Philadelphia, PA, 1989.

44. K. Vasilikos, and A. Beuter. Effects of noise on a delayed visual feedback system. *J. theor. Biol.*, **165**: 389–407, 1993.
45. E. M. Wright. A nonlinear difference-differential equation. *J. Reine Angew. Math.*, **194**: 66–87, 1955.
46. M. K. S. Yeung, and S. H. Strogatz. Time delay in the kuramoto model of coupled oscillators. *Phys. Rev. Lett.*, **82**: 648–651, 1999.
47. K. Zabrocki, S. Tatur, S. Trimper, and R. Mahnke. Relationship between a non-Markovian process and Fokker-Planck equation. *Phys. Lett. A*, 359: 349–356, 2006.

Chapter 7
Global Convergent Dynamics of Delayed Neural Networks

Wenlian Lu and Tianping Chen

7.1 Introduction

Artificial neural networks arise from the research of the configuration and function of the brain. As pointed out in [79], the brain can be regarded as a complex non-linear parallel information processing system with a concept of neuron as a basic functional unit. Compared with modern computer, the processing speed of a single neuron is 5–6 times slower than that of a single silicic logic gate but the brain has a processing speed 10^9 times faster than any computer due to a huge quantity of synapses that interconnect neurons. Based on this viewpoint, scientists proposed a network model to describe the function and state of the brain called neural networks. In short, a neural network is a computing network that accomplishes given tasks by connecting a large number of simple computing units. The most important characteristic of neural networks is the ability to learn. Reference [75] defined learning as the process through which the neural network adjusts its parameters using information of its circumstances via a simulating process. Many learning algorithms have been proposed in the past decades, for example, error-correction learning, Hebbian learning [52], competitive learning [47], and Boltzmann learning [1].

In particular, [2] proposes a definition of artificial neural network. Neural network is a large-scale parallel distributed processing system, which can learn and employ knowledge and satisfies that (1) knowledge is obtained by learning (learning algorithm); (2) knowledge is stored in the interconnection weights of the network. Since neural networks have many advantages, for instance, the ability to solve nonlinear problems, adaptability, fault tolerance, and mass computability, they have been one of the focal research topics for the last 50–60 years.

References [53, 36, 37] proposed multi-layered neuronal perceptron model which can approximate any continuous function. This model can be formulated as

W. Lu (✉)

Laboratory of Nonlinear Sciences, School of Mathematical Sciences, Fudan University, 200433, Shanghai, China; Max Planck Institute for Mathematics in the Sciences, 04103, Leipzig, Germany

e-mail: wenlian.lu@gmail.com

F.M. Atay (ed.), *Complex Time-Delay Systems,* Understanding Complex Systems,
DOI 10.1007/978-3-642-02329-3_7, © Springer-Verlag Berlin Heidelberg 2010

$$f(x_1, x_2, \ldots, x_n) \approx \sum_{i=1}^{n} C_i g \left(\sum_{j=1}^{m} \xi_{ij} x_j + \theta_i \right), \qquad (7.1)$$

where x_i denotes the state variable of neuron i and $g(\cdot)$ is a certain nonlinear activation function. Furthermore, [22, 23] proved that this model can approximate any nonlinear function and operator. This is the theoretical basis of neural networks.

References [35, 47] proposed a competitive and cooperative model to generate self-organized and self-adaptive neural networks, which can be modeled as an ODE system:

$$\frac{dx_i}{dt} = a_i(x_i) \left[-d_i(x_i) + \sum_{j=1}^{n} t_{ij} g_j(x_j) + I_i \right], \qquad i = 1, \ldots, n,$$

which is named Cohen–Grossberg neural network and widely used in pattern recognition, signal processing, and associative memory. Here, $x_i(t)$ denotes the state variable of the i-th neuron, $d_i(\cdot)$ represents the self-inhibition function with which the i-th neuron will reset its potential to the resting state in isolation when disconnected from the network, t_{ij} denotes the strength of j-th neuron on the i-th neuron, $g_i(\cdot)$ denotes the activation function of i-th neuron, I_i denotes the external input to the i-th neuron, and $a_i(\cdot)$ denotes the amplification function of the i-th neuron.

References [54, 57] developed a computing method using recurrent networks based on energy functions, which is called Hopfield neural network:

$$\frac{dx_i}{dt} = -d_i x_i + \sum_{j=1}^{n} t_{ij} g_j(x_j) + I_i, \qquad i = 1, \ldots, n,$$

which has been applied to solve some combinatorial optimization problems such as the traveling salesman problem.

As pointed out by [51], the common characteristic is that each neural network model can be regarded as a class of nonlinear signal-flow graphs. As indicated in Fig. 7.1, x_i denotes the state of neuron i, $y_i = \phi_i(x_i)$ denotes the output of neuron i by a nonlinear activation function $\phi_i(\cdot)$, t_{ij} denotes the weight of interconnection from neuron j to i, and I_i is the external input. Hence, neural networks are in fact a class of nonlinear dynamical systems due to the nonlinearity of the activations. The computation developed from neural networks is a self-adaptive distributed method based on a learning algorithm. The key point of success of an algorithm lies on whether the dynamical flow converges to a given equilibrium or manifold. So, dynamical analysis of neural networks is the first step for the expected applications.

In practice, time delays inevitably occur due to the finite switching speed of the amplifiers and communication time. Moreover, to process moving images, one must introduce time delays in the signals transmitted among the cells [25]. Neural networks with time delays have much more complicated dynamics due to the

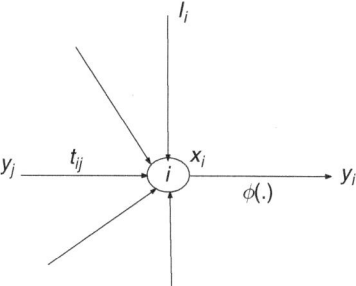

Fig. 7.1 Signal-flow graph

incorporation of delays. These neural networks can be modeled by the following delayed differential equations:

$$\frac{dx_i(t)}{dt} = -d_i x_i(t) + \sum_{j=1}^{n} a_{ij} g_j(x_j(t)) + \sum_{j=1}^{n} b_{ij} f_j\left(x_j(t - \tau_{ij})\right) + I_i,$$

$$i = 1, \ldots, n, \qquad (7.2)$$

where b_{ij} denotes the delayed feedback of the j-th neuron on the i-th neuron and τ_{ij} denotes the transmission delay from neuron j to i. If the activation functions concerned with delayed or without delayed terms are the same, i.e., $f_j = g_j$, $j = 1, \ldots, n$, then this model can be formulated as

$$\frac{dx_i(t)}{dt} = -d_i x_i(t) + \sum_{j=1}^{n} a_{ij} g_j(x_j(t)) + \sum_{j=1}^{n} b_{ij} g_j(x_j(t - \tau_{ij})) + I_i,$$

$$i = 1, \ldots, n.$$

One can see that this model contains cellular neural networks [32, 33] as a special case. If $\tau_{ij} = \tau$ is uniform, it has the following form:

$$\frac{dx_i(t)}{dt} = -d_i x_i(t) + \sum_{j=1}^{n} a_{ij} g_j(x_j(t)) + \sum_{j=1}^{n} b_{ij} g_j(x_j(t - \tau)) + I_i,$$

$$i = 1, \ldots, n.$$

Also, the delayed Cohen–Grossberg neural networks can be written as

$$\frac{dx_i(t)}{dt} = a_i(x_i)\left[-d_i x_i(t) + \sum_{j=1}^{n} a_{ij} g_j(x_j(t)) + \sum_{j=1}^{n} b_{ij} g_j(x_j(t - \tau)) + I_i \right],$$

$$i = 1, \ldots, n, \qquad (7.3)$$

a special form.

Research of delayed neural networks with varying self-inhibitions, interconnection weights, and inputs is an important issue, because in many real-world applications, self-inhibitions, interconnection weights, and inputs vary with time. Thus, we also study the delayed neural networks with a more general form, which is first introduced in [26]:

$$\frac{dx_i(t)}{dt} = -d_i(t)x_i(t) + \sum_{j=1}^{n} a_{ij}(t)g_j(x_j(t)) + \sum_{j=1}^{n} \int_0^{\infty} f_j(x_j(t-s))d_sK_{ij}(s)$$
$$+I_i(t), \quad i = 1,\ldots,n,$$

where $d_sK_{ij}(t,s)$, $i,j = 1,\ldots,n$, are Lebesgue–Stieltjes measures with respect to s, which denotes the delayed terms. For example, if $d_sK_{ij}(t,s)$ has the form $b_{ij}\delta_{\tau_{ij}}(t-s)ds$, one obtains (7.2). More details about the descriptions of the models will be discussed in the following sections.

In this chapter, we study the global convergent dynamics of a class of delayed neural networks. The models are rather general, including Hopfield neural networks, Cohen–Grossberg networks, cellular neural networks, as well as the case of discontinuous activation functions. The purpose of this chapter is not only to present the existing results but also to illustrate the methodologies used in obtaining and proving these results. These methodologies could be utilized or extended in analysis of global convergent dynamics of other models or general delayed differential systems.

Two mathematical problems must be solved. One is the existence of a static orbit: an equilibrium, a periodic orbit, or an almost periodic orbit. Ordinarily, this can be investigated by the fixed point theory. In addition, in this chapter we use novel methods. We study the system of the derivative of the delayed Hopfield neural networks instead and conclude that the global exponential stability of the derivative can lead the global exponential stability of the intrinsic neural networks. Moreover, the existence of periodic or almost periodic orbits can be handled by regarding it as a clustering orbit of any trajectory. The second problem is the stability of such a static orbit. This is investigated by designing a suitable Lyapunov functional. We should point out that it is not the theorems but the ideas of Lyapunov and Lyapunov–Krasovskii stability theory that is used to prove global stability. The main results and proofs in this chapter come from our recent literature [19–21, 30, 63, 65–68].

We organize this chapter as follows. In Sect. 7.2, we discuss the stability of delayed neural networks. We study the periodicity and almost periodicity in Sect. 7.3. In Sect. 7.4, we investigate the convergence analysis of delayed neural networks with discontinuous activation functions. We present reviews of literature on this topic and compare them with the results in Sect. 7.5.

We first present the notation used in this chapter. $\|\cdot\|$ denotes the norm of a vector in some sense. In particular, $\|v\|_2$ for a vector $v = (v_1,\ldots,v_n)^{\top}$ denotes the 2-norm, i.e., $\|v\|_2 = \sqrt{\sum_{i=1}^{n} |v_i|^2}$ and $\|v\|_1 = \sum_{i=1}^{n} |v_i|$. For some positive vector $\xi = (\xi_1,\ldots,\xi_n)^{\top}$, we denote $\|v\|_{\{\xi,\infty\}} = \max_i \xi_i^{-1}|v_i|$ and $\|v\|_{\{\xi,1\}} = \sum_{i=1}^{n} \xi_i|v_i|$. The norm of a matrix is induced by the definition of the norm of vectors. $C([a,b],\mathbb{R}^n)$

denotes the class of continuous functions from $[a, b]$ to $\mathbb{R}^n$. The norm of $x(\,\cdot\,) \in$ $C([a, b], \mathbb{R}^n)$ is denoted by $\|x(\,\cdot\,)\| = \max_{a \le t \le b} \|x(t)\|$ for some vector norm $\|\,\cdot\,\|$. We write $a^+ = \max\{a, 0\}$ for a real number a. The spectral set of a square matrix A is denoted by $\lambda(A)$. Among them, $\lambda_{\min}(A)$ and $\lambda_{\max}(A)$ denote the minimum and maximum one, respectively, if all eigenvalues of A are real. For a matrix A, $A^\top$ denotes its transpose and A^s denotes its symmetric part, i.e., $A^s = (A + A^\top)/2$. For a matrix $A \in \mathbb{R}^{n,n}$, $A > 0$ denotes that A is positive definite, with similar definitions for the notations $A \ge 0$, $A < 0$, and $A \le 0$. For two matrices $A, B \in \mathbb{R}^{n,n}$, $A > B$ denotes $A - B > 0$; similarly with $A \ge B$, $A < B$, and $A \le B$. $\mathbb{R}^n_+$ denotes the first orthant, $\mathbb{R}^n_+ = \{x = (x_1, \ldots, x_n)^\top : x_i > 0, \ \forall \ i = 1, \ldots, n\}$. For a matrix $A = (a_{ij})^n_{i,j=1} \in \mathbb{R}^{n,n}$, $|A|$ denotes the matrix $(|a_{ij}|)^n_{i,j=1}$. Finally, $\text{sign}(\,\cdot\,)$ denotes the signature function.

7.2 Stability of Delayed Neural Networks

In this section we will study the global stability of delayed neural networks. The basic mathematical method is the theory of functional differential equations. For more details, we refer interested readers to [49]. The study of stability of these differential systems contains two main contents: (1) existence of an equilibrium and (2) global attractivity of this equilibrium as done in previous literature. We study the delayed Hopfield neural network (7.2) and the delayed Cohen–Grossberg neural network (7.3) and prove that under several assumptions, diagonal dominant conditions can lead the global stability.

7.2.1 Preliminaries

Before presenting the main results, we provide a brief review of necessary theoretical preliminaries.

7.2.1.1 Functional Differential Equations (FDE)

Delayed neural networks can be modeled as a class of functional differential equations, which have the following general forms:

$$\frac{dx}{dt} = f(x_t). \tag{7.4}$$

Here, $x(t) \in \mathbb{R}^n$, $x_t(\theta) = x(t+\theta)$, $\theta \in [-\tau, 0]$, where $\tau > 0$ can even be infinite, $f(\,\cdot\,)$ is a function in $C([-\tau, 0], \mathbb{R}^n)$. A solution of the system (7.4) with initial condition $\phi \in C([-\tau, 0], \mathbb{R}^n)$ is a smooth $x(t)$ satisfying (1) $x(\theta) = \phi(\theta)$ for all $\theta \in [-\tau, 0]$ and (2) (7.4) holds for all $t \ge 0$. As pointed out in [49], local Lipschitz continuity of $f(\,\cdot\,)$ can guarantee the existence and uniqueness of the solution of the system (7.4). In addition, if the solution is bounded, then the solution exists for the whole time interval.

Stability of (7.4) is with respect to an equilibrium. An *equilibrium* $x^* \in \mathbb{R}^n$ is a solution of the equation

$$f(x^*) = 0, \tag{7.5}$$

i.e., $x_t(\cdot)$ is picked as a constant function. Stability is then defined as follows.

Definition 7.1 Equation (7.4) is said to be globally stable if for any initial condition $\phi \in C([-\tau, 0], \mathbb{R}^n)$, the corresponding solution $x(t)$ satisfies $\lim_{t \to \infty} x(t) = x^*$. Moreover, if there exist some $M > 0$ and $\epsilon > 0$ such that $\|x(t) - x^*\| \leq M \exp(-\epsilon t)$ for all $t \geq 0$, (7.4) is said to be globally exponentially stable. If there exist some $M > 0$ and $\gamma > 0$ such that $\|x(t) - x^*\| \leq M t^{-\gamma}$ for all $t \geq 0$, (7.4) is said to be globally stable in power rate.

In this chapter, we use Lyapunov functional methods to study the global stability of the equilibrium. Actually, we do not directly cite Lyapunov stability theorem for FDEs but use the underlying idea. We design a suitable functional which is zero if and only if $x_t = x^*$, give conditions to guarantee that it decreases through the system, and directly prove that the Lyapunov functional converges to zero. Also, we use the idea of Lyapunov–Krasovskii theory instead of the theorem, which can be cited as the following simple lemma:

Lemma 7.2 *Let $x(t)$ be a solution of the system (7.4) with the initial time $t_0 > 0$ and $\phi(t) = \|x_t(\cdot)\|$. If at each t^* with $\phi(t^*) = \|x(t^*)\|$, we have*

$$\frac{d\|x(t)\|}{dt} \Big|_{t=t^*} \leq -\eta\phi(t^*) + M(t^*) \tag{7.6}$$

for some positive continuous function $M(t^*)$, then $\phi(t) \leq \max\{M(t)/\eta, \phi(t_0)\}$ for all $t \geq t_0$.

Proof We prove it by discussing the following two cases.

Case 1: $\phi(t_0) \leq M(t_0)/\eta$. We can prove $\phi(t) \leq M(t)/\eta$ for all $t \geq t_0$. In fact, if there exists some $t_1 > t_0$ such that $\phi(t_1) = M(t_1)$ for the first time, then $\phi(t)$ is non-increasing at t_1. Otherwise, if $\phi(t)$ is strictly increasing at t_1, then $\phi(t_1) = \|x(t_1)\|$ and $\|x(t)\|$ is strictly increasing at t_1, which by (7.6) is impossible. Hence, $\phi(t)$ will never increase beyond $M(t)$.

Case 2: $\phi(t_0) > M(t_0)/\eta$. Then, $\phi(t)$ is decreasing in a small right neighborhood of t_0. If at some $t_1 > T_0, \phi(t_1) \leq M(t_1)/\eta$, then it reduces to Case 1. Otherwise, $\phi(t)$ keeps decreasing.

In both cases, it can be concluded that $\phi(t) \leq \max\{M(t)/\eta, \phi(t_0)\}$. □

7.2.1.2 Matrix Theory

A matrix $T \in \mathbb{R}^{n,n}$ is said to be *Lyapunov diagonally stable* (LDS) if there exists a positive definite diagonal matrix $D \in \mathbb{R}^{n,n}$ such that $DT + T^\top D$ is positive definite.

Lemma 7.3 (See Lemma 2 in [41]) *Let D and G be positive definite diagonal matrices and $T \in \mathbb{R}^{n,n}$. If $DG^{-1} - T$ is LDS, then for any positive definite diagonal matrix $\bar{D} \geq D$ and nonnegative definite diagonal matrix $0 \leq K \leq G$, we have $det(\bar{D} - TK) \neq 0$.*

A nonsingular matrix $C \in \mathbb{R}^{n,n}$ with $c_{ij} \leq 0$, $i,j = 1,\ldots,m$, $i \neq j$, is said to be an M-matrix if all elements of C^{-1} are nonnegative.

Lemma 7.4 ([11]) *Let $C = (c_{ij}) \in \mathbb{R}^{n,n}$ be a nonsingular matrix with $c_{ij} \leq 0$, $i,j = 1,\ldots,n$, $i \neq j$. Then the following statements are equivalent.*

1. *C is an M-matrix;*
2. *All the successive principal minors of C are positive;*
3. *$C^{\top}$ is an M-matrix;*
4. *The real parts of all eigenvalues are positive;*
5. *There exists a vector $\xi = (\xi_1, \xi_2, \ldots, \xi_n)^{\top}$ with $\xi_i > 0$, $i = 1,\ldots,n$, such that every component of $\xi^{\top} C$ is positive, or every component of $C\xi$ is positive;*
6. *C is LDS;*
7. *For any two diagonal matrices $P = \text{diag}\{p_1, p_2, \ldots, p_n\}$, $Q = \text{diag}\{q_1, q_2, \ldots, q_n\}$, where $p_i > 0$, $q_i > 0$, $i = 1,\ldots,n$, PCQ is an M-matrix.*

The following lemma states the Schur Complement.

Lemma 7.5 (Schur Complement [13]) *The following Linear Matrix Inequality (LMI)*

$$\begin{bmatrix} Q(x) & S(x) \\ S^{\top}(x) & R(x) \end{bmatrix} > 0,$$

where $Q(x) = Q^{\top}(x), R(x) = R^{\top}(x)$, and $S(x)$ depend affinely on x, is equivalent to

$$R(x) > 0 \text{ and } Q(x) - S(x)R^{-1}(x)S^{\top}(x) > 0.$$

7.2.1.3 Nonlinear Complimentary Problems

To discuss the existence and uniqueness of the equilibrium, we give a brief review on Nonlinear Complementarity Problem (NCP).

Definition 7.6 For a continuous function $f(x) = (f_1(x), \ldots, f_n(x))^{\top} : \mathbb{R}^n_+ \to \mathbb{R}^n$, an NCP is to find x_i, $i = 1,\ldots,n$, satisfying

$$x_i \geq 0, \quad f_i(x) - I_i \geq 0, \quad x_i(f_i(x) - I_i) = 0 \quad \text{for all } i = 1,\ldots,n. \quad (7.7)$$

Define a function $F(x) : \mathbb{R}^n \to \mathbb{R}^n$

$$F(x) = f(x^+) + x^-,$$

where

$$x_i^+ = \begin{cases} x_i, & x_i \geq 0 \\ 0, & \text{otherwise}, \end{cases} \qquad x_i^- = \begin{cases} x_i, & x_i \leq 0 \\ 0, & \text{otherwise} \end{cases} \qquad \text{for } i = 1, \ldots, n.$$

The following lemma gives a sufficient and necessary condition for the solvability of a NCP.

Lemma 7.7 (Theorem 2.3 in [76])   *The NCP (7.7) has a unique solution for every $I \in \mathbb{R}^n$ if and only if $F(x)$ is norm-coercive, i.e.,*

$$\lim_{\|x\| \to \infty} \|F(x)\| = \infty,$$

and $F(x)$ is locally one-to-one.

7.2.1.4 Descriptions of Activations

The activation functions in these models are assumed to be Lipschitz continuous.

Definition 7.8 A continuous function $g(x) = (g_1(x_1), \ldots, g_n(x_n))^\top : \mathbb{R}^n \to \mathbb{R}^n$ is said to belong to the function class $H_1\{G_1, \ldots, G_n\}$ for some positive numbers $G_1, \ldots, G_n$ if $|g_i(\xi) - g_i(\zeta)| \leq G_i|\xi - \zeta|$ for all $\xi, \zeta \in \mathbb{R}$ and $i = 1, \ldots, n$. If, in addition, each $g_i(\cdot)$ is monotonously increasing, then g is said to belong to the function class $H_2\{G_1, \ldots, G_n\}$.

7.2.2 Delayed Hopfield Neural Networks

In this section we study the following delayed differential system:

$$\frac{dx_i(t)}{dt} = -d_i x_i(t) + \sum_{j=1}^{n} a_{ij} g_j(x_j(t)) + \sum_{j=1}^{n} b_{ij} f_j(x_j(t - \tau_{ij})) + I_i,$$

$$i = 1, \ldots, n. \qquad (7.8)$$

Different from the ordinary way to handle this topic, we do not first prove the existence of the equilibrium but derive it with global stability. Instead of directly studying the system (7.8), we consider its derivative system with respect to $\dot{x}$ and prove that under several diagonal dominant conditions, $\dot{x}$ converges to zero exponentially. This in fact implies that $x(t)$ converges to some equilibrium globally exponentially. This idea comes from [18–20] and can be summarized in the following theorems.

Theorem 7.9 *Suppose that $g(x) = (g_1(x), \ldots, g_n(x))^\top \in H_2\{G_1, \ldots, G_n\}$ and $f(x) = (f_1(x), \ldots, f_n(x))^\top \in H_1\{F_1, \ldots, F_n\}$. If there are positive constants $\xi_1, \ldots, \xi_n$ such that*

$$- \xi_j d_j + [\xi_j a_{jj} + \sum_{i \neq j} \xi_i |a_{ij}|]^+ G_j + \sum_{i=1}^{n} \xi_i |b_{ij}| F_j < 0, \quad j = 1, \dots, n, \qquad (7.9)$$

then the system (7.8) is globally exponentially stable.

Proof According to the condition, there exists some $\alpha > 0$ such that

$$\xi_j(-d_j + \alpha) + [\xi_j a_{jj} + \sum_{i \neq j} \xi_i |a_{ij}|]^+ G_j + \sum_{i=1}^{n} \xi_i |b_{ij}| e^{\alpha \tau_{ij}} F_j \leq 0$$

for all $j = 1, \dots, n$. Let $v_i(t) = \dot{x}_i(t)$ and $y(t) = e^{\alpha t} v(t)$. Then, for almost every $t \geq 0$, we have

$$\frac{dy_i(t)}{dt} = (-d_i + \alpha)y_i(t) + \sum_{j=1}^{n} a_{ij} g_j'(x_j(t))y_j(t) \qquad (7.10)$$

$$+ \sum_{j=1}^{n} b_{ij} f_j'(x_j(t - \tau_{ij}))e^{\alpha \tau_{ij}} y_j(t - \tau_{ij}), \quad i = 1, \dots, n.$$

Define the following candidate Lyapunov functional

$$L(t) = \sum_{i=1}^{n} \xi_i |y_i(t)| + \sum_{i,j=1}^{n} \xi_i |b_{ij}| \int_{t-\tau_{ij}}^{t} e^{\alpha(s+\tau_{ij})} |f_j'(x_j(s))| |v_j(s)| ds. \qquad (7.11)$$

Differentiating $L(t)$ gives

$$\dot{L}(t) = \sum_{i=1}^{n} \xi_i \text{sign}\{y_i(t)\} \left\{ (-d_i + \alpha)y_i(t) + \sum_{j=1}^{n} a_{ij} g_j'(x_j(t))y_j(t) \right.$$

$$\left. + \sum_{j=1}^{n} b_{ij} f_j'(x_j(t - \tau_{ij}))e^{\alpha \tau_{ij}} y_j(t - \tau_{ij}) \right\} + \sum_{i,j=1}^{n} \xi_i |b_{ij}| e^{\alpha(t+\tau_{ij})}$$

$$|f_j'(x_j(t))||v_j(t)| - \sum_{i,j=1}^{n} \xi_i |b_{ij}| |f_j'(x_j(t - \tau_{ij}))|e^{\alpha t}|v_j(t - \tau_{ij})|$$

$$\leq \sum_{j=1}^{n} \left\{ \xi_j(-d_j + \alpha) + \left[\xi_j a_{jj} + \sum_{i \neq j} \xi_i |a_{ij}| \right]^+ G_j + \sum_{i=1}^{n} \xi_i |b_{ij}| e^{\alpha \tau_{ij}} F_j \right\} |y_j(t)|$$

$$\leq 0.$$

Therefore, $L(t)$ is bounded and $\sum_{i=1}^{n} \xi_i |\dot{x}_i(t)| = O(e^{-\alpha t})$. By Cauchy convergence principle, there exists an equilibrium point $x^* = (x_1^*, \dots, x_1^*)^\top$, such that

$$\sum_{i=1}^{n}\xi_i|x_i(t)-x_i^*|=O(e^{-\alpha t}). \tag{7.12}$$

Uniqueness of the equilibrium point can be proved by defining another candidate Lyapunov functional

$$L(t)=\sum_{i=1}^{n}\xi_i|x_i(t)-x_i^*|+\sum_{i,j=1}^{n}|b_{ij}|\xi_i\int_{t-\tau_{ij}}^{t}e^{\alpha(s+\tau_{ij})}|f_j(x_j(s))-f_j(x_j^*)|ds$$

and differentiating it similarly as done above. This completes the proof.   □

Another result comes from another Lyapunov functional for $y(t)$.

Theorem 7.10 *Suppose that* $g(x)=(g_1(x),\ldots,g_n(x))^\top \in H_2\{G_1,\ldots,G_n\}$ *and* $f(x)=(f_1(x),\ldots,f_n(x))^\top \in H_1\{F_1,\ldots,F_n\}$. *If there are positive constants* $\xi_1,\ldots,\xi_n$ *such that*

$$-\xi_i d_i+\xi_i a_{ii}^+ G_i+\sum_{j=1,j\neq i}^{N}\xi_j|a_{ij}|G_j+\sum_{j=1}^{n}\xi_j|b_{ij}|F_j<0,\quad j=1,\ldots,n, \tag{7.13}$$

then the system (7.8) is globally exponentially stable.

Proof Let $v_i(t)$ and $y(t)$ be defined in the same way as in the proof of Theorem 7.9. Define

$$\|y(t)\|_{\{\xi,\infty\}}=\max_{i=1,\ldots,n}\xi_i^{-1}\|y_i(t)\|,\quad \varphi(t)=\sup_{0\leq s<\tau}\|y(t-s)\|_{\{\xi,\infty\}}.$$

Denoting $i_0=i_0(t)$ by $\xi_{i_0}^{-1}|y_{i_0}(t)|=\|y(t)\|_{\{\xi,\infty\}}$, we have

$$\xi_{i_0}\frac{d\|y(t)\|_{\{\xi,\infty\}}}{dt}=\text{sign}(x_{i_0}(t))\frac{dy_{i_0}}{dt}$$

$$=\text{sign}(x_{i_0}(t))\left\{-(d_{i_0}-\alpha)y_{i_0}+a_{i_0 i_0}g_{i_0}'(x_{i_0}(t))y_{i_0}(t)\right.$$

$$\left.+\sum_{j=1,j\neq i_0}^{n}g_j'(y_j)(t)y_j(t)+\sum_{j=1}^{n}b_{ij}f_j'(x_j(t-\tau_{ij}))y_j(t-\tau_{ij})e^{\alpha\tau_{ij}}\right\}$$

$$\leq[-(d_{i_0}-\alpha)\xi_{i_0}+a_{i_0 i_0}^+ G_{i_0}\xi_{i_0}]\xi_{i_0}^{-1}|y_{i_0}(t)|+\sum_{j=1,j\neq i_0}^{N}|a_{i_0 j}|G_j\xi_j\xi_j^{-1}|y_j(t)|$$

$$+\sum_{j=1}^{N}F_j\xi_j|b_{i_0 j}|\xi_j^{-1}|y_j(t-\tau_{i_0 j})|e^{\alpha\tau_{ij}}.$$

If $\varphi(t)$ is strictly monotone increasing at $t = t^*$, then $\varphi(t^*) = \|x(t^*)\|_{\{\xi,\infty\}}$ and we have

$$
\xi_{i_0} \frac{d\|y(t)\|_{\{\xi,\infty\}}}{dt} \leq \left\{ -(d_{i_0} - \alpha)\xi_{i_0} + a_{i_0 i_0}^+ G_{i_0}\xi_{i_0} + \sum_{j=1, j\neq i_0}^{n} |a_{i_0 j}| G_j \xi_j \right.
$$
$$
\left. + \sum_{j=1}^{n} F_j \xi_j |b_{i_0 j}| e^{\alpha\tau_{i_0 j}} \right\} \|y(t)\|_{\{\xi,\infty\}} \leq 0,
$$

which implies that $\|y(t)\|_{\{\xi,\infty\}}$ is bounded according to Lemma 7.2, i.e., $\max_i \xi_i^{-1}$ $|\dot{x}_i(t)| = O(e^{-\alpha t})$. By the Cauchy convergence principle, there exists an equilibrium point $x^* = (x_1^*, \ldots, x_n^*)^\top$ such that $\max_i \xi_i^{-1} |x_i(t) - x_i^*| = O(e^{-\alpha t})$. The uniqueness of the equilibrium point can be proved by arguments similar to those used in the proof of the previous theorem. ☐

A direct corollary can be obtained in the M-matrix term.

Corollary 7.11 *Suppose that* $g(x) = (g_1(x), \ldots, g_n(x))^\top \in H_1\{G_1, \ldots, G_n\}$ *and* $f(x) = (f_1(x), \ldots, f_n(x))^\top \in H_1\{F_1, \ldots, F_n\}$. *Let* $G = diag\{G_1, \ldots, G_n\}$ *and* $F = diag\{F_1, \ldots, F_n\}$. *If* $-D + |A|G + |B|F$ *is a M-matrix, then the system (7.8) is globally exponentially stable.*

So far we have studied the exponential stability of delayed Hopfield neural networks with constant delays. However, in many cases the time delays are temporally variant. Then the delayed system can be formulated as

$$
\frac{dx_i(t)}{dt} = -d_i x_i(t) + \sum_{j=1}^{n} a_{ij} g_j(x_j(t))
$$
$$
+ \sum_{j=1}^{n} b_{ij} f_j(x_j(t - \tau_{ij}(t))) + I_i, \quad i = 1, \ldots, n. \tag{7.14}
$$

For the case of bounded delays, i.e., $\tau_{ij}(t) \leq \tau$ for all $i, j = 1, \ldots, n$ and $t \geq 0$, the method in the proof of Theorem 7.10 can be used and the same results can be obtained. However, the case of unbounded delays needs further investigation. It should be pointed out that most of the literature is concerned with stability of delayed neural networks with unbounded delays, which always assumes $\dot{\tau}_{ij}(t) < 1$. Reference [30] presented a novel analysis with a weaker assumption $\tau_{ij}(t) < t$, which includes $\dot{\tau}_{ij} < 1$ as a special case. The result can be summarized as follows.

Theorem 7.12 *Suppose* $\tau_{ij}(t) \leq \mu t$ *for some* $0 < \mu < 1$ *and all* $t \geq 0$, $g(\cdot) \in H_1\{G_1, \ldots, G_n\}$, *and* $f(\cdot) \in H_1\{F_1, \ldots, F_n\}$. *If there are positive constants* $\xi_1, \ldots, \xi_n$ *such that*

$$
-\xi_i d_i + \sum_{j=1}^{n} \xi_j |a_{ij}| G_j + \sum_{j=1}^{n} \xi_j |b_{ij}| F_j < 0, \quad i = 1, \ldots, n, \tag{7.15}
$$

then the system (7.14) has a unique equilibrium x^* which is globally stable in power rate, i.e., there exists some $\gamma > 0$ such that

$$\|x(t) - x^*\| = O(t^{-\gamma})$$

Proof Under the condition (7.15), according to the results in Theorem 7.10, there exists an equilibrium point $x^* = (x_1^*, \ldots, x_n^*)^\top$ for the system (7.14). Moreover, there exists a scalar $\gamma > 0$ and a sufficiently large T, such that for all $t > T$,

$$\left(-d_i + \frac{\gamma}{t}\right)\xi_i + \sum_{j=1}^{n} \xi_j |a_{ij}| G_j + (1-\mu)^{-\gamma} \sum_{j=1}^{n} \xi_j |b_{ij}| F_j < 0,$$

$$i = 1, \ldots, n. \qquad (7.16)$$

We always assume $t > T$ afterward.

Let $x(t)$ be a solution of the system (7.8). Define $z(t) = t^\gamma (x(t) - x^*)$ and

$$M_2(t) = \sup_{s \le t} \|z(s)\|_{\{\xi, \infty\}}. \qquad (7.17)$$

We will prove that $M_2(t)$ is bounded. For any t_0 with $\|z(t_0)\|_{\{\xi, \infty\}} = M_2(t_0)$, letting $i_{t_0} = i_{t_0}(t_0)$ be such an index that $|\xi_{i_{t_0}}^{-1} z_{i_{t_0}}(t_0)| = \|z(t_0)\|_{\{\xi, \infty\}}$, we have

$$\left\{\frac{d|z_{i_{t_0}}(t)|}{dt}\right\}_{t=t_0} = \text{sign}\{z_{i_{t_0}}(t_0)\}\xi_{i_{t_0}}\left(-d_{i_{t_0}} + \frac{\gamma}{t_0}\right)\xi_{i_{t_0}}^{-1}z_{i_{t_0}}(t_0)$$

$$+ \text{sign}\{z_{i_{t_0}}(t_0)\}t_0^\gamma\left\{\sum_{j=1}^{n} a_{i_{t_0}j}\left[g_j(u_j(t_0)) - g_j(v_j^*)\right]\right.$$

$$\left. + \sum_{j=1}^{n} b_{i_{t_0}j}\left[f_j\left(u_j(t_0 - \tau_{i_{t_0}j}(t_0))\right) - f_j(v_j^*)\right]\right\}$$

$$\le \left\{\xi_{i_{t_0}}\left(-d_{i_{t_0}} + \frac{\gamma}{t_0}\right) + \sum_{j=1}^{n} \xi_j|a_{i_{t_0}j}|G_j\right\}\|z(t_0)\|_{\{\xi, \infty\}}$$

$$+ \sum_{j=1}^{n} \xi_j|b_{i_{t_0}j}|F_j\left[\frac{t_0}{t_0 - \tau_{i_{t_0}j}(t_0)}\right]^\gamma \xi_j^{-1}|z_j\left(t_0 - \tau_{i_{t_0}j}(t_0)\right)|$$

$$\le \left\{\xi_{i_{t_0}}\left(-d_{i_{t_0}} + \frac{\gamma}{t_0}\right) + \sum_{j=1}^{n} \xi_j|a_{i_{t_0}j}|G_j\right.$$

$$\left. + \sum_{j=1}^{n} \xi_j|b_{i_{t_0}j}|F_j\left\{\frac{t_0}{t_0 - \tau_{i_{t_0}j}(t_0)}\right\}^\gamma\right\}M_2(t_0)$$

$$\leq \left\{ \xi_{i_{t_0}} \left(-d_{i_{t_0}} + \frac{\gamma}{t_0} \right) \right.$$
$$\left. + \sum_{j=1}^{n} \xi_j |a_{i_{t_0}j}| G_j + (1-\mu)^{-\gamma} \sum_{j=1}^{n} \xi_j |b_{i_{t_0}j}| F_j \right\} M_2(t_0)$$
$$< 0.$$

By Lemma 7.2, we can conclude that $M_2(t)$ is bounded, which implies that $\|u(t) - v^*\|_{\{\xi,\infty\}} = O(t^{-\gamma})$, which completes the proof. $\square$

We give a numerical example to verify the theoretical results. We consider the following system

$$\dot{x}(t) = -5x(t) + x(t - \tau(t)), \qquad (7.18)$$

where $\tau(t) \leq \mu t$, with $\mu = 0.5$. The power convergence is shown in Fig. 7.2. The slope of the straight line is approximately -2.3221, which means that $x(t) \approx O(t^{-2.3217})$. The theoretical result is $x(t) \approx O(t^{-\gamma})$, where $\gamma \approx -\frac{\log 5}{\log (1-\mu)} = -\frac{\log 5}{\log (0.5)} \approx 2.3219$, which agrees well with the numerical result.

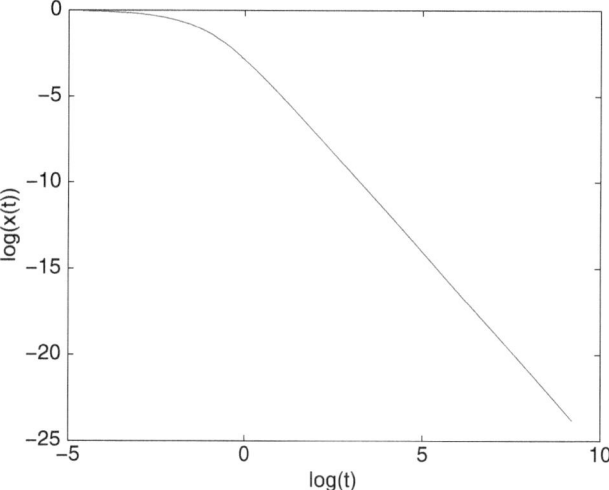

Fig. 7.2 Illustration of power stability. Slope of the straight line is -2.3221

7.2.3 Delayed Cohen–Grossberg Competitive and Cooperative Networks

We consider delayed Cohen–Grossberg neural networks with a uniform delay, which can be formalized as follows:

$$\frac{dx_i(t)}{dt} = a_i(x_i(t))\left[-d_i(x_i) + \sum_{j=1}^{n} a_{ij}g_j(x(t)) + \sum_{j=1}^{n} b_{ij}g_j(x_j(t-\tau)) + I_i\right],$$

$$i = 1,\ldots,n. \qquad (7.19)$$

This model is very general, and includes a large class of existing neural field and evolution models. For instance, assuming that $a_i(\rho) = 1$ for all $\rho \in \mathbb{R}$ and $i = 1,\ldots,n$, then it is the famous Hopfield neural network, which can be written as

$$\frac{dx_i(t)}{dt} = -d_ix_i(t) + \sum_{j=1}^{n} a_{ij}g_j(x_j(t)) + \sum_{j=1}^{n} b_{ij}g_j(x_j(t-\tau)) + I_i, \ i = 1,\ldots,n$$

with $d_i(\rho) = d_i\rho$ for given $d_i > 0$, $i = 1,\ldots,n$. It also includes the famous Volterra–Lotka competitive-cooperation equations:

$$\frac{dx_i}{dt} = A_ix_i\left(I_i - \sum_{j=1}^{n} a_{ij}x_j\right), \ i = 1,\ldots,n$$

with $a_i(\rho) = A_i\rho$, for all $\rho > 0$ and given $A_i > 0$, and $g_i(\rho) = \rho$, $i = 1,\ldots,n$. Most existing results in the literature are based on the assumption that the amplifier function $a_i(\cdot)$ is always *positive* (see [29, 71, 82]). But in the original papers [35, 46, 47] this model was proposed as a kind of competitive-cooperation dynamical system for decision rules, pattern formation, and parallel memory storage. Here, the state of the neuron x_i might be the population size, activity, or concentration, etc., of the i-th species in the system, which is always nonnegative for all time. To guarantee the positivity of the states, one should assume $a_i(\rho) > 0$ for all $\rho > 0$ and $a_i(0) = 0$ for all $i = 1,\ldots,n$.

The purpose of this section is to study the convergent dynamics of the delayed Cohen–Grossberg neural networks without assuming the strict positivity of $a_i(\cdot)$, symmetry of the connection matrix, or boundedness of the activation functions, but with considering a time delay. Hereby, we focus our study of the dynamical behavior on the first orthant: $\mathbb{R}_+^n = \{(x_1,\ldots,x_n)^\top \in \mathbb{R}^n: x_i \geq 0, \ i = 1,\ldots,n\}$ and introduce the concept of $\mathbb{R}_+^n$-global stability, which means that all trajectories are initiated in the first orthant $\mathbb{R}_+^n$ instead of the whole space $\mathbb{R}^n$. We point out that an asymptotically stable nonnegative equilibrium is closely related to the solution of a Nonlinear Complementary Problem (NCP). Based on the Linear Matrix Inequality (LMI) technique (for more details on LMI, see [13]) and NCP theory (for more details on NCP, we refer to [76]), we give a sufficient condition for existence and uniqueness of nonnegative equilibrium. Moreover, the $\mathbb{R}_+^n$-global asymptotic stability and exponential stability of the equilibrium are investigated. The main results of this section comes from [67].

Let $x(t) = (x_i(t), x_2(t),\ldots,x_n(t))^\top$, $d(x) = (d_i(x_i), d_2(x_2),\ldots,d_n(x_n))^\top$, $g(x) = (g_1(x_1), g_2(x_2),\ldots,g_n(x_n))^\top$, $a(x) = \text{diag}\{a_1(x_1), a_2(x_2),\ldots,a_n(x_n)\}$, $A = (a_{ij})_{i,j=1}^n$,

$B = (b_{ij})_{i,j=1}^{n} \in \mathbb{R}^{n,n}$, and $I = (I_1, I_2, \ldots, I_n)^{\top}$. Then, the system (7.19) can be rewritten in the matrix form:

$$\frac{dx(t)}{dt} = a(x)\left[-d(x) + Ag(x(t)) + Bg(x(t-\tau)) + I \right]. \tag{7.20}$$

For the amplifier and activation functions, we give the following assumptions:

(i) $a(\cdot) \in \mathcal{A}_1$; that is, every $a_i(\rho)$ is continuous with $a_i(0) = 0$, and $a_i(\rho) > 0$, whenever $\rho > 0$;

(ii) $a(\cdot) \in \mathcal{A}_2$; that is, $a(\cdot) \in \mathcal{A}_1$, and for any $\epsilon > 0$, $\int_0^{\epsilon} d\rho/a_i(\rho) = +\infty$ for all $i = 1, \ldots, n$;

(iii) $a(\cdot) \in \mathcal{A}_3$; that is, $a(\cdot) \in \mathcal{A}_1$, and for any $\epsilon > 0$, $\int_{\epsilon}^{\infty} \rho \, d\rho/a_i(\rho) = +\infty$ for all $i = 1, \ldots, n$;

(iv) $a(\cdot) \in \mathcal{A}_4$; that is, $a(\cdot) \in \mathcal{A}_1$, and for any $\epsilon > 0$, $\int_0^{\epsilon} \rho \, d\rho/a_i(\rho) < +\infty$ for all $i = 1, \ldots, n$;

(v) $d(\cdot) \in \mathcal{D}$; that is, $d_i(\cdot)$ is continuous and satisfies $[d_i(\xi) - d_i(\zeta)]/(\xi - \zeta) \geq D_i$, for all $\xi \neq \zeta$, where D_i are positive constants, $i = 1, \ldots, n$, and $g(\cdot)$ belongs to $H_2\{G_1, \ldots, G_n\}$ for some $G_i > 0$, $i = 1, \ldots, n$.

First, we define positive solutions componentwise.

Definition 7.13 A solution $x(t)$ of the system (7.20) is said to be a positive solution if for every positive initial condition $\phi(t) > 0$, $t \in [-\tau, 0]$, the trajectory $x(t) = (x_1(t), \ldots, x_n(t))^{\top}$ satisfies $x_i(t) > 0$ for all $t \geq 0$ and $i = 1, \ldots, n$.

Lemma 7.14 (Positive Solution) *If $a(\cdot) \in \mathcal{A}_2$, then the solution of the system (7.20) is a positive solution.*

Proof Assume that the initial value $\phi(t) = (\phi_1(t), \ldots, \phi_n(t))^{\top}$ satisfies $\phi_i(t) > 0$ for $i = 1, \ldots, n$ and $t \in [-\tau, 0]$. Suppose for some $t_0 > 0$ and some index i_0, $x_{i_0}(t_0) = 0$. Then, the assumption $a(\cdot) \in \mathcal{A}_2$ leads

$$\int_0^{t_0}\left[-d_i(x_i(t)) + \sum_{j=1}^{n} a_{ij}g_j(x_j(t)) + \sum_{j=1}^{n} b_{ij}g_j(x_j(t-\tau)) + I_i \right]dt$$

$$= \int_0^{t_0} \frac{\dot{x}_i(t)dt}{a_i(x_i(t))} = -\int_0^{\phi_i(0)} \frac{d\rho}{a_i(\rho)} = -\infty,$$

which is impossible due to the continuity of $x_i(\cdot)$ on $[0, t_0]$. Hence, $x_i(t) \neq 0$ for all $t \geq 0$ and $i = 1, \ldots, n$. This implies that $x_i(t) > 0$ for all $t \geq 0$ and $i = 1, \ldots, n$. $\square$

By this lemma we can actually concentrate on the first orthant $\mathbb{R}_+^n$. If $a(\cdot) \in \mathcal{A}_1$, then any equilibrium in $\mathbb{R}_+^n$ of the system (7.19) is a solution of the equations

$$x_i[f_i(x) - I_i] = 0, \ i = 1, \ldots, n, \tag{7.21}$$

where $f_i(x) = d_i(x_i) - \sum_{j=1}^{n}(a_{ij} + b_{ij})g_j(x_j)$, $i = 1, \ldots, n$. Even though (7.21) might possess multiple solutions, we can show that an asymptotically stable nonnegative equilibrium is just a solution of a nonlinear complementary problem (NCP).

Proposition 7.15 *Suppose* $a(\,\cdot\,) \in \mathcal{A}_1$. *If* $x^* = (x_1^*, \ldots, x_n^*)^\top \in \mathbb{R}_+^n$ *is an asymptotically stable equilibrium of the system (7.20), then it must be a solution of the following nonlinear complementary problem (NCP):*

$$x_i^* \geq 0, \quad f_i(x^*) - I_i \geq 0, \quad x_i^*(f_i(x^*) - I_i) = 0, \quad i = 1, \ldots, n, \qquad (7.22)$$

where $f_i(x) = d_i(x_i) - \sum_{j=1}^{n}(a_{ij} + b_{ij})g_j(x_j)$, $i = 1, \ldots, n$.

Proof Suppose that $x^* \in \mathbb{R}_+^n$ is an asymptotically stable equilibrium of the system (7.20). Then $x_i^* > 0$ or $x_i^* = 0$. In case $x_i^* > 0$, we have $f_i(x_i^*) - I_i = 0$. If $x_i^* = 0$, we claim that $f_i(x_i^*) - I_i \geq 0$. Otherwise, if $f_{i_0}(x^*) - I_{i_0} < 0$ for some index i_0, then $\dot{x}_{i_0}(t) = a_i(x_{i_0}(t))[-f_{i_0}(x_{i_0}(t)) + I_{i_0}] > (1/2)a_i(x_{i_0}(t))[-f_{i_0}(x^*) + I_{i_0}] > 0$ when $x_{i_0}(t)$ is sufficiently close to x^*, which implies that $x_{i_0}(t)$ will never converge to 0. Therefore, x^* is unstable. $\qquad\square$

Thus, we can propose a definition of a nonnegative equilibrium of the system (7.20).

Definition 7.16 x^* is said to be a nonnegative equilibrium of the system (7.20) in the NCP sense, if x^* is the solution of the Nonlinear Complementarity Problem (NCP) (7.22); moreover, if $x_i^* > 0$, for all $i = 1, \ldots, n$, then x^* is said to be a positive equilibrium of system (7.20). In this case, x^* must satisfy

$$d(x^*) - (A + B)g(x^*) + I = 0, \quad x_i^* > 0, \quad i = 1, \ldots, n,$$

where $\mathbf{0} = (0, \ldots, 0)^\top \in \mathbb{R}^n$.

Definition 7.17 A nonnegative equilibrium x^* of the system (7.20) in the NCP sense is said to be $\mathbb{R}_+^n$-globally asymptotically stable if for any positive initial condition $\phi_i(t) > 0$, $t \in [-\tau, 0]$ and $i = 1, \ldots, n$, the trajectory $x(t)$ of the system (7.20) satisfies $\lim_{t \to \infty} x(t) = x^*$. Moreover, if there exist constants $M > 0$ and $\epsilon > 0$ such that

$$\|x(t) - x^*\| \leq Me^{-\epsilon t}, \quad t \geq 0,$$

then x^* is said to be $\mathbb{R}_+^n$-exponentially stable.

So, we discuss the existence and uniqueness of the nonnegative equilibrium in the NCP sense.

Theorem 7.18 (Existence and Uniqueness of Nonnegative Equilibrium) *Suppose* $a(\,\cdot\,) \in \mathcal{A}_2$, $d(\,\cdot\,) \in \mathcal{D}$, *and* $g(\,\cdot\,) \in H_2\{G_1, \ldots, G_n\}$ *for* $G_i > 0$, $i = 1, \ldots, n$.

Let $D = diag\{D_1, \ldots, D_n\}$ and $G = diag\{G_1, \ldots, G_n\}$. If there exists a positive definite diagonal matrix $P = diag\{P_1, P_2, \ldots, P_n\}$ such that

$$\left\{P[DG^{-1} - (A + B)]\right\}^s > 0, \qquad (7.23)$$

then for each $I \in \mathbb{R}^n$, there exists a unique nonnegative equilibrium of the system (7.20) in the NCP sense.

The proof is given in the Appendix.

The following corollary is a direct consequence of Theorem 7.18.

Corollary 7.19 Suppose $a(\cdot) \in \mathcal{A}_2$, $d(\cdot) \in \mathcal{D}$, and $g(\cdot) \in H_2\{G_1, \ldots, G_n\}$ for $G_i > 0$, $i = 1, \ldots, n$. Let $D = diag\{D_1, \ldots, D_n\}$ and $G = diag\{G_1, \ldots, G_n\}$. If there exist a positive definite diagonal matrix P and a positive definite symmetric matrix Q such that

$$\begin{bmatrix} 2PDG^{-1} - PA - A^{\top}P - Q & -PB \\ -B^{\top}P & Q \end{bmatrix} > 0, \qquad (7.24)$$

then for each $I \in \mathbb{R}^n$, there exists a unique nonnegative equilibrium for the system (7.20) in the NCP sense.

Let x^* be the nonnegative equilibrium of the system (7.20) in the NCP sense and $y(t) = x(t) - x^*$. Thus, the system (7.19) can be rewritten as

$$\frac{dy_i(t)}{dt} = a_i^*(y_i(t))\left[-d_i^*(y_i(t)) + \sum_{j=1}^{n} a_{ij}g_j^*(y_j(t)) + \sum_{j=1}^{n} b_{ij}g_j^*(y_j(t-\tau)) + J_i\right]$$

or in matrix form

$$\frac{dy(t)}{dt} = a^*(y(t))\left[-d^*(y(t)) + Ag^*(y(t)) + Bg^*(y(t-\tau)) + J\right], \qquad (7.25)$$

where for $i = 1, \ldots, n$, $a_i^*(s) = a_i(s + x_i^*)$, $a^*(y) = diag\{a_1^*(y_1), \ldots, a_n^*(y_n)\}$, $d_i^*(s) = d_i^*(s + x_i^*) - d_i^*(x_i^*)$, $d^*(y) = [d_1^*(y_1), \ldots, d_n^*(y_n)]^{\top}$, $g_i^*(s) = g_i^*(s + x_i^*) - g_i^*(x_i^*)$, $g^*(y) = [g_1^*(y_1), \ldots, g_n^*(y_n)]^{\top}$, and

$$J_i = \begin{cases} -d_i(x_i^*) + \sum_{j=1}^{n}(a_{ij} + b_{ij})g_j(x_j^*) + I_i \ x_i^* = 0 \\ 0 \qquad\qquad\qquad\qquad\qquad\qquad x_i^* > 0 \end{cases} \qquad J = (J_1, \ldots, J_n)^{\top}.$$

Since x^* is the nonnegative equilibrium of (7.20) in the NCP sense, i.e., the solution of NCP (7.7), $J_i \leq 0$ holds for all $i = 1, \ldots, n$ which implies that $g_i^*(y_i(t))J_i \leq 0$ for all $i = 1, \ldots, n$ and $t \geq 0$.

Theorem 7.20 ($\mathbb{R}_+^n$-Global Asymptotic Stability of the Nonnegative Equilibrium)
Suppose $a(\cdot) \in \mathcal{A}_2 \bigcap \mathcal{A}_3 \bigcap \mathcal{A}_4$, $d(\cdot) \in \mathcal{D}$, and $g(\cdot) \in H_2\{G_1, \ldots, G_n\}$ for $G_i > 0$,
$i = 1, \ldots, n$. Let $D = diag\{D_1, \ldots, D_n\}$ and $G = diag\{G_1, \ldots, G_n\}$. If there exist
a positive definite diagonal matrix $P = diag\{P_1, \ldots, P_n\}$ and a positive definite
symmetric matrix Q such that

$$\begin{bmatrix} 2PDG^{-1} - PA - A^\top P - Q & -PB \\ -B^\top P & Q \end{bmatrix} > 0, \tag{7.26}$$

then the unique nonnegative equilibrium x^* for the system (7.20) in the NCP sense
is $\mathbb{R}_+^n$-globally asymptotically stable.

Proof Without loss of generality, we assume $x_i^* = 0$, $i = 1, 2, \ldots, p$ and $x_i^* > 0$,
$i = p + 1, \ldots, n$ for some integer p. By the assumptions $\mathcal{A}_3$ and $\mathcal{A}_4$, it can be seen
that

$$\int_0^{y_i(t)} \frac{\rho d\rho}{a_i^*(\rho)} < +\infty, \quad \int_0^{+\infty} \frac{\rho d\rho}{a_i^*(\rho)} = +\infty, \quad \int_0^{y_i(t)} \frac{g_i^*(\rho) d\rho}{a_i^*(\rho)} < +\infty$$

for $i = 1, \ldots, n$ and $t \geq 0$. By inequality (7.26), there exists $\beta > 0$ such that

$$Z = \begin{bmatrix} 2\beta D & -\beta A & -\beta B \\ -\beta A^\top & 2PDG^{-1} - PA - A^\top P - Q & -PB \\ -\beta B^\top & -B^\top P & Q \end{bmatrix} > 0.$$

Let

$$V(t) = 2\beta \sum_{i=1}^n \int_0^{y_i(t)} \frac{\rho d\rho}{a_i^*(\rho)} + 2 \sum_{i=1}^n P_i \int_0^{y_i(t)} \frac{g_i^*(\rho) d\rho}{a_i^*(\rho)} + \int_{t-\tau}^t g^{*\top}(y(s))Qg^*(y(s))ds.$$

It is easy to see that $V(t)$ is positive definite and radially unbounded. Noting
$g_i^*(y_i(t))J_i \leq 0$, we have

$$\frac{dV(t)}{dt} = 2\beta \sum_{i=1}^n y_i(t)\left[-d_i^*(y_i(t)) + \sum_{j=1}^n a_{ij}g_j^*(y_j(t)) + \sum_{j=1}^n b_{ij}g_j^*(y_j(t-\tau)) + J_j\right]$$

$$+2 \sum_{i=1}^n P_i g_i^*(y_i(t))\left[-d_i^*(y_i(t)) + \sum_{j=1}^n a_{ij}g_j^*(y_j(t)) + \sum_{j=1}^n g_j^*(y_j(t-\tau)) + J_j\right]$$

$$+g^{*\top}(y(t))Qg^*(y(t)) - g^{*\top}(y(t-\tau))Qg^*(y(t-\tau))$$

$$\leq -2\beta\left[y^\top(t)Dy(t) - y^\top(t)Ag^*(y(t)) - y^\top(t)Bg^*(y(t-\tau))\right]$$

$$-2\left[g^{*\top}(y(t))PDG^{-1}g^*(y(t)) - g^{*\top}(y(t))PBg^*(y(t))\right]$$

$$-g^{*\top}(y(t))PBg^{*}(y(t-\tau))\Big]$$

$$+g^{*\top}(y(t))Qg^{*}(y(t))-g^{*\top}(y(t-\tau))Qg^{*}(y(t-\tau))$$

$$= -[y^{\top}(t),g^{*\top}(y(t)),g^{*\top}(y(t-\tau))]Z\begin{bmatrix} y(t) \\ g^{*}(y(t)) \\ g^{*}(y(t-\tau)) \end{bmatrix} \le -\delta y^{\top}(t)y(t),$$

where $\delta = \lambda_{\min}(Z) > 0$. Therefore, $\lim_{t\to\infty} \|y(t)\|_2 = 0$. This completes the proof. □

In the following, we present a numerical example to verify the theoretical results obtained above and compare the convergent dynamics of the Cohen–Grossberg neural systems with an amplification function which is always positive versus an amplification function which is only positive in the first orthant. A result for positive amplification function was provided in [28, 69].

Theorem 7.21 *Suppose that $g \in H_2\{G_1, G_2, ..., G_n\}$ and there exists $\alpha > 0$ such that $a_i(\rho) > \alpha$ for any $\rho \in \mathbb{R}$ and $i = 1, \dots, n$. If there exist a positive definite diagonal matrix P and a positive definite matrix Q such that inequality (7.26) holds, then for each $I \in \mathbb{R}^n$, the system (7.20) has a unique equilibrium point that is globally exponentially stable.*

Consider the dynamical behavior of the following two systems:

$$\begin{cases} \frac{dx_1(t)}{dt} = x_1(t)\Big[-6x_1(t) + 2g(x_1(t)) - g(x_2(t)) \\ \qquad\qquad +3g(x_1(t-2)) + g(x_2(t-2)) + I_1\Big] \\ \frac{dx_2(t)}{dt} = x_2(t)\Big[-6x_2(t) - 2g(x_1(t)) \\ \qquad\qquad +3g(x_2(t)) + \frac{1}{2}g(x_1(t-2)) + 2g(x_2(t-2)) + I_2\Big], \end{cases} \tag{7.27}$$

$$\begin{cases} \frac{du_1(t)}{dt} = \frac{1}{|u_1(t)|+1}\Big[-6u_1(t) + 2g(u_1(t)) \\ \qquad\qquad -g(u_2(t)) + 3g(u_1(t-2)) + g(u_2(t-2)) + I_1\Big] \\ \frac{du_2(t)}{dt} = \frac{1}{|u_2(t)|+1}\Big[-6u_2(t) - 2g(u_1(t)) \\ \qquad\qquad +3g(u_2(t)) + \frac{1}{2}g(u_1(t-2)) + 2g(u_2(t-2)) + I_2\Big], \end{cases} \tag{7.28}$$

where $g(\rho) = (1/2)(\rho + \arctan(\rho))$ and $I = (I_1, I_2)^{\top}$ is the constant input that will be determined below. Furthermore,

$$D = 6 \times \begin{bmatrix} 1 & 0 \\ 0 & 1 \end{bmatrix}, \quad G = \begin{bmatrix} 1 & 0 \\ 0 & 1 \end{bmatrix}, \quad A = \begin{bmatrix} 2 & -1 \\ -2 & 3 \end{bmatrix}, \quad B = \begin{bmatrix} 3 & 1 \\ \frac{1}{2} & 2 \end{bmatrix}.$$

By the Matlab LMI and Control Toolbox, we obtain

$$P = \begin{bmatrix} 0.2995 & 0 \\ 0 & 0.3298 \end{bmatrix}, \quad Q = \begin{bmatrix} 1.0507 & 0.3258 \\ 0.3258 & 0.9430 \end{bmatrix}.$$

The eigenvalues of

$$Z = \begin{bmatrix} 2PDG^{-1} - PA - A^{\top}P - Q & -PB \\ -B^{\top}P & Q \end{bmatrix}$$

are 2.6490, 1.1343, 0.5302, and 0.0559, which implies that Z is positive definite. By Theorem 7.20, for any $I \in \mathbb{R}^2$, the system (7.27) has a unique nonnegative equilibrium x^* in the NCP sense which is $\mathbb{R}_+^2$-globally asymptotically stable. By Theorem 7.21, for any $I \in \mathbb{R}^2$, system (7.28) has a unique equilibrium x^0, which is globally asymptotically stable in $\mathbb{R}^2$.

In case $I = (1, 0.1)^{\top}$, the equilibria of the system (7.27) are $(0, 0)^{\top}$, $(0.7414, 0)^{\top}$, $(0, 0.0992)^{\top}$, and $(0.7414, -0.7062)^{\top}$. Among them, $x^* = (0.7414, 0)^{\top}$ is the nonnegative equilibrium of the system (7.27) in the NCP sense and $x^0 = (0.7414, -0.7062)$ is the unique equilibrium of the system (7.28). Pick initial condition $\phi_1(t) = (7/2)(\cos(t) + 1)$ and $\phi_2(t) = e^{-t}$, for $t \in [-2, 0]$. Figure 7.3 shows that the solution of the system (7.27) converges to $x^* = (0.7414, 0)^{\top}$, while the solution of the system (7.28) converges to $x^0 = (0.7414, -0.7062)$

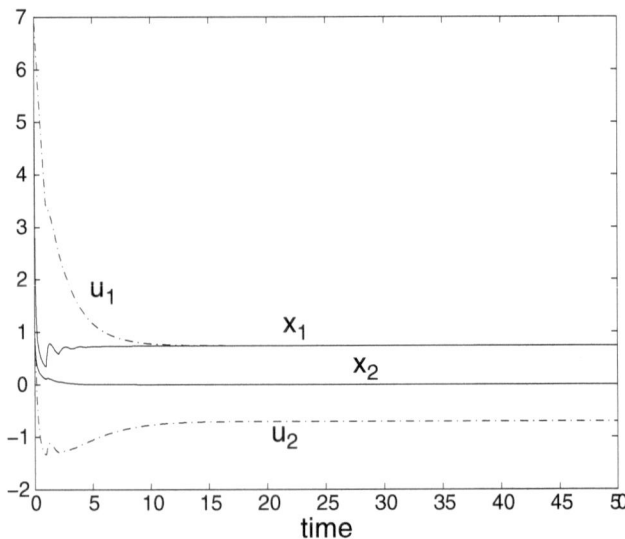

Fig. 7.3 Dynamical behavior of systems (7.27) and (7.28) with $I = (1, 0.1)^{\top}$

7.3 Periodicity and Almost Periodicity of Delayed Neural Networks

In this section, we discuss a large class of delayed neural networks with time-varying inhibitions, interconnection weights, and inputs which can be periodic or almost periodic. We will prove that under several diagonal dominant conditions, the periodic or almost periodic system has at least one periodic or almost periodic solution, respectively, which is globally stable. Moreover, the equilibrium of the delayed neural networks with constant coefficients can be regarded as a periodic orbit with arbitrary period.

We consider a rather general delayed system,

$$
\begin{aligned}
\frac{du_i}{dt} = -d_i(t)u_i(t) + \sum_{j=1}^{n} a_{ij}(t)g_j(u_j(t)) \\
+ \sum_{j=1}^{n} \int_0^\infty f_j(u_j(t - \tau_{ij} - s))d_sK_{ij}(t, s) + I_i(t), \quad i = 1, \ldots, n
\end{aligned} \tag{7.29}
$$

or

$$
\begin{aligned}
\frac{du_i}{dt} = -d_i(t)u_i(t) + \sum_{j=1}^{n} a_{ij}(t)g_j(u_j(t)) \\
+ \sum_{j=1}^{n} \int_0^\infty f_j(u_j(t - \tau_{ij}(t) - s))d_sK_{ij}(t, s) + I_i(t), \quad i = 1, \ldots, n,
\end{aligned} \tag{7.30}
$$

where for any fixed t, $d_sK_{ij}(t, s)$ are Lebesgue–Stieltjes measures with respect to s.

This model contains many delayed recurrent neural network models as special cases. For example, if $d_sK_{ij}(t, t_k) = b_{ij}^k(t)$ for $0 < t_1 < \cdots t_m < \infty$ and $d_sK_{ij}(t, s) = 0$ for $s \neq t_k$, we obtain the following system with multi-discrete delays,

$$
\begin{aligned}
\frac{du_i}{dt} = -d_i(t)u_i(t) + \sum_{j=1}^{n} a_{ij}(t)g_j(u_j(t)) \\
+ \sum_{j=1}^{n} b_{ij}^k(t)f_j(u_j(t - \tau_{ij}(t) - t_k)) + I_i(t), \quad i = 1, 2, \ldots, n.
\end{aligned} \tag{7.31}
$$

Instead, if $d_sK_{ij}(t, s) = b_{ij}(t)k_{ij}(s)ds$, then we have the following system with distributed delays,

$$\frac{du_i(t)}{dt} = -d_i(t)u_i(t) + \sum_{j=1}^{n} a_{ij}(t)g_j(u_j(t))$$

$$+ \sum_{j=1}^{n} b_{ij}(t) \int_0^{\infty} k_{ij}(s)f_j(u_j(t - \tau_{ij}(t) - s))ds + I_i(t), \ i = 1, 2, \ldots, n.$$

$$(7.32)$$

From [49], one can see that if the activation functions $g_i(\cdot)$ and $f_i(\cdot)$ are Lipschitz continuous, then the system has a unique solution for any bounded continuous initial condition.

Periodicity and almost periodicity are defined as follows.

Definition 7.22 A vector-valued function $x(t): \mathbb{R} \to \mathbb{R}^n$ is said to be periodic if there exists $\omega > 0$ such that $x(t + \omega) = x(t)$ for all $t \in \mathbb{R}$. In this case, ω is called the period of $x(t)$. The function $x(t)$ is said to be almost periodic on $\mathbb{R}$ if for any $\epsilon > 0$, it is possible to find a real number $l = l(\epsilon) > 0$, such that for any interval with length $l(\epsilon)$, there exists a number $\omega = \omega(\epsilon)$ in this interval such that $\|x(t + \omega) - x(t)\| < \epsilon$ for all $t \in \mathbb{R}$.

The key problem of this section is to prove the existence of a periodic or almost periodic solution. Different from the existing literature, which uses Mawhin coincidence degree theory (see [44]), we use two methods to prove existence. The first method is to regard the periodic solution as a fixed point of a Poincaré–Andronov map [63]. A basic result is the famous Brouwer fixed point theorem [60].

Lemma 7.23 *A continuous map T over a compact subset Ω of a Banach space such that $T(\Omega) \subset \Omega$ has at least one fixed point, namely, there exists $\omega^* \in \Omega$ such that $T(\omega^*) = \omega^*$.*

The second method is to regard the periodic or almost periodic solution as a limit of a solution of (7.29). See [27].

The global stability of such periodic or almost periodic solutions is studied by Lyapunov and Lyapunov–Krasovskii methods.

7.3.1 Delayed Periodic Hopfield Neural Networks

Considering the system (7.30), we give the following hypotheses.

$\mathcal{B}_1$:

(1) $d_i(t), a_{ij}(t), b_{ij}(t), I_i(t), \tau_{ij}(t): \mathbb{R}^+ \to \mathbb{R}$ are continuous functions, and $d_s K_{ij}(t, s)$ is continuous in the sense that $\lim_{h \to \infty} \int_0^{\infty} |d_s K_{ij}(t + h, s) - d_s K_{ij}(t, s)| = 0$ for all $i, j = 1, \ldots, n$; they are all periodic functions with period $\omega > 0$, i.e., $d_i(t + \omega) = d_i(t), a_{ij}(t) = a_{ij}(t + \omega), b_{ij}(t) = b_{ij}(t + \omega), I_i(t) = I_i(t + \omega), \tau_{ij}(t + \omega) = \tau_{ij}(t)$, and $dK_{ij}(t + \omega, s) = d_s K_{ij}(t, s)$ for all $t > 0$ and $i, j = 1, \ldots, n$.
(2) $g(\cdot) \in H_2\{G_1, \ldots, G_n\}$ and $f(\cdot) \in H_1\{F_1, \ldots, F_n\}$ for some positive constants G_i and $F_i, i = 1, \ldots, n$;

(3) the initial condition $x(\theta) = \phi(\theta)$, $\theta \in (-\infty, 0]$ satisfies that $\phi \in C((-\infty, 0], \mathbb{R}^n)$ is bounded.

The following result is concerned with the Poincaré–Andronov map.

Theorem 7.24 *Suppose that the hypotheses $\mathcal{B}_1$ above are satisfied. If there exist positive constants $\xi_1, \xi_2, \ldots, \xi_n$ such that for all $\omega \geq t > 0$,*

$$- \xi_i d_i(t) + \sum_{j=1}^{n} \xi_j G_j |a_{ij}(t)| + \sum_{j=1}^{n} \xi_j F_j \int_0^\infty |d_s K_{ij}(t, s)| < 0,$$

$$i = 1, 2, \ldots, n, \qquad (7.33)$$

then the system (7.30) has at least one ω-periodic solution $v(t)$. In addition, if there exists a constant α such that for all $\omega \geq t > 0$,

$$-\xi_i(d_i(t) - \alpha) + \sum_{j=1}^{n} \xi_j G_j |a_{ij}(t)|$$

$$+ \sum_{j=1}^{n} \xi_j F_j e^{\alpha \tau_{ij}} \int_0^\infty e^{\alpha s} |d_s K_{ij}(t, s)| \leq 0, \quad i = 1, 2, \ldots, n, \qquad (7.34)$$

then for any solution $x(t) = (x_1(t), \ldots, x_n(t))$ of (7.30),

$$\|x(t) - v(t)\| = O(e^{-\alpha t}) \quad t \to \infty. \qquad (7.35)$$

Proof Pick a constant M satisfying $M > J/\eta$, where

$$J = \max_i \max_t \left\{ \sum_{j=1}^{n} |a_{ij}(t)| C_j + \sum_{j=1}^{n} D_j \int_0^\infty |d_s K_{ij}(t, s)| + |I_i(t)| \right\}$$

and let $C = C((-\infty, 0], \mathbb{R}^n)$ be the Banach space with norm

$$\|\phi\| = \sup_{\{-\infty < \theta \leq \omega\}} \|\phi(\theta)\|_{\{\xi, \infty\}}.$$

Denote

$$\Omega = \{x(\theta) \in C : \|x(\theta)\| \leq M, \|\dot{x}(\theta)\| \leq N\},$$

where $N = (\alpha + \beta + \gamma)M + c$, $\alpha = \max_i \sup_t |d_i(t)|\xi_i^{-1}$, $\beta = \max_{i,j} \sup_t |a_{ij}(t)|\xi_i^{-1} G_j$, $\gamma = \max_{i,j} \sup_t \int_0^\infty |d_s K_{ij}(t, s)| F_j \xi_i^{-1}$, and $c = \max_i \sup_{0 \leq t < \omega} |I_i(t)|\xi_i^{-1}$. It is easy to check that Ω is a convex compact set.

Now, define a map T from Ω to C by

$$T:\phi(\theta) \to x(\theta + \omega, \phi),$$

where $x(t) = x(t, \phi)$ is the solution of the system (7.29) with the initial condition $x_i(\theta) = \phi_i(\theta)$, for $\theta \in (-\infty, 0]$ and $i = 1, \ldots, n$.

In the following, we will prove that $T\Omega \subset \Omega$, i.e., if $\phi \in \Omega$, then $x \in \Omega$. To do that, we define the following function

$$M(t) = \sup_{s \in (-\infty, 0]} \|x(t + s)\|_{\{\xi, \infty\}}.$$

It is easy to see that $\|x(t)\|_{\{\xi, \infty\}} \le M(t)$. Therefore, what we need to do is to prove $M(t) \le M$ for all $t > 0$.

Assume that $t_0 \ge 0$ is the smallest value such that $\|x(t_0)\|_{\{\xi, \infty\}} = M(t_0) = M$, and $\|x(t)\|_{\{\xi, \infty\}} \le M$ if $t < t_0$. Let i_0 be an index such that $\xi_{i_0}^{-1} |x_{i_0}(t)| = \|x(t_0)\|_{\{\xi, \infty\}}$. Then, direct calculations give

$$\left\{ \frac{d|x_{i_0}(t)|}{dt} \right\}_{t=t_0} \le \operatorname{sign}(x_{i_0}(t_0)) \left\{ - d_{i_0}(t_0) x_{i_0}(t_0) + \sum_{j=1}^{n} a_{i_0 j}(t_0) g_j(x_j) \right.$$

$$+ \sum_{j=1}^{n} \int_0^{\infty} f_j(x_j(t_0 - \tau_{i_0 j}(t_0) - s)) d_s K_{i_0 j}(t_0, s) + I_{i_0}(t_0) \Big\}$$

$$\le \left[- d_{i_0} \xi_{i_0} + \sum_{j=1}^{n} |a_{i_0 j}(t_0)| G_j \xi_j \right] \|x(t_0)\|_{\{\xi, \infty\}}$$

$$+ \sum_{j=1}^{n} F_j \xi_j \int_0^{\infty} \|x(t_0 - \tau_{i_0 j}(t_0) - s)\|_{\{\xi, \infty\}} |d_s K_{i_0 j}(t_0, s)| + J$$

$$\le \left[- d_{i_0} \xi_{i_0} + \sum_{j=1}^{n} |a_{i_0 j}(t_0)| G_j \xi_j \right.$$

$$+ \sum_{j=1}^{n} F_j \xi_j \int_0^{\infty} |d_s K_{i_0 j}(t_0, s)| \right] M(t_0) + J$$

$$\le -\eta M(t_0) + J = -\eta M + J < 0,$$

which implies that $\|x(t)\|_{\{\xi, \infty\}}$ can never exceed M. Thus, $\|x(t)\|_{\{\xi, \infty\}} \le M(t) \le M$ for all $t > t_0$. Moreover, it is easy to see that $\|\dot{x}(\theta + \omega)\| \le N$. Therefore, $T\Omega \subset \Omega$. By Lemma 7.23, there exists $\phi^* \in \Omega$ such that $T\phi^* = \phi^*$. Hence $x(t, \phi^*) = x(t, T\phi^*)$, i.e., $x(t, \phi^*) = x(t + \omega, \phi^*)$, which is an ω-periodic solution of the system (7.30).

Now, we prove that inequality (7.34) leads to the global attractivity of the periodic solution. Let $\bar{x}(t) = [x(t) - v(t)]$ and $z(t) = e^{\alpha t} \bar{x}(t)$. We have

$$\frac{dz_i(t)}{dt} = -(d_i(t) - \alpha)z_i(t) + e^{\alpha t}\left\{\sum_{j=1}^{n} a_{ij}(t)\Big[g_j(x_j(t)) - g_j(v_j(t))\Big]\right.$$

$$+ \sum_{j=1}^{n}\int_0^{\infty}\left[f_j(x_j(t - \tau_{ij}(t) - s)) - f_j(v_j(t - \tau_{ij}(t) - s))\right]d_s K_{ij}(t, s)\Bigg\}.$$

Therefore,

$$\frac{d|z_i(t)|}{dt} \leq -(d_i(t) - \alpha)|z_i(t)| + \sum_{j=1}^{n}|a_{ij}(t)|G_j|z_j(t)|$$

$$+ \sum_{j=1}^{n} F_j e^{\alpha \tau_{ij}(t)}\int_0^{\infty} e^{\alpha s}|z_j(t - \tau_{ij}(t) - s)||d_s K_{ij}(t, s)|$$

$$\leq \left[-\xi_i(d_i(t) - \alpha) + \sum_{j=1}^{n}\xi_j|a_{ij}(t)|G_j\right]\|z(t)\|_{\{\xi, \infty\}}$$

$$+ \sum_{j=1}^{n}\xi_j F_j e^{\alpha \tau_{ij}(t)}\int_0^{\infty} e^{\alpha s}|z_j(t - \tau_{ij}(t) - s)||d_s K_{ij}(t, s)|.$$

By the same approach used before, we can prove that $z(t)$ is bounded. That is, $\bar{x}(t) = O(e^{-\alpha t})$. This completes the proof of the theorem.   □

7.3.2 Delayed Periodic Cohen–Grossberg Competitive and Cooperative Neural Networks

In this section, we investigate the following delayed Cohen–Grossberg neural network:

$$\frac{dx_i(t)}{dt} = -a_i(x_i(t))\Bigg[d_i(x_i(t)) - \sum_{j=1}^{n} c_{ij}(t)g_j(x_j(t))$$

$$- \sum_{j=1}^{n}\int_0^{\infty} f_j(x_j(t - s))d_s K_{ij}(t, s) + I_i(t)\Bigg], \quad i = 1, \ldots, n, \quad (7.36)$$

where $x_i(t)$ denotes the state variable of neuron i, all coefficients satisfy the condition $\mathcal{B}_1$, $d(x) = (d_1(x_1), \ldots, d_n(x_n))^\top \in \mathcal{D}$ as defined in Sect. 7.2.3, and the amplification functions $a_i(\cdot)$, $i = 1, \ldots, n$, might satisfy some of the assumptions $\mathcal{A}_{1-4}$ defined in Sect. 7.2.3. The initial condition is $x_i(\theta) = \phi_i(\theta)$, $\theta \in (-\infty, 0]$ for some continuous bounded positive functions $\phi_i(\cdot) \in C(-\infty, 0]$. The main results come from [21].

By Lemma 7.14, the assumption $\mathcal{A}_2$ implies that for positive, bounded, and continuous initial conditions, the trajectory of the system (7.36) is always positive. Moreover, we can obtain its boundedness.

Lemma 7.25 *Assume the hypotheses $\mathcal{B}_1$, and suppose further that $a(\cdot) \in \mathcal{A}_2$ and $d(\cdot) \in \mathcal{D}$. If there exist constants $\xi_i > 0$, $i = 1, \ldots, n$, such that for all $i = 1, 2, \ldots, n$ and $0 \le t < \omega$,*

$$- \gamma_i\xi_i + \sum_{j=1}^{n} |c_{ij}(t)|G_j\xi_j + \sum_{j=1}^{n} F_j\xi_j \int_0^\infty |d_sK_{ij}(t,s)| < 0, \qquad (7.37)$$

then any solution $x(t)$ of the system (7.36) is bounded.

Proof First, by Lemma 7.14, any solution of (7.36) under positive initial conditions is globally positive. Since $c_{ij}(t)$ are continuous and periodic with period ω, $d_sK_{ij}(t,s)$ are ω-periodic with respect to t, and there exists a constant $\eta > 0$ with

$$\eta = \min_i \min_{0 \le t < \omega} \left\{ \gamma_i\xi_i - \sum_{j=1}^{n} |c_{ij}(t)|G_j\xi_j - \sum_{j=1}^{n} F_j\xi_j \int_0^\infty |d_sK_{ij}(t,s)| \right\}.$$

So, we have

$$-\gamma_i\xi_i + \sum_{j=1}^{n} |c_{ij}(t)|G_j\xi_j + \sum_{j=1}^{n} F_j\xi_j \int_0^\infty |d_sK_{ij}(t,s)| \le -\eta < 0.$$

Let $M(t) = \max_{s \le t} \|x(s)\|_{\{\xi,\infty\}}$. Clearly, $M(t)$ is nondecreasing and $\|x(t)\|_{\{\xi,\infty\}} \le M(t)$. Denote

$$H = \sup_{0 < t \le \omega} \max_i \left\{ |d_i(0)| + |I_i^*| + \sum_{j=1}^{n} c_{ij}^*|g_j(0)| + \sum_{j=1}^{n} |f_j(0)| \int_0^\infty |d_sK_{ij}(t,s)| \right\}.$$

Now, we can prove that $M(t) \le \max\{M(0), H/\eta\}$. For any $t_0 \ge 0$ with $M(t_0) = \|x(t_0)\|_{\{\xi,\infty\}}$, let i_0 be the index with $\|x(t_0)\|_{\{\xi,\infty\}} = |x_{i_0}(t_0)|\xi_{i_0}^{-1}$. Note that the assumptions imply that, for $i = 1, \ldots, n$,

$$|g_i(s)| \le G_i|s| + |g_i(0)|, \quad |f_i(s)| \le F_i|s| + |f_i(0)|, \quad s \in \mathbb{R},$$

and

$$\text{sign}(s)d_i(s) \ge \gamma_i|s| + \text{sign}(s)d_i(0), \quad s \in \mathbb{R}.$$

Hence, we have

$$\left\{ \frac{d}{dt} |x_{i_0}(t)| \right\}_{t=t_0} = a_{i_0}(x_{i_0}(t_0))\mathrm{sign}(x_{i_0}(t_0))[- b_{i_0}(x_{i_0}(t_0))$$

$$+ \sum_{j=1}^{n} c_{i_0 j}(t_0) g_j(x_j(t_0))$$

$$+ \sum_{j=1}^{n} \int_0^\infty f_j(x_j(t_0 - s)) d_s K_{i_0 j}(t_0, s) + I_{i_0}(t_0)]$$

$$\leq a_{i_0}(x_{i_0}(t_0)) \left[-\gamma_{i_0}\xi_{i_0}|x_{i_0}(t_0)|\xi_{i_0}^{-1} + \sum_{j=1}^{n} |c_{i_0 j}(t_0)| G_j \xi_j |x_j(t_0)| \xi_j^{-1} \right.$$

$$+ |b_{i_0}(0)| + |I_{i_0}(t_0)| + \sum_{j=1}^{n} F_j \xi_j \int_0^\infty |x_j(t_0 - s)| \xi_j^{-1} |d_s K_{i_0 j}(t_0, s)|$$

$$\left. + \sum_{j=1}^{n} |c_{i_0 j}(t)||g_j(0)| + \sum_{j=1}^{n} |f_j(0)| \int_0^\infty |d_s K_{i_0 j}(t_0, s)| \right]$$

$$\leq a_{i_0}(x_{i_0}(t_0)) \left\{ \left[-\gamma_{i_0}\xi_{i_0} + \sum_{j=1}^{n} |c_{i_0 j}(t_0)| G_j \xi_j \right.\right.$$

$$\left.\left. + \sum_{j=1}^{n} F_j \xi_j \int_0^\infty |d_s K_{i_0 j}(t_0, s)| \right] \|x(t_0)\|_{\{\xi, \infty\}} + H \right\}$$

$$\leq a_{i_0}(x_{i_0}(t_0))(- \eta\|x(t_0)\|_{\{\xi, \infty\}} + H) = a_{i_0}(x_{i_0}(t_0))(- \eta M(t_0) + H).$$

This implies $M(t) \leq \max\{M(t_0), H/\eta\}$ according to Lemma 7.2. So, $x(t)$ is bounded. This completes the proof. $\qquad\square$

Thus, we can give the main result of this section.

Theorem 7.26 *Assume the hypotheses* $\mathcal{B}_1$, *and suppose further that* $a(\,\cdot\,) \in \mathcal{A}_2$ *and* $d(\,\cdot\,) \in \mathcal{D}$. *If there exist constants* $\zeta_i > 0$, $i = 1, 2, \ldots, n$, *such that for all* $i = 1, 2, \ldots, n$ *and* $0 \leq t < \omega$,

$$- \zeta_i\gamma_i + \sum_{j=1}^{n} |c_{ji}(t)|\zeta_j G_i + \sum_{j=1}^{n} \zeta_j F_i \int_0^\infty |dK_{ji}(s)| < 0, \qquad (7.38)$$

then the system (7.36) has a nonnegative periodic solution with period ω *which is globally asymptotically stable.*

Proof First, inequality (7.38) implies that inequality (7.37) holds owing to the M-matrix theory (see Lemma 7.4). By Lemma 7.14, any solution of the system (7.36) with a positive, bounded, and continuous initial condition is globally positive. Let

$$-\lambda = \max_i \sup_{0 \le t < \omega} \left\{ -\zeta_i \gamma_i + \sum_{j=1}^n |c_{ji}(t)| \zeta_j G_i + \sum_{j=1}^n \zeta_j F_i \int_0^\infty |dK_{ji}(s)| \right\}.$$

The conditions stated in the theorem implies that $\lambda > 0$.

For a specific positive solution $x(t)$ of system (7.36), let $u_i(t) = x_i(t + \omega) - x_i(t)$, and $v_i(t) = \int_{x_i(t)}^{x_i(t+\omega)} 1/a_i(\rho)\,d\rho$, $i = 1, 2, \ldots, n$. Note that $a_i(\cdot)$ is continuous, $a_i(\rho) > 0$ when $\rho > 0$, and x_i is positive and bounded, thus $\int_{x_i(t)}^{x_i(t+\omega)} 1/a_i(\rho)\,d\rho$ exists. By the mean-value theorem for integrals, $v_i(t) = 1/a_i(\xi)(x_i(t + \omega) - x_i(t)) = (1/a_i(\xi))u_i(t)$, where $\xi \in [\min\{x_i(t), x_i(t+\omega)\}, \max\{x_i(t), x_i(t+\omega)\}]$. Since $a_i(x) > 0$ when $x > 0$, we have $\mathrm{sign}(v_i(t)) = \mathrm{sign}(u_i(t))$.

Direct calculations give

$$\frac{dv_i(t)}{dt} = \frac{1}{a_i(x_i(t + \omega))} \left\{ \frac{dx_i(s)}{ds} \right\}_{s=t+\omega} - \frac{1}{a_i(x_i(t))} \left\{ \frac{dx_i(s)}{ds} \right\}_{s=t}$$

$$= -d_i(x_i(t + \omega)) + \sum_{j=1}^n c_{ij}(t + \omega)g_j(x_j(t + \omega))$$

$$+ \sum_{j=1}^n \int_0^\infty f_j(x_j(t + \omega - s))d_s K_{ij}(t + \omega, s) - I_i(t + \omega)$$

$$- \left[-d_i(x_i(t)) + \sum_{j=1}^n c_{ij}(t)g_j(x_j(t)) + \sum_{j=1}^n \int_0^\infty f_j(x_j(t - s))d_s K_{ij}(t, s) - I_i(t) \right]$$

$$= - \left(d_i(x_i(t + \omega)) - d_i(x_i(t)) + \sum_{j=1}^n c_{ij}(t)(g_j(x_j(t + \omega)) - g_j(x_j(t))) \right)$$

$$+ \sum_{j=1}^n \int_0^\infty (f_j(x_j(t + \omega)) - f_j(x_j(t)))d_s K_{ij}(t, s),$$

and

$$\frac{d}{dt}|v_i(t)| = \mathrm{sign}(v_i(t)) \left\{ -(d_i(x_i(t + \omega)) - d_i(t)) \right.$$

$$+ \sum_{j=1}^n c_{ij}(t)(g_j(x_j(t + \omega)) - g_j(x_j(t)))$$

$$+ \left. \sum_{j=1}^n \int_0^\infty (f_j(x_j(t + \omega)) - f_j(x_j(t)))d_s K_{ij}(t, s) \right\}$$

$$\le -\gamma_i |u_i(t)| + \sum_{j=1}^n |c_{ij}(t)|G_j|u_j(t)| + \sum_{j=1}^n \int_0^\infty F_j |u_j(t - s)||dK_{ij}(t, s)|.$$

Define

$$L(t) = \sum_{i=1}^{n} \zeta_i |v_i(t)| + \sum_{i,j=1}^{n} \zeta_i F_j \int_0^\infty \int_{t-s}^t |u_j(\rho)| d\rho \, |dK_{ij}(s)|.$$

Differentiating $L(t)$ along the trajectory $x(t)$ of system (7.36) gives

$$\frac{dL(t)}{dt} \le \sum_{i=1}^{n} \zeta_i \left[-\gamma_i |u_i(t)| + \sum_{j=1}^{n} |c_{ij}(t)| G_j |u_j(t)| + \sum_{j=1}^{n} \int_0^\infty F_j |u_j(t-s)| |dK_{ij}(s)| \right]$$

$$+ \sum_{i,j=1}^{n} \zeta_i F_j \left[\int_0^\infty |u_j(t)| |dK_{ij}(s)| - \int_0^\infty |u_j(t-s)| |dK_{ij}(s)| \right]$$

$$= \sum_{i=1}^{n} \left[-\zeta_i \gamma_i + \sum_{j=1}^{n} |c_{ji}(t)| \zeta_j G_i \right] |u_i(t)| + \sum_{i,j=1}^{n} \zeta_j F_i \int_0^\infty |u_i(t-s)| |dK_{ji}(s)|$$

$$+ \sum_{i,j=1}^{n} \zeta_j F_i \int_0^\infty |u_i(t)| |dK_{ji}(s)| - \sum_{i,j=1}^{n} \zeta_j F_i \int_0^\infty |u_i(t-s)| |dK_{ji}(s)|$$

$$= \sum_{i=1}^{n} \left[-\zeta_i \gamma_i + \sum_{j=1}^{n} |c_{ji}(t)| \zeta_j G_i + \sum_{j=1}^{n} \zeta_j F_i \int_0^\infty |dK_{ji}(s)| \right] |u_i(t)| \le -\lambda \|u(t)\|_1.$$

Since $L(t) \ge 0$, integrating both sides of (7.39) from 0 to ∞ gives

$$\int_0^\infty \sum_{i=1}^{n} |u_i(t)| dt \le \frac{1}{\lambda} L(0) < +\infty, \tag{7.39}$$

which implies

$$\sum_{n=1}^\infty \int_0^\omega \|x(t+n\omega) - x(t+(n-1)\omega)\|_1 dt < +\infty.$$

By the Cauchy convergence principle, we have that $x(t+n\omega)$ converges in $L^1[0, \omega]$ as $n \to \infty$. Since $x(t)$ is bounded, $a_i(x_i(t))$, $i = 1, 2, \ldots, n$, are also bounded and $x(t)$ is uniformly continuous. Then, the sequence $\{x(t+n\omega)\}$ is uniformly bounded and equicontinuous. Thus, by the Arzéla–Ascoli theorem, there exists a subsequence $\{x(t+n_k\omega)\}$ converging on any compact set of $\mathbb{R}$. Denote its limit by $x^*(t)$. We have that $x^*(t)$ is also the limit of $\{x(t+n\omega)\}$ in $L^1[0, \omega]$, i.e.,

$$\lim_{n \to \infty} \int_0^\infty \|x(t+n\omega) - x^*(t)\| dt = 0.$$

Then, we have that $\|x(t + n\omega) - x^*(t)\| \to 0$ uniformly on $[0, \omega]$. Similarly, $\|x(t + n\omega) - x^*(t)\| \to 0$ uniformly on any compact set of $\mathbb{R}$.

We next prove that $x^*(t)$ is a nonnegative periodic solution with period ω. Since

$$x^*(t + \omega) = \lim_{n \to \infty} x(t + (n + 1)\omega) = \lim_{n \to \infty} x(t + n\omega) = x^*(t),$$

we have that $x^*(t)$ is periodic with period ω. Then, replacing $x(t)$ with $x(t + n_k\omega)$ in system (7.36) and letting $k \to \infty$ give

$$\frac{dx_i^*(t)}{dt} = -a_i(x_i^*(t)) \left[d_i(x_i^*(t)) - \sum_{j=1}^{n} c_{ij}(t)g_j(x_j^*(t)) \right.$$
$$\left. - \sum_{j=1}^{n} \int_0^{\infty} f_j(x_j^*(t - s))d_s K_{ij}(t, s) + I_i(t) \right], \quad i = 1, \ldots, n.$$

Hence, $x^*(t)$ is a solution of the system (7.36). Let $t = t_1 + n\omega$, where $0 \le t_1 < \omega$. Then, $\|x(t) - x^*(t)\| = \|x(t_1 + n\omega) - x^*(t_1)\|$. The uniform convergence of $\{x(t + n\omega)\}$ on $[0, \omega]$ implies that

$$\lim_{t \to \infty} \|x(t) - x^*(t)\| = 0. \tag{7.40}$$

Finally, we prove that any positive solution of the system (7.36) converges to $x^*(t)$. Suppose that $y(t)$ is another positive solution of system (7.36) and let $u_i(t) = y_i(t) - x_i(t)$, $v_i(t) = \int_{x_i(t)}^{y_i(t)} 1/a_i(\rho)d\rho$, $i = 1, \ldots, n$. The same arguments above yield $\lim_{t \to \infty} \|y(t) - x(t)\| = 0$. In conjunction with (7.40), we conclude that $\lim_{t \to \infty} \|x(t) - x^*(t)\| = 0$, completing the proof. □

7.3.3 Delayed Almost Periodic Hopfield Neural Networks

In this section, we investigate the dynamical system (7.30) with almost periodic coefficients. The main results come from [65]. At this stage, we give the following set of hypotheses.

$\mathcal{B}_2$:

(1) The activation functions g, f satisfy $g(\cdot) \in H_2\{G_1, G_2, \ldots, G_n\}$ and $f(\cdot) \in H_1\{F_1, F_2, \ldots, F_n\}$ for some positive constants;
(2) $d_i(t)$, $a_{ij}(t)$, $\tau_{ij}(t)$, and $I_i(t)$ are continuous, $d_i(t) \ge d_i > 0$ and $\tau_{ij} \ge 0$ for $i, j = 1, 2, \ldots, n$;
(3) For any $s \in \mathbb{R}$, $K_{ij}(t, s): t \mapsto K_{ij}(t, s)$ is continuous in the same sense as in $\mathcal{B}_1$, and for any $t \in \mathbb{R}$, $dK_{ij}(t, s): s \mapsto dK_{ij}(t, s)$ is a Lebesgue–Stieltjes measure, for all $i, j = 1, 2, \ldots, n$;

(4) For any $\epsilon > 0$, there exists $l = l(\epsilon) > 0$, such that every interval $[\alpha, \alpha + l]$ contains at least one number ω for which $|d_i(t + \omega) - d_i(t)| < \epsilon$, $|a_{ij}(t + \omega) - a_{ij}(t)| < \epsilon$, $|I_i(t + \omega) - I_i(t)| < \epsilon$, $|\tau_{ij}(t + \omega) - \tau_{ij}(t)| < \epsilon$, and $\int_0^\infty |dK_{ij}(t + \omega, s) - dK_{ij}(t, s)| < \epsilon$ for all $i, j = 1, 2, \ldots, n$ and $t \in \mathbb{R}$.
(5) $|dK_{ij}(t, s)| \leq |dK_{ij}(s)|$, and for some $\epsilon > 0$, $\int_0^\infty e^{\epsilon s} |dK_{ij}(s)| < \infty$.

It can be seen that under Item 4 in this assumption, $d_i(t)$, $a_{ij}(t)$, $I_i(t)$, and $\tau_{ij}(t)$ are almost periodic functions. Therefore, they are all bounded. We also denote $|a_{ij}^*| = \sup_{\{t \in \mathbb{R}\}} |a_{ij}(t)|$, $|b_{ij}^*| = \sup_{\{t \in \mathbb{R}\}} |b_{ij}(t)|$, $|I_i^*| = \sup_{\{t \in \mathbb{R}\}} |I_i(t)|$, $\tau_{ij}^* = \sup_{\{t \in \mathbb{R}\}} \tau_{ij}(t)$, $i, j = 1, \ldots, n$, which are surely finite due their almost periodicity.
Before stating the main result, we need several lemmas for the proof of the main theorem.

Lemma 7.27 *Suppose that the hypotheses $\mathcal{B}_2$ are satisfied. If there exist $\xi_i > 0$, $i = 1, \ldots, n$, and $\eta > 0$ such that*

$$- d_i(t)\xi_i + \sum_{j=1}^n |a_{ij}(t)| G_j \xi_j + \sum_{j=1}^n F_j \xi_j \int_0^\infty |dK_{ij}(t, s)| < -\eta < 0 \qquad (7.41)$$

for all $t > 0$ and $i = 1, \ldots, n$, then any solution $x(t)$ of the system (7.29) is bounded.

Proof Define $M(t) = \max_{s \leq t} \|x(s)\|_{\{\xi, \infty\}}$. It is obvious that $\|x(t)\|_{\{\xi, \infty\}} \leq M(t)$, and $M(t)$ is nondecreasing. We will prove that $M(t) \leq \max\{M(0), (2/\eta)\hat{I}\}$, where

$$\hat{I} = \max_i \left\{ |I_i^*| + \sum_{j=1}^n \left[|a_{ij}^*| |g_j(0)| + |b_{ij}^*| |f_j(0)| \right] \right\}.$$

Fix t_0 such that $\|x(t_0)\|_{\{\xi, \infty\}} = M(t_0) = \max_{s \leq t_0} \|x(s)\|_{\{\xi, \infty\}}$. In this case, let i_{t_0} be such an index that $\xi_{i_{t_0}}^{-1} |x_{i_{t_0}}(t_0)| = \|x(t_0)\|_{\{\xi, \infty\}}$. Then, noting that $|g_j(s)| \leq G_j|s| + |g_j(0)|$ and $|f_j(s)| \leq F_j|s| + |f_j(0)|$ for $j = 1, \ldots, n$ and $s \in \mathbb{R}$, we have

$$\left\{ \frac{d}{dt} |x_{i_{t_0}}(t)| \right\}\Big|_{t-t_0} = \text{sign}(x_{i_{t_0}}(t_0)) \Bigg[- d_{i_{t_0}}(t_0)x_{i_{t_0}}(t_0) + \sum_{j=1}^n a_{i_{t_0}j}(t_0)g_j(x_j(t_0))$$

$$+ \sum_{j=1}^n \int_0^\infty f_j(x_j(t_0 - \tau_{i_{t_0}j}(t_0) - s))dK_{i_{t_0}j}(t_0, s) + I_{i_{t_0}}(t_0) \Bigg]$$

$$\leq -d_{i_{t_0}}(t_0)|x_{i_{t_0}}(t_0)|\xi_{i_{t_0}}^{-1}\xi_{i_{t_0}} + \sum_{j=1}^n |a_{i_{t_0}j}(t_0)| G_j |x_j(t_0)| \xi_j^{-1} \xi_j$$

$$+ \sum_{j=1}^n F_j \xi_j \int_0^\infty |x_j(t_0 - \tau_{i_{t_0}j}(t_0) - s)| \xi_j^{-1} |dK_{i_{t_0}j}(t_0, s)| + |I_{i_{t_0}}(t_0)|$$

$$+ \sum_{j=1}^{n} |a_{i_0j}(t)||g_j(0)| + |b_{i_0j}(t)||f_j(0)|$$

$$\leq -d_{i_{t_0}}(t_0)\xi_{i_{t_0}} + \sum_{j=1}^{n} \left[|a_{i_{t_0}j}(t_0)|G_j\xi_j + F_j\xi_j \int_0^\infty |dK_{i_{t_0}j}(t_0,s)| \|x(t_0)\|_{\{\xi,\infty\}} \right] + \hat{I}$$

$$\leq -\eta \|x(t_0)\|_{\{\xi,\infty\}} + \hat{I} = -\eta M(t_0) + \hat{I}, \tag{7.42}$$

which implies $M(t) \leq \max\{M(0), (2/\eta)\hat{I}\}$ for all $t > 0$ according to Lemma 7.2. This proves that $x(t)$ is bounded. The lemma is proved. □

Lemma 7.28 *Suppose that the hypotheses $\mathcal{B}_2$ are satisfied. If there exist $\xi_i > 0$, $i = 1, 2, \ldots, n$, $\beta > 0$, and $\eta > 0$ such that for all $t > 0$,*

$$- d_i(t)\xi_i + \sum_{j=1i}^{n} |a_{ij}(t)|G_j\xi_j + \sum_{j=1}^{n} F_j\xi_j e^{\beta \tau_{ij}^*} \int_0^\infty e^{\beta s}|dK_{ij}(t,s)| < -\eta, \quad (7.43)$$

then for any $\epsilon > 0$, there exist $T > 0$ and $l = l(\epsilon) > 0$, such that every interval $[\alpha, \alpha + l]$ contains at least one number ω for which the solution $x(t)$ of system (7.30) satisfies

$$\|x(t + \omega) - x(t)\|_{\{\xi,\infty\}} \leq \epsilon \quad \text{for all } t > T. \tag{7.44}$$

Proof Let

$$\epsilon_i(\omega, t) = -[d_i(t + \omega) - d_i(t)]x_i(t + \omega) + \sum_{j=1}^{n} [a_{ij}(t + \omega) - a_{ij}(t)]g_j(x_j(t + \omega))$$

$$+ \sum_{j=1}^{n} \int_0^\infty [f_j(x_j(t - \tau_{ij}(t + \omega) + \omega - s)) - f_j(x_j(t - \tau_{ij}(t) + \omega - s))]dK_{ij}(t + \omega, s)$$

$$+ \sum_{j=1}^{n} \int_0^\infty f_j(x_j(t - \tau_{ij}(t) + \omega - s))d[K_{ij}(t + \omega, s) - K_{ij}(t, s)] + [I_i(t + \omega) - I_i(t)].$$

Lemma 7.27 tells that $x(t)$ is bounded. Thus, the right side of (7.30) is also bounded, which implies that $x(t)$ is uniformly continuous. Therefore, by the fourth item in assumption $\mathcal{B}_2$, for any $\epsilon > 0$, there exists $l = l(\epsilon) > 0$ such that every interval $[\alpha, \alpha + l]$, $\alpha \in \mathbb{R}$, contains an ω for which $|\epsilon_i(\omega, t)| \leq (1/2)\eta\epsilon$, for all $t \in \mathbb{R}$ and $i = 1, 2, \ldots, n$.

Denote $z_i(t) = x_i(t + \omega) - x_i(t)$. We have

$$\frac{dz_i(t)}{dt} = -d_i(t)z_i(t) + \sum_{j=1}^{n} a_{ij}(t)[g_j(x_j(t + \omega)) - g_j(x_j(t))]$$

$$+ \sum_{j=1}^{n} \int_0^{\infty} \left[f_j(x_j(t + \omega - \tau_{ij}(t) - s)) - f_j(x_j(t - \tau_{ij}(t) - s)) \right] dK_{ij}(t, s)$$

$$+ \epsilon_i(\omega, t).$$

Let i_t be such an index that $\xi_{i_t}^{-1}|z_{i_t}(t)| = \|z(t)\|_{\{\xi, \infty\}}$. Differentiating $e^{\beta s}|z_{i_t}(s)|$ gives

$$\frac{d}{ds}\left\{ e^{\beta s}|z_{i_t}(s)| \right\}\bigg|_{s=t} = \beta e^{\beta t}|z_{i_t}(t)| + e^{\beta t}\text{sign}(z_{i_t}(t))\left\{ -d_{i_t}(t)z_{i_t}(t) \right.$$

$$+ \sum_{j=1}^{n} a_{i_t j}(t)\left[g_j(x_j(t + \omega)) - g_j(x_j(t)) \right]$$

$$+ \sum_{j=1}^{n} \int_0^{\infty} \left[f_j(x_j(t + \omega - \tau_{i_t j}(t) - s)) - f_j(x_j(t - \tau_{i_t j}(t) - s)) \right] dK_{i_t j}(t, s)$$

$$\left. + \epsilon_{i_t}(\omega, t) \right\}$$

$$\le e^{\beta t}\left\{ -[d_{i_t}(t) - \beta]|z_{i_t}(t)|\xi_{i_t}^{-1}\xi_{i_t} + \sum_{j=1}^{n} |a_{i_t j}(t)||G_j|z_j(t)|\xi_j^{-1}\xi_j \right.$$

$$\left. + \sum_{j=1}^{n} F_j\xi_j \int_0^{\infty} |z_j(t - \tau_{i_t j}(t) - s)|\xi_j^{-1} e^{-\beta(\tau_{i_t j}(t) + s)} e^{\beta(s + \tau_{ij}^*)} |dK_{i_t j}(t, s)| \right\}$$

$$+ \frac{1}{2}\eta\epsilon e^{\beta t}.$$

Using arguments similar to those in the proof of Lemma 7.27, let

$$\Psi(t) = \max_{s \le t}\left\{ e^{\beta s}\|z(s)\|_{\{\xi, \infty\}} \right\}. \tag{7.45}$$

For any $t_0 > 0$ with $\Psi(t_0) = e^{\beta t_0}\|z(t_0)\|_{\{\xi, \infty\}}$, we have $d\{e^{\beta t}|z_{i_t}(t)|\}/dt|_{t=t_0} \le -\eta\Psi(t_0) + \frac{1}{2}\eta\epsilon e^{\beta t}$. From Lemma 7.2, this implies that there must exist $T > 0$ such that $\|z(t)\|_{\{\xi, \infty\}} \le \epsilon$ for all $t > T$. $\qquad\square$

Thus, we obtain the main theorem.

Theorem 7.29 Suppose that the hypotheses $\mathcal{B}_2$ are satisfied. If there exist $\xi_i > 0$, $i = 1, 2, \ldots, n$, $\beta > 0$, and $\eta > 0$ such that the inequality

$$- [d_i(t) - \beta]\xi_i + \sum_{j=1}^{n} |a_{ij}(t)|G_j\xi_j + \sum_{j=1}^{n} F_j\xi_j e^{\beta\tau_{ij}^*} \int_0^\infty e^{\beta s}|dK_{ij}(t,s)| < -\eta,$$

$$i = 1,\ldots,n \tag{7.46}$$

holds for all $t > 0$, then the system (7.30) has a unique almost periodic solution $v(t) = (v_1(t),\ldots,v_n(t))^\top$, and for any solution $x(t) = (x_1(t),\ldots,x_n(t))^\top$ of (7.30), one has

$$\|x(t) - v(t)\| = O(e^{-\beta t}). \tag{7.47}$$

Proof $\epsilon_{i,k}(t)$ is defined as in the proof of Lemma 7.28. From the hypotheses $\mathcal{B}_2$ and the boundedness of $u(t)$, we can select a sequence $\{t_k\} \to \infty$ such that $|\epsilon_{i,k}(t)| \le 1/k$ for all i,t. Since $\{x(t+t_k)\}_{k=1}^\infty$ are uniformly bounded and equicontinuous, by the Arzela–Ascoli lemma and the diagonal selection principle, we can select a subsequence t_{k_j} of t_k, such that $x(t+t_{k_j})$ (for convenience, we still denote by $x(t+t_k)$) uniformly converges to a continuous function $v(t) = [v_1(t), v_2(t), \ldots, v_n(t)]^\top$ on any compact subset of $\mathbb{R}$.

Now, we prove $v(t)$ is a solution of system (7.30). In fact, by Lebesgue dominated convergence theorem, for any $t > 0$ and $\delta t \in \mathbb{R}$, we have

$$v_i(t + \delta t) - v_i(t) = \lim_{k \to \infty} \left[u_i(t + \delta t + t_k) - u_i(t + t_k) \right]$$

$$= \lim_{k \to \infty} \int_t^{t+\delta t} \Bigg\{ -d_i(\sigma + t_k)u_i(\sigma + t_k) + \sum_{j=1}^{n} a_{ij}(\sigma + t_k)g_j(u_j(\sigma + t_k))$$

$$+ \sum_{j=1}^{n} \int_0^\infty f_j(u_j(\sigma + t_k - \tau_{ij}(\sigma + t_k) - s))dK_{ij}(\sigma + t_k, s) + I_i(\sigma + t_k) \Bigg\} d\sigma$$

$$= \int_t^{t+\delta t} \Bigg\{ -d_i(\sigma)v_i(\sigma) + \sum_{j=1}^{n} a_{ij}(\sigma)g_j(v_j(\sigma))$$

$$+ \sum_{j=1}^{n} \int_0^\infty f_j(v_j(\sigma - \tau_{ij}(\sigma) - s))dK_{ij}(\sigma, s) + I_i(\sigma) \Bigg\} d\sigma + \lim_{k \to \infty} \int_t^{t+\delta t} \epsilon_{i,k}(s)d\sigma$$

$$= \int_t^{t+\delta t} \Bigg\{ -d_i(\sigma)v_i(\sigma) + \sum_{j=1}^{n} a_{ij}(\sigma)g_j(v_j(\sigma))$$

$$+ \sum_{j=1}^{n} \int_0^\infty f_j(v_j(\sigma - \tau_{ij}(\sigma) - s))dK_{ij}(\sigma, s) + I_i(\sigma) \Bigg\} d\sigma,$$

which implies

$$\frac{dv_i}{dt} = -d_i(t)v_i(t) + \sum_{j=1}^{n} a_{ij}(t)g_j(u_j(t))$$

$$+ \int_0^{\infty} f_j(u_j(t - \tau_{ij}(t) - s))dK_{ij}(t,s) + I_i(t),$$

i.e., $v(t)$ is a solution of the system (7.30).

Second, we prove that $v(t)$ is an almost periodic function. By Lemma 7.28, for any $\epsilon > 0$, there exist $T > 0$ and $l = l(\epsilon) > 0$, such that every interval $[\alpha, \alpha + l]$ contains at least one number ω for which $|x_i(t + \omega) - x_i(t)| \leq \epsilon$, for all $t > T$. Then we can find a sufficient large $K \in N$ such that for any $k > K$ and all $t > 0$, we have $|x_i(t + t_k + \omega) - x_i(t + t_k)| \leq \epsilon$. Let $k \to \infty$, we have $|v_i(t + \omega) - v_i(t)| \leq \epsilon$, for all $t > 0$. In other words, $v(t)$ is an almost periodic function.

Finally, we prove that every solution $x(t)$ of the system (7.30) converges to $v(t)$ exponentially with rate β.

Denote $y(t) = x(t) - v(t)$. We have

$$\frac{dy_i(t)}{dt} = -d_i(t)y_i(t) + \sum_{j=1}^{n} a_{ij}(t)\left[g_j(x_j(t)) - g_j(v_j(t))\right]$$

$$+ \sum_{j=1}^{n} \int_0^{\infty} \left[f_j(x_j(t - \tau_{ij}(t) - s)) - f_j(v_j(t - \tau_{ij}(t) - s))\right] dK_{ij}(t,s).$$

Let i_t be an index such that $|y_{i_t}(t)| = \xi_{i_t}\|y(t)\|_{\{\xi,\infty\}}$. Differentiating $e^{\beta s}|y_{i_t}(s)|$, we have

$$\frac{d}{ds}\left\{e^{\beta s}|y_{i_t}(s)|\right\}\Bigg|_{s=t} = \beta e^{\beta t}|y_{i_t}(t)| + e^{\beta t}\text{sign}(y_{i_t}(t))\Bigg\{ -d_{i_t}(t)y_{i_t}(t)$$

$$+ \sum_{j=1}^{n} a_{i_t j}(t)[g_j(x_j(t)) - g_j(v_j(t))]$$

$$+ \sum_{j=1}^{n} [f_j(x_j(t - \tau_{i_t j}(t) - s)) - f_j(v_j(t - \tau_{i_t j}(t) - s))]dK_{i_t j}(t,s)\Bigg\}$$

$$\leq e^{\beta t}\Bigg\{ -[d_{i_t} - \beta]|y_{i_t}(t)|\xi_{i_t}^{-1}\xi_{i_t} + \sum_{j=1}^{n} |a_{i_t j}(t)|G_j|y_j(t)|\xi_j^{-1}\xi_j$$

$$+ \sum_{j=1}^{n} F_j\xi_j \int_0^{\infty} |y_j(t - \tau_{i_t j}(t) - s)|\xi_j^{-1}e^{-\beta(s+\tau_{i_t j}(t))}e^{\beta(s+\tau_{i_t j}^*)}|dK_{i_t j}(t,s)|\Bigg\}$$

$$(7.48)$$

Define $\Delta(t) = \max_{s \leq t}\{e^{\beta s}\|y(s)\|_{\{\xi,\infty\}}\}$. Fix t_0 such that $\Delta(t_0) = e^{\beta t_0}\|y(t_0)\|_{\{\xi,\infty\}}$. Inequality (7.48) becomes $d\{e^{\beta t}|y_{i_0}(t)|\}/dt_{t=t_0} \leq -\eta\Delta(t_0) \leq 0$. By Lemma 7.2, this implies that $\Delta(t) \leq \Delta(0)$ for all $t \geq 0$ and $\|y(t)\|_{\{\xi,\infty\}} \leq \Delta(0)e^{-\beta t}$. In other words, $\|x(t) - v(t)\|_{\{\xi,\infty\}} \leq \Delta(0)e^{-\beta t}$. The theorem is proved. □

Since periodic functions are a special case of almost periodic functions, the results in this section can easily be used to obtain the criterion guaranteeing the existence of a periodic trajectory and its global stability for the case when coefficients are all periodic with a uniform period. Hence, the following theorem is a direct consequence of Theorem 7.29.

Theorem 7.30 *Suppose that the hypotheses $\mathcal{B}_1$ are satisfied. If there exist positive constants $\xi_1, \ldots, \xi_n$ and $\beta > 0$ such that*

$$- \xi_i[d_i - \beta] + \sum_{j=1}^n |a_{ij}(t)|\xi_j G_j + \sum_{j=1}^n \xi_i F_j e^{\beta \tau_{ij}^*} \int_0^\infty e^{\beta s} |d\bar{K}_{ij}(s)| < 0,$$

$$i = 1, \ldots, n, \tag{7.49}$$

then the system (7.29) has a unique periodic solution $v(t) = (v_1(t), \ldots, v_n(t))^\top$, and for any solution $x(t) = (x_1(t), \ldots, x_n(t))^\top$ of (7.29), one has $|x(t) - v(t)| = O(e^{-\beta t})$ as $t \to \infty$.

Moreover, consider the following system with constant coefficients:

$$\frac{du_i}{dt} = -d_i u_i(t) + \sum_{j=1}^n a_{ij} g_j(u_j(t)) + \sum_{j=1}^n \int_0^\infty f_j(u_j(t - \tau_{ij} - s)) d_s K_{ij}(s)$$

$$+ I_i, \quad i = 1, \ldots, n, \tag{7.50}$$

where $d_s K_{ij}(s)$ denotes the Lebesgue–Stieltjes measures, $i, j = 1, \ldots, n$. Since a constant can be regarded as a function with arbitrary period, we have the following result.

Theorem 7.31 *Suppose $g(\cdot) \in H_1\{G_1, \ldots, G_n\}$ and $f(\cdot) \in H_1\{F_1, \ldots, F_n\}$. If there are positive constants $\xi_1, \ldots, \xi_n$ and $\beta > 0$ such that*

$$- \xi_i[d_i - \beta] + \sum_{j=1}^n |a_{ij}|\xi_j G_j + \sum_{j=1}^n \xi_j F_j e^{\beta \tau_{ij}} \int_0^\infty e^{\beta s} |dK_{ij}(s)| F_j < 0, \quad i = 1, \ldots, n,$$

$$\tag{7.51}$$

then the system (7.50) is globally exponentially stable.

7.4 Delayed Neural Network with Discontinuous Activations

So far, all discussions and results have been based on the assumption that the activation functions are Lipschitz continuous. As pointed in [40], a brief review on some common neural network models reveals that neural networks with discontinuous activation functions are of importance and arise frequently in practice. For example, consider the classical Hopfield neural networks with graded response neurons (see [54]). The standard assumption is that the activations are used in the high-gain limit where they closely approach discontinuous and comparator functions. As shown in [54, 57], the high-gain hypothesis is crucial to make negligible the connection to the neural network energy function of the term depending on neuron self-inhibitions, and to favor binary output formation, as in a hard comparator function like sign(s).

A conceptually analogous model based on hard comparators are discrete-time neural networks discussed in [50]. Another important example concerns the class of neural networks introduced in [59] to solve linear and nonlinear programming problems. Those networks exploit constraint neurons with diode-like input–output activations. Again, in order to guarantee satisfaction of the constraints, the diodes are required to possess a very high slope in the conducting region, i.e., they should approximate the discontinuous characteristic of an ideal diode (see [31]). When dealing with dynamical systems possessing high-slope nonlinear elements, it is often advantageous to model them with a system of differential equations with discontinuous right-hand side, rather than studying the case where the slope is high but of finite value (see [85]).

In this section, we consider the following delayed dynamical system:

$$\frac{dx_i(t)}{dt} = -d_i(t)x_i(t) + \sum_{j=1}^{n} a_{ij}(t)g_j(x_j(t))$$

$$+ \sum_{j=1}^{n} \int_0^\infty g_j(x_j(t-s))d_s K_{ij}(t,s) + I_i(t), \ i = 1,\dots,n, \quad (7.52)$$

with discontinuous activations g_j for both delayed and undelayed terms. A special form with a uniform discrete delay is

$$\frac{dx(t)}{dt} = -Dx(t) + Ag(x(t)) + Bg(x(t-\tau)) + I, \quad (7.53)$$

when rewritten in matrix form. We introduce the concept of a solution in the Filippov sense for the system (7.52) and prove its existence by the idea introduced in [48]. We construct a sequence of delayed systems in which the activations have high slope and converge to the discontinuous activations. First, we prove that under diagonal dominance conditions, the sequence of solutions has at least a subsequence converging to a solution of the system (7.52) with discontinuous activations by a well-known diagonal selection argument. Second, we consider the system (7.53). Without assuming the boundedness and the continuity of the neuron activations, we present sufficient conditions for the global stability of neural networks with time delay based on linear matrix inequalities and discuss their convergence. Third, we discuss the system (7.52) with almost periodic coefficients. We use the Lyapunov functional method to obtain an asymptotically almost periodic solution which leads to the existence of an almost periodic solution [84]. We also use the Lyapunov functional to obtain the global exponential stability of this almost periodic solution. Furthermore, from the proof of the existence and uniqueness of the solution, we can conclude that each solution sequence of the system with high-slope activations which converge to the discontinuous activations will actually converge to the unique solution of the system (7.52) with discontinuous activations in the Filippov sense. The main results come from [64, 66].

7.4.1 Preliminaries

In this section, we introduce the definitions and lemmas on nonsmooth and variational analysis, report some definitions and existing results on differential inclusions, and based on those results, give the mathematical description for the generalized neural network model to be studied.

7.4.1.1 Nonsmooth Analysis of Single-Valued Functions

Here, we introduce some necessary definitions and lemma on nonsmooth and variational analysis. We refer interested readers to [34, 80] for more details on these topics.

A single-valued function $f: \mathbb{R}^n \to \mathbb{R}$ is said to be *strictly continuous* at $\bar{x} \in \mathbb{R}^n$ if the value $\text{lip} f(\bar{x}) := \lim \sup_{x,x' \to \bar{x}, \, x \neq x'} |f(x) - f(x')| / \|x - x'\|$ is finite. If f is strictly continuous at each $\bar{x} \in \mathbb{R}^n$, then f is said to be *strictly continuous in $\mathbb{R}^n$*. A strictly continuous function $f: \mathbb{R}^n \to \mathbb{R}$ is said to be *(Clarke) regular* at $x \in \mathbb{R}^n$ if there exists the usual *one-sided directional derivative* $f'(x, v) = \lim_{\rho \searrow 0} [f(x + \rho v) - f(x)] / \rho$ for all $v \in \mathbb{R}^n$ and it equals to the *generalized directional derivative* $f^o(x, v) = \lim \sup_{y \to x, t \searrow 0} [f(y + tv) - f(y)]/t$. f is said to be *regular in $\mathbb{R}^n$* if f is regular on each $x \in \mathbb{R}^n$. For a strictly continuous function $f: \mathbb{R}^n \to \mathbb{R}$, the *Clarke's generalized gradient* of f at $x \in \mathbb{R}^n$, which can be used to handle gradient flow on nonsmooth functions, can be written as

$$\partial f = \{p \in \mathbb{R}^n : f^o(x, v) \geq \langle p, v \rangle, \ \forall \, v \in \mathbb{R}^n\}.$$

A point $x_0 \in \mathbb{R}^n$ is said to be a *critical point* of f if $0 \in \partial f(x_0)$, and crit(f) denotes the set of critical points of f.

The following chain rule for nonsmooth functions is very important for later arguments.

Lemma 7.32 (Chain Rule, Theorem 2.3.9 in [34]) *If $x(t): \mathbb{R}^+ \to \mathbb{R}^n$ is locally absolutely continuous and a single-valued function $f: \mathbb{R}^n \to \mathbb{R}$ is strictly continuous and regular in $\mathbb{R}^n$, then the derivative $\frac{d}{dt} f(x(t))$ exists for almost all $t \geq 0$ and*

$$\frac{d}{dt} f(x(t)) = \langle p, \dot{x}(t) \rangle, \ \text{for all } p \in \partial f(x(t)),$$

for almost all $t \geq 0$.

7.4.1.2 Set-Valued Map

We introduce some definitions and lemmas for set-valued and variational analysis. We refer the interested readers to [5, 80] for more details.

Suppose $E \subset \mathbb{R}^n$. Then $x \mapsto F(x)$ is called a set-valued map from $E \hookrightarrow \mathbb{R}^n$, if to each point x of a set $E \subset \mathbb{R}^n$, there corresponds a non-empty set $F(x) \subset \mathbb{R}^n$. A set-valued map F with non-empty values is said to be *upper semicontinuous* (u.s.c. for short) at $x_0 \in E$, if for any open set N containing $F(x_0)$, there exists a neighborhood M of x_0 such that $F(M) \subset N$. $F(x)$ is said to have closed (convex, compact) image, if for each $x \in E$, $F(x)$ is closed (convex, compact).

7.4.1.3 Description of the Solution of the Model

Consider the following system:

$$\frac{dx}{dt} = f(x), \tag{7.54}$$

where $f(\,\cdot\,)$ is not continuous. Reference [39] proposed the following definition of the solution for the system (7.54).

Definition 7.33 Let ϕ be a set-valued map given by

$$\phi(x) = \bigcap_{\delta>0} \bigcap_{\mu(N)=0} \overline{co}\Big[f(\overline{\mathcal{O}}(x, \delta) - N) \Big], \tag{7.55}$$

where $\overline{co}(E)$ is the closure of the convex hull of some set E, $\overline{\mathcal{O}}(x, \delta) = \{y \in \mathbb{R}^n : \|y - x\| \le \delta\}$, and $\mu(N)$ is the Lebesgue measure of the set N. A solution of the Cauchy problem for (7.54) with initial condition $x(0) = x_0$ is an absolutely continuous function $x(t)$, $t \in [0, T)$, which satisfies $x(0) = x_0$, and the differential inclusion

$$\frac{dx}{dt} \in \phi(x), \qquad \text{a.e. } t \in [0, T). \tag{7.56}$$

Furthermore, [4, 6, 48] have proposed the following functional differential inclusion with memory:

$$\frac{dx}{dt}(t) \in F(t, A(t)x), \tag{7.57}$$

where $F : \mathbb{R} \times C([-\tau, 0], \mathbb{R}^n) \mapsto \mathbb{R}^n$ is a given set-valued map, and

$$[A(t)x](\theta) = x_t(\theta) = x(t + \theta). \tag{7.58}$$

Inspired by these works, we denote $\overline{co}[g_i(s)] = [g_i^-(s), g_i^+(s)]$ and $\overline{co}[g(x)] = \overline{co}[g_1(x_1)] \times \overline{co}[g_2(x_2)] \times \cdots \times \overline{co}[g_n(x_n)]$, where $\times$ denotes the Cartesian product.

The set-valued map $\overline{co}[g(x)]$ is always u.s.c., convex, and compact. Thus, we can define solution of the system (7.52) in the Filippov sense as follows.

Definition 7.34 For a continuous function $\phi(\theta) = (\phi_1(\theta), \ldots, \phi_n(\theta))^\top$ and a measurable function $\lambda(\theta) = (\lambda_1(\theta), \ldots, \lambda_n(\theta))^\top \in \overline{co}[g(\phi(\theta))]$ for almost all $\theta \in (-\infty, 0]$, an absolute continuous function $x(t) = x(t, \phi, \lambda) = (x_1(t), \ldots, x_n(t))^\top$ associated with a measurable function $\gamma(t) = (\gamma_1(t), \ldots, \gamma_n(t))^\top$ is said to be a solution of the Cauchy problem for the system (7.52) on $[0, T)$ (T might be ∞) with initial value $(\phi(\theta), \lambda(\theta))$, $\theta \in (-\infty, 0]$, if

$$
\begin{cases}
\dfrac{dx_i(t)}{dt} = -d_i(t)x_i(t) + \displaystyle\sum_{j=1}^{n} a_{ij}(t)\gamma_j(t) \\
\quad + \int_0^\infty \gamma_j(t-s)d_s K_{ij}(t,s) + I_i(t) & \text{a.e. } t \in [0, T), \\
\gamma_i(t) \in \overline{co}[g_i(x_i(t))] & \text{a.e. } t \in [0, T), \\
x_i(\theta) = \phi_i(\theta) & \theta \in (-\infty, 0], \\
\gamma_i(\theta) = \lambda_i(\theta) & \text{a.e. } \theta \in (-\infty, 0],
\end{cases}
\tag{7.59}
$$

for all $i = 1, \ldots, n$.

The solution of the system (7.53) can be defined in the same way.

7.4.1.4 Set-up of Discontinuous Activations

We summarize the set-up of the model with the following assumptions.

C_1: Every $g_i(\cdot)$ is nondecreasing and local Lipschizian, except on a set of isolated points $\{\rho_k^i\}$. More precisely, for each $i = 1, \ldots, n$, $g_i(\cdot)$ is nondecreasing and continuous except on a set of isolated points $\{\rho_k^i\}$, where the right and left limits $g_i^+(\rho_k^i)$ and $g_i^-(\rho_k^i)$ satisfy $g_i^+(\rho_k^i) > g_i^-(\rho_k^i)$. In each compact set of $\mathbb{R}$, $g_i(\cdot)$ has only finite number of discontinuities. Moreover, ordering the set of discontinuities as $\{\rho_k^i: \rho_{k+1}^i > \rho_k^i, k \in \mathbb{Z}\}$, there exist positive constants $G_{i,k} > 0$, $i = 1, \ldots, n$, $k \in \mathbb{Z}$, such that $|g_i(\xi) - g_i(\zeta)| \le G_{i,k}|\xi - \zeta|$ for all $\xi, \zeta \in (\rho_k^i, \rho_{k+1}^i)$.

C_2: The initial condition $\phi(\theta) \in C((-\infty, 0], \mathbb{R}^n)$ is bounded, and $\lambda(\theta)$ is measurable and essentially bounded.

7.4.1.5 Viability

Here, we give the conditions guaranteeing the existence of Filippov solution in the sense (7.59) for the system (7.52). Similar to the idea proposed in [48], the solution of the system (7.52) in the sense (7.59) can be regarded as an approximation of the solutions of delayed neural networks with high-slope activations. This is the main idea of proving the existence and almost periodicity of the solution. More precisely, define a family of functions Ξ containing $f(x) = [f_1(x_1), f_2(x_2), \ldots, f_n(x_n)]^\top \in C(\mathbb{R}^n, \mathbb{R}^n)$ and satisfying the following properties: (1) every $f_i(\cdot)$ is monotonically nondecreasing, for $i = 1, 2, \ldots, n$; (2) every $f_i(\cdot)$ is uniformly locally bounded, i.e., for any compact set $Z \subset \mathbb{R}^n$, there exists a constant $M > 0$ independent of

f such that $|f_i(x)| \leq M$ for all $x \in Z$ and $i = 1, \ldots, n$; (3) every $f_i(\cdot)$ is locally Lipschitz continuous, i.e., for any compact set $Z \subset \mathbb{R}^n$, there exists $\lambda > 0$ such that $|f_i(\xi) - f_i(\zeta)| \leq \lambda |\xi - \zeta|$ for all $\xi, \zeta \in Z$, and $i = 1, 2, \ldots, n$. For any $f \in \Xi$, by the theory given in [49], the following system:

$$
\begin{cases}
\frac{du_i^f}{dt}(t) = -d_i(t)u_i^f(t) + \sum_{j=1}^{n} a_{ij}(t)\sigma_j^f(t) + \sum_{j=1}^{n} \int_0^\infty \sigma_j^f(t-s) d_s K_{ij}(t,s) + I_i(t) \\
u_i^f(\theta) = \phi_i(\theta), \; \theta \in (-\infty, 0] \\
\sigma_i^f(\theta) = \begin{cases} \lambda_i(\theta), & \theta \leq 0 \\ f_i(u_i^f(\theta)), & \theta \geq 0 \end{cases} \quad i = 1, \ldots, n
\end{cases}
$$

$$(7.60)$$

admits a unique solution $u_f(t) = (u_1(t), u_2(t), \ldots, u_n(t))^\top$ on $[0, T)$, where T might be ∞.

First, we prove that the solutions $u^f(t)$ are uniformly bounded with respect to $f \in \Xi$.

Lemma 7.35 *Suppose that the assumptions $C_{1,2}$ and B_2 hold. If there exist constants $\xi_i > 0$, $i = 1, \ldots, n$, and $\delta > 0$ such that $d_i(t) \geq \delta$ and*

$$
\xi_i a_{ii}(t) + \sum_{j=1, j \neq i}^{n} \xi_j |a_{ji}(t)| + \sum_{j=1}^{n} \xi_j \int_0^\infty e^{\delta s} |d\bar{K}_{ji}(s)| < 0 \qquad (7.61)
$$

for all $t \geq 0$ and $i = 1, \ldots, n$, then the solutions $u^f(t)$ are uniformly bounded with respect to $f \in \Xi$. That is, there exists $M = M(\phi, \lambda) > 0$, which is independent of $f \in \Xi$, such that $\|u^f(t)\|_{\{\xi, 1\}} \leq M$ for all $f \in \Xi$ and $t \geq 0$. Consequently, the existence interval of $u^f(t)$ can be extended to $[0, \infty)$.

Proof Let

$$
V^f(t) = \sum_{i=1}^{n} \xi_i \left| u_i^f(t) \right| e^{\delta t} + \sum_{i,j=1}^{n} \xi_i \int_0^\infty \int_{t-s}^t \left| \sigma_j^f(\theta) \right| e^{\delta(s+\theta)} d\theta |d\bar{K}_{ij}(s)|.
$$

Differentiating yields

$$
\frac{d}{dt} V^f(t) = \sum_{i=1}^{n} \delta e^{\delta t} \xi_i \left| u_i^f(t) \right| + \sum_{i=1}^{n} \xi_i e^{\delta t} \text{sign}\left(u_i^f(t) \right) \left\{ -d_i(t)u_i^f(t) \right.
$$

$$
+ a_{ii}(t) f_i \left(u_i^f(t) \right) + \sum_{j=1, j \neq i}^{n} a_{ij}(t) f_j \left(u_j^f(t) \right) + \left. \sum_{j=1}^{n} \int_0^\infty \sigma_j^f(t-s) d_s K_{ij}(t,s) \right\}
$$

$$
+ \sum_{i=1}^{n} \xi_i e^{\delta t} \text{sign}\left(u_i^f(t) \right) I_i(t) + \sum_{i,j=1}^{n} \xi_i \left| f_j \left(u_j^f(t) \right) \right| e^{\delta t} \int_0^\infty e^{\delta s} |d\bar{K}_{ij}(s)|
$$

$$-\sum_{i,j=1}^{n} \xi_j e^{\delta t} \int_0^\infty \left| \sigma_j^f(t-s) \right| |\bar{K}_{ij}(s)|$$

$$\leq \sum_{i=1}^{n} \xi_i \left| u_i^f(t) \right| e^{\delta t} (-d_i(t) + \delta) + \sum_{i=1}^{n} e^{\delta t} \left| f_i \left(u_i^f(t) \right) \right| \left\{ a_{ii}(t)\xi_i \right.$$

$$+ \sum_{j=1, j\neq i}^{n} |a_{ji}(t)|\xi_j + \sum_{j=1}^{n} \xi_j \int_0^\infty e^{\delta s} |d\bar{K}_{ji}(s)| \left. \right\} + e^{\delta t}\hat{I} \leq e^{\delta t}\hat{I},$$

where $\hat{I} = \sup\limits_{t\geq 0} \|I(t)\|_{\{\xi,1\}} < +\infty$. It follows that

$$\|u^f(t)\|_{\{\xi,1\}} \leq e^{-\delta t} V^f(t) = e^{-\delta t} \left[\int_0^t \dot{V}^f(s)ds + V^f(0) \right]$$

$$\leq e^{-\delta t} \int_0^t e^{\delta s}\hat{I}ds + e^{-\delta t}V^f(0)$$

$$\leq \frac{\hat{I}}{\delta}(1 - e^{-\delta t}) + e^{-\delta t}V^f(0) < \frac{\hat{I}}{\delta} + V^f(0).$$

Noting that $V^f(0)$ is independent of $f \in \Xi$, we obtain the uniform boundedness of the solutions $u^f(t)$ by letting $M = \hat{I}/\delta + V^f(0)$. Moreover, $f(\cdot)$ is locally Lipschitz continuous, and we conclude that the existence interval of the solution $u^f(t)$ can be extended to the infinite interval $[0, +\infty)$ according to the results given in [49]. This lemma is proved. $\qquad\square$

Now, for any sequence $\{g^m(x) = (g_1^m(x_1), \ldots, g_n^m(x_n))^\top\}_{m\in\mathbb{N}} \in \Xi$ satisfying

$$\lim_{m\to\infty} d_H(\text{Graph}(g^m(K)), \overline{co}[g(K)]) = 0, \quad \text{for all } K \subset \mathbb{R}^n, \tag{7.62}$$

where $d_H(\cdot,\cdot)$ denotes the Hausdorff metric on $\mathbb{R}^n$; we construct a sequence of delayed systems with high-slope continuous activations as follows:

$$\frac{du_i^m(t)}{dt} = -d_i(t)u_i^m(t) + \sum_{j=1}^{n} a_{ij}(t)\sigma_j^m(t)$$

$$+ \sum_{j=1}^{n} \int_0^\infty \sigma_j^m(t-s)d_s K_{ij}(t,s) + I_i(t), \ i = 1,\ldots,n, \tag{7.63}$$

where $u_i^m(\theta) = \phi_i(\theta), \theta \in (-\infty, 0]$, and

$$\sigma_j^m(\theta) = \begin{cases} \lambda_j(\theta), & \theta \leq 0 \\ g_j^m(u_j(\theta)), & \theta > 0 \end{cases}.$$

For instance, let $\{\rho_{k,i}\}$ be the set of discontinuous points of $g_i(\cdot)$. Pick a strictly decreasing sequence $\{\delta_{k,i,m}\}$ with $\lim_{m\to\infty}\delta_{k,i,m} = 0$ and define $I_{k,i,m} = [\rho_{k,i} - \delta_{k.i.m}, \rho_{k,i} + \delta_{k,i,m}]$ such that $I_{k_1,i,m} \cap I_{k_2,i,m} = \emptyset$ for every $k_1 \neq k_2$. Then, define functions $g_i^m(\cdot)$ as follows:

$$
g_i^m(s) = \begin{cases} g_i(s) & s \notin \bigcup_{k\in\mathbb{Z}} I_{k,i,m}, \\ \dfrac{g_i(\rho_{k,i} + \delta_{k,i,m}) - g_i(\rho_{k,i} - \delta_{k,i,m})}{2\delta_{k,i,m}}[s - \rho_{k,i} - \delta_{k,i,m}] \\ +g_i(\rho_{k,i} + \delta_{k,i,m}) & s \in I_{k,i,m}. \end{cases}
$$

It can be seen that the sequence $\{g^m(\cdot)\}_{m\in\mathbb{N}} \subset \Xi$ satisfies condition (7.62).

We point out that the solution sequence of the system sequence (7.63) converges to a solution of the system (7.52) in the sense (7.59).

Lemma 7.36 *Suppose the assumptions $\mathcal{C}_{1,2}$ and $\mathcal{B}_2$ are satisfied. If the condition (7.61) holds, then for each initial value pair (ϕ, λ), the system (7.52) has a solution in the sense of (7.59) on the whole time interval $[0, \infty)$.*

Proof Lemma 7.35 states that all solutions $\{u^m(t)\}_{m\in\mathbb{N}}$ are uniformly bounded, which implies that $\{\dot{u}^m(t)\}_{m\in\mathbb{N}}$ is uniformly essentially bounded. By the Arzela–Ascoli lemma and the diagonal selection principle, we can select a subsequence of $\{u^m(t)\}_{m\in\mathbb{N}}$ (still denoted by $u^m(t)$) such that $u^m(t)$ converges uniformly to a continuous function $u(t)$ on any compact interval of $\mathbb{R}$. Since $\{\dot{u}^m(t)\}_{m\in\mathbb{N}}$ is uniformly essentially bounded, $u(t)$ is Lipschitz continuous on $[0, T]$ for any $T > 0$. This implies that $\dot{u}(t)$ exists for almost all $t \in [0, T]$ and is bounded almost everywhere in $[0, T]$.

We claim that $\{\dot{u}^m(t)\}_{m\in\mathbb{N}}$ weakly converges to $\dot{u}(t)$ on the space $L^\infty([0, T], \mathbb{R}^n)$.

In fact, since $C_0^\infty([0, T], \mathbb{R}^n$ is dense in the Banach space $L^1([0, T], \mathbb{R}^n)$ and is the conjugate space $L^\infty([0, T], \mathbb{R}^n)$, for each $p(t) \in C_0^\infty([0, T], \mathbb{R}^n)$, we have

$$
\int_0^T \langle \dot{u}^m(t) - \dot{u}(t), p(t)\rangle dt = - \int_0^T \langle \dot{p}(t), u^m(t) - u(t)\rangle dt.
$$

By the uniform essential boundedness of $\{\dot{u}^m(t)\}_{m\in\mathbb{N}}$ and the Lebesgue dominated convergence theorem, we conclude that $\{\dot{u}^m(t)\}_{m\in\mathbb{N}}$ weakly converges to $\dot{u}(t)$ on the space $L^\infty([0, T], \mathbb{R}^n)$.

By Mazur's convexity theorem (see p. 120–123 in [83]), for any m, we can find a finite number of constants $\alpha_l^m \geq 0$ satisfying $\sum_{l=m}^\infty \alpha_l^m = 1$, such that $\lim_{m\to\infty} y^m(t) = u(t)$, uniformly on $[0, T]$, $\lim_{m\to\infty} \dot{y}^m(t) = \dot{u}(t)$, a.e. $t \in [0, T]$, where $y^m(t) = \sum_{l=m}^\infty \alpha_l^m u^l(t)$. Let $\eta_j^m(t) = \sum_{l=m}^\infty \alpha_l^m \sigma_j^l(u_j(t))$. Then,

$$
\dot{y}_i^m(t) = -d_i(t)y_i^m(t) + \sum_{j=1}^n a_{ij}(t)\eta_j^m(t) + \sum_{j=1}^n \int_0^\infty \eta_j^m(t-s)d_s K_{ij}(t,s) + I_i(t)
$$

for $i = 1, \ldots, n$.

Let $\varphi^m(t) = \int_0^t \eta^m(s)\,ds$, which is absolutely continuous and has uniformly essentially bounded derivative. By the same arguments, we can find $\gamma^m(t) = \sum_{l=m}^{\infty} \beta_l^m \eta^l(t)$ such that $\lim_{m\to\infty} \gamma^m(t) = \gamma(t)$ for almost every $t \in (-\infty, T]$ and $\gamma(t)$ is measurable.

Now, denoting $z^m(t) = \sum_{l=m}^{\infty} \beta_l^m y^m(t)$, we have

$$\dot{z}_i^m(t) = -d_i(t)z_i^m(t) + \sum_{j=1}^{n} a_{ij}(t)\gamma_j^m(t)$$

$$+ \sum_{j=1}^{n} \int_0^{\infty} \gamma_j^m(t-s)d_sK_{ij}(t,s) + I_i(t), \quad i = 1, \ldots, n. \qquad (7.64)$$

Letting $m \to \infty$, by the Lebesgue dominated convergence theorem, we obtain

$$\dot{u}_i(t) = -d_i(t)u_i(t) + \sum_{j=1}^{n} a_{ij}(t)\gamma_j(t)$$

$$+ \sum_{j=1}^{n} \int_0^{\infty} \gamma_j(t-s)d_sK_{ij}(t,s) + I_i(t), \quad i = 1, \ldots, n,$$

for a.e. $t \in [0, T]$. It remains to prove $\gamma(t) \in \overline{co}[g(u(t))]$ on $t \in [0, T]$. Since $u^m(t)$ converges to $u(t)$ uniformly with respect to $t \in [0, T]$ and $\overline{co}[g(\cdot)]$ is an upper-semi-continuous set-valued map, for any $\epsilon > 0$, there exists $N > 0$ such that $g^m(u^m(t)) \in \mathcal{O}(\overline{co}[g(u(t))], \epsilon)$ for all $m > N$ and $t \in [0, T]$. Noting that $\overline{co}[g(u(t))]$ is convex and compact, we conclude that $\gamma^m(t) \in \mathcal{O}(\overline{co}[g(u(t))], \epsilon)$, which implies $\gamma(t) \in \mathcal{O}(\overline{co}[g(u(t))], \epsilon)$ for any $t \in [0, T]$. Because of the arbitrariness of ϵ, we conclude that $\gamma(t) \in \overline{co}[g(u(t))]$, $t \in [0, T]$. Since T is also arbitrary, the solution can be extended to $[0, \infty)$. This completes the proof. $\qquad\square$

Similar arguments yield existence of solutions for the system (7.53). The Filippov solution of the system (7.53) with discontinuous activation functions can be described as

$$\frac{dx}{dt}(t) = -Dx(t) + A\alpha(t) + B\alpha(t-\tau) + I, \quad \text{for almost all } t, \qquad (7.65)$$

where the output $\alpha(t)$ is measurable and satisfies $\alpha(t) \in \overline{co}[g(x(t))]$ for almost all t.

Lemma 7.37 *Suppose the assumptions $\mathcal{C}_{1,2}$ satisfied. If there exist $P = \text{diag}\{P_1, P_2, \ldots, P_n\}$ with $P_i > 0$, and a positive definite symmetric matrix Q such that*

$$Z = \begin{bmatrix} -PA - A^{\top}P - Q & -PB \\ -B^{\top}P & Q \end{bmatrix} > 0, \qquad (7.66)$$

then the system (7.53) has a solution $x(t) = (x_1(t) \ldots, x_n(t))^\top$ *for* $t \in [0, \infty)$.

The details of the proof can be found in [64].

7.4.2 Stability of Equilibrium

In this section, we study the global stability of the system (7.53) in the sense (7.65). The main results come from [64]. Here, the equilibrium of such system is defined as follows:

Definition 7.38 (Equilibrium) x^* is said to be an equilibrium of the system (7.53) if there exists $\alpha^* \in \overline{co}[g(x^*)]$ such that

$$0 = -Dx^* + A\alpha^* + B\alpha^* + I.$$

Definition 7.39 An equilibrium x^* of the system (7.53) is said to be globally asymptotically stable if for any solution $x(t)$ of (7.65), whose existence interval is $[0, +\infty)$, we have

$$\lim_{t \to \infty} x(t) = x^*.$$

Moreover, $x(t)$ is said to be globally exponentially asymptotically stable, if there exist constants $\epsilon > 0$ and $M > 0$, such that

$$\|x(t) - x^*\| \le Me^{-\epsilon t}.$$

We first investigate the existence of an equilibrium point. For this purpose, consider the differential inclusion

$$\frac{dy}{dt} \in -Dy(t) + T\overline{co}[g(y(t))] + I, \tag{7.67}$$

where $y(t) = (y_1(t), y_2(t), \ldots, y_n(t))^\top$, D, $\overline{co}[g(\,\cdot\,)]$, and I are the same as those in the system (7.53). We have the following result.

Lemma 7.40 (Theorem 2 in [64]) *Suppose that* $g(\,\cdot\,)$ *satisfies the assumption* C_1. *If there exists a positive definite diagonal matrix* P *such that* $-PT - T^\top P$ *is positive definite, then there exists an equilibrium point of system (7.67), i.e., there exist* $y^* \in \mathbb{R}^n$ *and* $\alpha^* \in \overline{co}[g(y^*)]$, *such that*

$$0 = -Dy^* + T\alpha^* + I.$$

See Appendix B for the proof.

By Lemma 7.40, we can prove the following theorem.

Theorem 7.41 *If there exist a positive definite diagonal matrix* $P = \text{diag}\{P_1,$
$P_2, \ldots, P_n\}$ *and a positive definite symmetric matrix* Q *such that*

$$\begin{bmatrix} -PA - A^\top P - Q & -PB \\ -B^\top P & Q \end{bmatrix} > 0, \tag{7.68}$$

then there exists an equilibrium point of system (7.65).

Proof By the Schur Complement Theorem (Lemma 7.5), inequality (7.68) is equiv-
alent to $-(PA + A^\top P) > PBQ^{-1}B^\top P + Q$. By the inequality $[Q^{-\frac{1}{2}}B^\top P - Q^{\frac{1}{2}}]^\top [Q^{-\frac{1}{2}}$
$B^\top P - Q^{\frac{1}{2}}] \geq 0$, one has $PBQ^{-1}B^\top P + Q \geq PB + B^\top P$. Then, the inequality (7.68)
becomes $-P(A + B) - (A + B)^\top P > 0$. By Lemma 7.40, there exist an equilibrium
point $x^* \in \mathbb{R}^n$ and $\alpha^* \in \overline{co}[g(x^*)]$ such that

$$0 = -Dx^* + (A + B)\alpha^* + I, \tag{7.69}$$

which implies that α^* is an equilibrium point of system (7.65).   □

Suppose that $x^* = (x_1^*, x_2^*, \ldots, x_n^*)^\top$ is an equilibrium point of the system (7.65),
i.e., there exists $\alpha^* = (\alpha_1^*, \alpha_2^*, \ldots, \alpha_n^*)^\top \in \overline{co}[g(x)]$ such that (7.69) is satisfied. Let
$u(t) = x(t) - x^*$ be a translation of $x(t)$ and $\gamma(t) = \alpha(t) - \alpha^*$ be a translation of $\alpha(t)$.
Then $u(t) = (u_1(t), u_2(t), \ldots, u_n(t))^\top$ satisfies

$$\frac{du(t)}{dt} = -Du(t) + A\gamma(t) + B\gamma(t - \tau), \quad \text{a.e. } t \in \mathbb{R},$$

where $\gamma(t) \in \overline{co}[g^*(u(t))]$, $g_i^*(s) = g_i(s + x_i^*) - \gamma_i^*$, $i = 1, 2, \ldots, n$. To simplify,
we still use $g_i(s)$ to denote $g_i^*(s)$. Therefore, in the following, instead of the system
(7.65), we will investigate

$$\frac{du(t)}{dt} = -Du(t) + A\gamma(t) + B\gamma(t - \tau), \quad \text{a.e. } t \in \mathbb{R}, \tag{7.70}$$

where $\gamma(t) \in \overline{co}[g(u(t))]$, $g(\cdot) \in \bar{G}$, and $0 \in \overline{co}[g_i(0)]$, for all $i = 1, 2, \ldots, n$. It
can be seen that the dynamical behavior of (7.65) is equivalent to that of (7.70).
Namely, if there exists a solution $u(t)$ for (7.70), then $x(t) = u(t) + x^*$ must be a
solution for (7.65); moreover, if all trajectories of (7.70) converge to the origin, then
the equilibrium x^* must be globally stable for system (7.65) as defined in Definition
7.39.

Theorem 7.42 (Global Exponential Asymptotic Stability) *If the matrix inequality*
(7.68) and the assumptions $\mathcal{C}_{1,2}$ *hold, then the system (7.53) is globally exponentially*
stable.

Proof From the condition (7.68), we can find a sufficiently small $\epsilon > 0$ such that the matrix

$$Z_1 = \begin{bmatrix} -2D + \epsilon I & \epsilon A & \epsilon B \\ \epsilon A^\top & PA + A^\top P + Q e^{\epsilon \tau} & PB \\ \epsilon B^\top & B^\top P & -Q \end{bmatrix}$$

is negative definite. Let

$$V_3(t) = e^{\epsilon t} u^\top(t) u(t) + 2 \sum_{i=1}^n e^{\epsilon t} P_i \int_0^{u_i(t)} g_i(\rho)\, d\rho + \int_{t-\tau}^t \gamma(s)^\top Q \gamma(s) e^{\epsilon(s+\tau)}\, ds,$$

with $\gamma(t) = \alpha(t) - \alpha^*$. Notice that for $p_i(s) = \int_0^s g_i(\rho)\, d\rho$, we have $\partial_c p_i(s) = \{v \in \mathbb{R} : g_i^-(s) \le v \le g_i^+(s)\}$. Differentiating $V_3(t)$ by the chain rule (Lemma 7.32) gives

$$\frac{dV_3(t)}{dt} = \varepsilon e^{\varepsilon t} u(t)^\top u(t) + 2 e^{\varepsilon t} u^\top \left[-Du + A\gamma(t) + B\gamma(t-\tau) \right]$$

$$+ 2 e^{\varepsilon t} \gamma(t) P \left[-Du(t) + A\gamma(t) + B\gamma(t-\tau) \right]$$

$$+ \varepsilon e^{\varepsilon t} \sum_{i=1}^n P_i \int_0^{u_i} g_i(\rho)\, d\rho - e^{\varepsilon t} \gamma^\top(t-\tau) Q \gamma(t-\tau)$$

$$+ e^{\varepsilon(t+\tau)} \gamma^\top(t) Q \gamma(t). \tag{7.71}$$

Since $\varepsilon < \min_i d_i$, we have $\varepsilon \int_0^{u_i} g_i(\rho)\, d\rho \le \varepsilon u_i(t) \gamma_i(t) \le d_i u_i(t) \gamma_i(t)$ and

$$\frac{dV_3(t)}{dt} \le e^{\varepsilon t} [u^\top(t), \gamma^\top(t), \gamma^\top(t-\tau)] Z_1 \begin{bmatrix} u(t) \\ \gamma(t) \\ \gamma(t-\tau) \end{bmatrix} \le 0.$$

Then, $u(t)^\top u(t) \le V_3(0) e^{-\varepsilon t}$ and $\|u(t)\|_2 \le \sqrt{V_3(0)} e^{-\frac{\varepsilon}{2} t}$. That is, $\|x(t) - x^*\|_2 \le \sqrt{V_3(0)} e^{-\frac{\varepsilon}{2} t}$. This proves the theorem. $\qquad\square$

In case $g(\,\cdot\,)$ is continuous, we have the following consequence.

Corollary 7.43 *If the condition (7.68) holds and $g_i(\,\cdot\,)$ is locally Lipschitz continuous, then there exist $\varepsilon > 0$ and $x^* \in \mathbb{R}^n$ such that for any solution $x(t)$ on $[0, \infty)$ of the system (7.53), there exist $M = M(\phi) > 0$ and $\epsilon > 0$ such that*

$$\|x(t) - x^*\| \le M e^{-\frac{\varepsilon}{2} t} \quad \text{for all } t > 0.$$

If every x_i^* is a continuous point of the activation functions $g_i(\,\cdot\,)$, $i = 1, \ldots, n$, for the outputs we have $\lim_{t \to \infty} g_i(x_i(t)) = g_i(x_i^*)$. Instead, if for some i, x_i^* is a discontinuous point of the activation function $g_i(\cdot)$, we can prove that the outputs converge in measure.

Theorem 7.44 (Convergence in measure of output) *If the condition (7.68) holds and* $g(\cdot) \in \bar{G}$, *then the output* $\alpha(t)$ *of the system (7.65) converges to* α^* *in measure, i.e., for all* $\epsilon > 0$ *we have* $\lim_{t\to\infty} \mu\{t: |\alpha(t) - \alpha^*| \geq \epsilon\} = 0$

Proof The condition (7.68) implies that there exists $\epsilon > 0$ such that the matrix

$$Z_2 = \begin{bmatrix} -2D\ \epsilon A & & \epsilon B \\ \epsilon A^\top\ PA + A^\top P + \epsilon I\ PB \\ \epsilon B^\top\ B^\top P & & -Q \end{bmatrix} \tag{7.72}$$

is negative definite. Let

$$V_5(t) = u^\top(t)u(t) + 2\sum_{i=1}^n P_i \int_0^{u_i} g_i(\rho)\,d\rho + \int_{t-\tau}^t \gamma(s)^\top Q\gamma(s)\,ds,$$

with $\gamma(t) = \alpha(t) - \alpha^*$, and P, Q, and ϵ are those in the matrix inequality (7.72). Differentiate $V_5(t)$:

$$\frac{dV_5(t)}{dt} = 2u^\top(t)\Big[-Du(t) + A\gamma(t) + B\gamma(t-\tau)\Big] + 2\gamma^\top(t)P\Big[-Du(t) + A\gamma(t)$$

$$+ B\gamma(t-\tau)\Big] + \gamma^\top(t)Q\gamma(t) - \gamma^\top(t-\tau)Q\gamma(t-\tau) + \epsilon\gamma(t)^\top\gamma(t)$$

$$- \epsilon\gamma(t)^\top\gamma(t)$$

$$= [u^\top(t), \gamma^\top(t), \gamma^\top(t-\tau)]Z_2 \begin{bmatrix} u(t) \\ \gamma(t) \\ \gamma(t-\tau) \end{bmatrix} - \epsilon\gamma^\top(t)\gamma(t)$$

$$\leq -\epsilon\gamma^\top(t)\gamma(t) \tag{7.73}$$

Then, $V_5(t) - V_5(0) \leq -\epsilon\int_0^t \gamma^\top(s)\gamma(s)\,ds$. Since $\lim_{t\to\infty} V_5(t) = 0$, we have $\int_0^\infty \gamma^\top(s)\gamma(s)\,ds \leq -(1/\epsilon)V_5(0)$. For any $\epsilon_1 > 0$, let $E_{\epsilon_1} = \{t \in [0,\infty): \|\gamma(t)\| > \epsilon_1\}$. Then,

$$\frac{V_5(0)}{\epsilon} \geq \int_0^\infty \gamma^\top(s)\gamma(s)\,ds \geq \int_{E_{\epsilon_1}} \gamma^\top(s)\gamma(s) \geq \epsilon_1^2\mu(E_\epsilon)$$

Hence, $\mu(E_{\epsilon_1}) < \infty$. From Proposition 2 in [40], one can see that $\gamma(t)$, i.e., $\alpha(t) - \alpha^*$, converges to zero in measure. □

7.4.3 Convergence of Periodic and Almost Periodic Orbits

Consider the system (7.52)

$$\frac{dx_i(t)}{dt} = -d_i(t)x_i(t) + \sum_{j=1}^{n} a_{ij}(t)g_j(x_j(t))$$

$$+ \sum_{j=1}^{n} \int_0^{\infty} g_j(x_j(t-s))d_s K_{ij}(t,s) + I_i(t), \ i = 1, \ldots, n, \quad (7.74)$$

with the almost periodic assumption $\mathcal{B}_2$. We study the almost periodicity of delayed neural networks. The main result stated below comes from [68].

Theorem 7.45 *Suppose the assumptions $\mathcal{C}_{1,2}$ and $\mathcal{B}_2$ are satisfied. Suppose further that there exist constants $\xi_i > 0$, $i = 1, \ldots, n$, and $\delta > 0$ such that $d_i(t) \geq \delta$ and*

$$\xi_i a_{ii}(t) + \sum_{j=1, j \neq i}^{n} \xi_j |a_{ji}(t)| + \sum_{j=1}^{n} \xi_j \int_0^{\infty} e^{\delta s} |d\bar{K}_{ji}(s)| < 0 \quad (7.75)$$

for all $t \geq 0$ and $i = 1, \ldots, n$. Then, (1) for every initial value (ϕ, λ), the system (7.74) has a unique solution in the sense of (7.59); (2) there exists a unique almost periodic solution $x^*(t)$ for the system (7.74), which is globally exponentially stable, that is, for any other solution $x(t)$ with initial condition (ϕ, λ), there exists a constant $M = M(\phi, \lambda) > 0$ such that

$$\|x(t) - x^*(t)\|_{\{\xi, 1\}} \leq M e^{-\delta t}$$

for all $t \geq 0$.

Besides the viability proved in Lemma 7.36, we prove this theorem step by step.
Step 1. We show that any solution of the system (7.74) in the sense (7.59) is asymptotically stable.

Lemma 7.46 *Suppose that the assumptions of Theorem 7.45 are satisfied. For any two solutions $x(t) = x(t, \phi, \lambda)$ and $v(t) = v(t, \psi, \chi)$ of the system (7.74) in the sense of (7.59) associated with the outputs $\gamma(t)$ and $\mu(t)$ and initial value pairs (ϕ, λ) and (ψ, χ), respectively, there exists a constant $M = M(\phi, \psi, \lambda, \chi)$ satisfying $M(\phi, \phi, \lambda, \lambda) = 0$ for all (ϕ, λ) such that*

$$\|x(t) \quad v(t)\|_{\{\xi, 1\}} \leq M e^{-\delta t}, \quad t \geq 0.$$

Moreover, the solution of the system (7.74) in the sense (7.59) is unique.

Proof Let $x(t) = (x_1(t), \ldots, x_n(t))^{\top}$ be a solution of

$$\frac{d}{dt}x_i(t) = -d_i(t)x_i(t) + \sum_{j=1}^{n} a_{ij}(t)\gamma_j(t) + \sum_{j=1}^{n} \int_0^\infty \gamma_j(t-s)d_s K_{ij}(t,s) + I_i(t),$$

and $v(t) = (v_1(t), \ldots, v_n(t))^\top$ be a solution of

$$\frac{d}{dt}v_i(t) = -d_i(t)v_i(t) + \sum_{j=1}^{n} a_{ij}(t)\mu_j(t) + \sum_{j=1}^{n} \int_0^\infty \mu_j(t-s)d_s K_{ij}(t,s) + I_i(t).$$

Then,

$$\frac{d}{dt}\Big[x_i(t) - v_i(t)\Big] = -d_i(t)\Big[x_i(t) - v_i(t)\Big] + \sum_{j=1}^{n} a_{ij}(t)\Big[\gamma_j(t) - \mu_j(t)\Big]$$

$$+ \sum_{j=1}^{n} \int_0^\infty \Big[\gamma_j(t-s) - \mu_j(t-s)\Big]d_s K_{ij}(t,s), \quad i = 1, \ldots, n.$$

Let

$$L_1(t) = \sum_{i=1}^{n} \xi_i |x_i(t) - v_i(t)| e^{\delta t}$$

$$+ \sum_{i,j=1}^{n} \xi_j \int_0^\infty \int_{t-s}^{t} |\gamma_j(\theta) - \mu_j(\theta)| e^{\delta(s+\theta)} d\theta \, |d\bar{K}_{ij}(s)|$$

and $M = M(\phi, \psi, \lambda, \chi) = L_1(0)$. By the chain rule (Lemma 7.32), differentiating the above expression gives

$$\frac{d}{dt}L_1(t) = \sum_{i=1}^{n} \delta e^{\delta t}\xi_i |x_i(t) - v_i(t)| + \sum_{i=1}^{n} \xi_i e^{\delta t}\text{sign}(x_i(t) - v_i(t))$$

$$\Big\{ -d_i(t)[x_i(t) - v_i(t)] + a_{ii}(t)[\gamma_i(t) - \mu_i(t)]$$

$$+ \sum_{j=1,j\neq i}^{n} a_{ij}(t)[\gamma_j(t) - \mu_j(t)]$$

$$+ \sum_{j=1}^{n} \int_0^\infty [\gamma_j(t-s) - \mu_j(t-s)]d_s K_{ij}(t,s) \Big\} + \sum_{i,j=1}^{n} \xi_i |\gamma_j(t) - \mu_j(t)|$$

$$e^{\delta t}\int_0^\infty e^{\delta s}|d\bar{K}_{ij}(s)| - \sum_{i,j=1}^{n} \xi_j e^{\delta t}\int_0^\infty |\gamma_j(t-s) - \mu_j(t-s)||\bar{K}_{ij}(s)|$$

$$\leq \sum_{i=1}^{n} \xi_i |x_j(t) - v_j(t)| e^{\delta t} (- d_i(t) + \delta) + \sum_{i=1}^{n} e^{\delta t} |\gamma_j(t) - \mu_j(t)| \left\{ a_{ii}(t) \xi_i \right.$$

$$\left. + \sum_{j=1,j\neq i}^{n} |a_{ji}(t)| \xi_j + \sum_{j=1}^{n} \xi_j \int_0^\infty e^{\delta s} |d\bar{K}_{ji}(s)| \right\} \leq 0,$$

which implies $\|x(t) - v(t)\|_{\{\xi,1\}} \leq L_1(0) e^{-\delta t} = M(\phi, \psi, \lambda, \chi) e^{-\delta t}$. It is clear that $M(\phi, \phi, \lambda, \lambda) = 0$. Therefore, the solution in unique. $\qquad\square$

In Lemma 7.36, we have proved that some subsequence of $u^m(t)$ converges to the solution $u(t)$. In fact, we can prove that $u^m(t)$ itself converges to the solution $u(t)$.

Proposition 7.47 *Suppose that the assumptions of the Main Theorem 7.45 are satisfied. For any function sequence $\{\tilde{g}^m(x) = (\tilde{g}_1^m(x_1), \dots, \tilde{g}_n^m(x_n))^\top : m = 1, 2, \dots\} \subset \Xi$ satisfying the condition (7.62) on any compact set in R^n, let $\tilde{u}^m(t) = [\tilde{u}_1^m(t), \dots, \tilde{u}_n^m(t)]^\top$ be the solution of the following system:*

$$\frac{d\tilde{u}_i^m}{dt} = -d_i(t)\tilde{u}_i^m(t) + \sum_{j=1}^{n} a_{ij}(t)\tilde{g}_j(\tilde{u}_j^m(t))$$

$$+ \sum_{j=1}^{n} \int_0^\infty \tilde{\sigma}_j^m(t - s) d_s K_{ij}(t, s) + I_i(t),$$

$$\tilde{u}_i^m(\theta) = \phi_i(\theta), \ \theta \in [-\infty, 0], \ \tilde{\sigma}_i^m(\theta) = \begin{cases} \lambda_i(\theta), & \theta \leq 0 \\ \tilde{g}_i^m(\tilde{u}_i^m(\theta)), & \theta \geq 0 \end{cases}, \quad (7.76)$$

for $i = 1, \dots, n$, and $u(t) = u(t, \phi, \lambda)$ be the solution of the delayed system (7.74) in the sense (7.59) with initial value (ϕ, λ). Then, $\tilde{u}^m(t)$ uniformly converges to $u(t)$ on any finite time interval $[0, T]$.

Proof First, we prove that $u^m(t)$ converges to the solution of the delayed system (7.74) in the sense (7.59) by reduction to absurdity. Assume that there exist $T > 0$, $\epsilon_0 \geq 0$, and a subsequence of integers $\{m_k\}_{k\in\mathbb{N}}$ such that

$$\max_{t\in[0,T]} \|u^{m_k}(t) - u(t)\| \geq \epsilon_0. \quad (7.77)$$

By the same arguments used in the proof of Lemma 7.36, we can select a subsequence $\{u^{m_{kl}}\}_{l\geq 0}$ of $\{u^{m_k}\}_{k\geq 0}$, which converges to a solution $v(t) = v(t, \phi, \lambda)$ of the delayed system (7.74) in the sense (7.59) uniformly in any finite interval $[0, T]$ with the initial value (ϕ, λ). By Lemma 7.46, $u(t) = v(t)$, which leads a contradiction with inequality (7.77). This completes the proof. $\qquad\square$

Remark 7.48 Proposition 7.47 indicates that the solution $v(t) = v(t, \phi, \lambda)$ of the delayed system (7.74) in the sense (7.59) does not depend on the choice of the sequence $\{g^m(x)\}_{m\in\mathbb{N}} \subset \Xi$ satisfying the condition (7.62).

The following lemma points out that any solution is asymptotically almost periodic [84].

Lemma 7.49 *Suppose that the assumptions of Theorem 7.45 are satisfied. Let* $u(t, \phi, \lambda)$ *be a solution of the system (7.74) in the sense of (7.59). For any* $\epsilon > 0$, *there exist* $T > 0$ *and* $l = l(\epsilon)$ *such that any interval* $[\alpha, \alpha + l]$ *contains an* ω *such that*

$$\|x(t + \omega) - x(t)\|_\xi \leq \epsilon \quad \text{for all } t \geq T.$$

Proof We introduce the following auxiliary functions

$$\epsilon_i(t, \omega) = x_i(t + \omega)[d_i(t + \omega) - d_i(t)] + \sum_{j=1}^{n} \gamma_j(t + \omega)[a_{ij}(t + \omega) - a_{ij}(t)]$$

$$+ \int_0^\infty \sum_{j=1}^{n} \gamma_j(t + \omega - s) d[K_{ij}(t + \omega, s) - K_{ij}(t, s)]$$

$$+ I_i(t + \omega) - I_i(t) \tag{7.78}$$

for $i = 1, \ldots, n$. From the assumption $\mathcal{C}_2$ and the boundedness of $x(t)$ and $\gamma(t)$, one can see that for any $\epsilon > 0$, there exists $l = l(\epsilon) > 0$ such that every interval $[\alpha, \alpha + l]$ contains at least one number ω with $\sum_{i=1}^{n} \xi_i |\epsilon_i(t, \omega)| < \delta\epsilon/2$ for all $t \geq 0$. Denote $z(t) = x(t + \omega) - x(t)$. Then,

$$\frac{dz_i(t)}{dt} = -d_i(t)z_i(t) + \sum_{j=1}^{n} a_{ij}(t)[\gamma_j(t + \omega) - \gamma_j(t)]$$

$$+ \sum_{j=1}^{n} \int_0^\infty [\gamma_j(t + \omega - s) - \gamma_j(t - s)] d_s K_{ij}(t, s) + \epsilon_i(t, \omega).$$

Let

$$L_2(t) = \sum_{i=1}^{n} \xi_i |z_i(t)| e^{\delta t} + \sum_{i,j=1}^{n} \xi_i \int_0^\infty \int_{t-s}^{t} |\gamma_j(\theta + \omega) - \gamma_j(\theta)| e^{\delta(\theta + s)} d\theta |d\bar{K}_{ij}(s)|.$$

Pick a sufficiently large T such that $e^{-\delta t} L_2(0) < \epsilon/2$ for all $t \geq T$. Differentiating $L_2(t)$ gives

$$\frac{dL_2(t)}{dt} = \sum_{i=1}^{n} \xi_i \delta e^{\delta t} |z_i(t)| + \sum_{i=1}^{n} \xi_i e^{\delta t} \text{sign}(z_i(t)) \left\{ - d_i(t)z_i(t) \right.$$

$$+ a_{ii}(t)[\gamma_i(t + \omega) - \gamma_i(t)] + \sum_{j=1, j \neq i} a_{ij}(t)[\gamma_j(t + \omega) - \gamma_j(t)]$$

$$+ \sum_{j=1}^{n} \int_{0}^{\infty} [\gamma_j(t + \omega - s) - \gamma_j(t - s)] d_s K_{ij}(t, s) + \epsilon_i(t, \omega) \Big\}$$

$$+ \sum_{i,j=1}^{n} \xi_i e^{\delta t} |\gamma_j(t + \omega) - \gamma_j(t)| \int_{0}^{\infty} e^{\delta s} |d\bar{K}_{ij}(s)|$$

$$- \sum_{i,j=1}^{n} \xi_i e^{\delta t} \int_{0}^{\infty} |\gamma_j(t + \omega - s) - \gamma_j(t - s)| |d\bar{K}_{ij}(s)|$$

$$\leq \sum_{i}^{n} \xi_i e^{\delta t} |z_i(t)| (- d_i(t) + \delta) + \sum_{i=1}^{n} |\gamma_j(t + \omega) - \gamma_j(t)| e^{\delta t} \Big\{ \xi_i a_{ii}(t)$$

$$+ \sum_{j=1, j\neq i}^{n} \xi_j |a_{ji}(t)| + \sum_{j=1}^{n} \xi_j \int_{0}^{\infty} e^{\delta s} |d\bar{K}_{ji}(s)| \Big\} + \sum_{i=1}^{n} \xi_i e^{\delta t} |\epsilon_i(t, \omega)|$$

$$\leq e^{\delta t} \frac{\delta}{2} \epsilon, \quad \text{a.e. } t \geq T.$$

Therefore,

$$\sum_{i=1}^{n} \xi_i |z_i(t)| \leq e^{-\delta} L_2(t) = e^{-\delta} \left[L_2(0) + \int_{0}^{t} \dot{L}_2(s) ds \right]$$

$$\leq e^{-\delta t} L_2(0) + e^{-\delta t} \int_{0}^{t} e^{\delta s} ds \frac{\delta}{2} \epsilon < \frac{\epsilon}{2} + \frac{\epsilon}{2} = \epsilon$$

for all $t \geq T$, which completes the proof.   □

Step 2. Now, we are to prove that the system (7.74) has at least one almost periodic solution in the sense of (7.59).

Lemma 7.50 *Under the assumptions of Theorem 7.45, the system (7.74) has at least one almost periodic solution in the sense of (7.59).*

Proof Let $x(t) = x(t, \phi, \lambda)$ be a solution of system (7.59). Pick a sequence $\{t_k\}_{k\in\mathbb{N}}$ satisfying $\lim_{k\to\infty} t_k = \infty$ and $\sup_{t\geq 0} \sum_{i=1}^{n} \xi_i |\epsilon_i(t, t_k)| \leq 1/k$, where $\epsilon_i(t, t_k)$, $i = 1, \ldots, n$, are the auxiliary functions (7.78) defined in the proof of Lemma 7.49.

Let $x^k(t) = x(t + t_k)$ and $\gamma^k(t) = \gamma(t + t_k)$. It is clear that the sequence $\{x(t + t_k)\}_{k\in\mathbb{N}}$ is uniformly continuous and bounded. By the Arzela–Ascoli lemma and the diagonal selection principle, we can select a subsequence of $x(t + t_k)$ (still denoted by $x(t + t_k)$), which converges to some absolutely continuous function $x^*(t)$ uniformly on any compact interval $[0, T]$.

In the following, we will prove that $x^*(t)$ is an almost periodic solution of the system (7.74) in the sense of (7.59). First, we prove that $x^*(t)$ is a solution of the system (7.74) in the sense of (7.59). With the notations above, we have

$$\frac{dx_i(t+t_k)}{dt} = -d_i(t)x_i(t+t_k) + \sum_{j=1}^{n} a_{ij}(t)\gamma_j(t+t_k)$$

$$+ \sum_{j=1}^{n} \int_0^\infty \gamma_j(t+t_k-s)d_s K_{ij}(t,s) + I_i(t) + \epsilon_i(t,t_k), \ i = 1,\dots,n.$$

With the method used in the proof of Lemma 7.36, we can select a subsequence from $x(t+t_k)$ (still denoted by $x(t+t_k)$) and constants $v_l^k \geq 0$ with finite $v_l^k > 0$ satisfying $\sum_{l=k}^{\infty} v_l^k = 1$ such that (i) $v^k(t) = \sum_{l=k}^{\infty} v_l^k x(t+t_l)$ converges to a Lipschitz continuous function $x^*(t)$ uniformly on $[0,T]$, and $\{\dot{v}^k(t)\}$ converges to $\dot{v}^*(t)$ for almost all $t \in [0,T]$ and (ii) $\zeta^k(t) = \sum_{l=k}^{\infty} v_l^k \gamma(t+t_l)$ converges to a measurable function $\zeta(t)$ for almost all $t \in [0,T]$.

Moreover, for each k, we have

$$\frac{dv_i^k(t)}{dt} = -d_i(t)v_i^k(t) + \sum_{j=1}^{n} a_{ij}(t)\zeta_j^k(t)$$

$$+ \sum_{j=1}^{n} \int_0^\infty \zeta_j^k(t-s)d_s K_{ij}(t,s) + I_i(t) + \bar{\epsilon}_i(t,k), \ i = 1,\dots,n,$$

where $\bar{\epsilon}_i(t,k) = \sum_{l=k}^{\infty} v_l^k \epsilon_i(t,t_k)$. Letting $k \to \infty$, we obtain

$$\frac{dx_i^*(t)}{dt} = -d_i(t)x_i^*(t) + \sum_{j=1}^{n} a_{ij}(t)\zeta_j(t)$$

$$+ \sum_{j=1}^{n} \int_0^\infty \zeta_j(t-s)d_s K_{ij}(t,s) + I_i(t), \ i = 1,\dots,n.$$

Repeating the proof of Lemma 7.36, we can prove $\zeta(t) \in \overline{co}[g(x^*(t))]$, which means that $x^*(t)$ is a solution of the system (7.74) in the sense of (7.59).

Second, we prove that $x^*(t)$ is almost periodic. By Lemma 7.49, for any $\epsilon > 0$, there exist $K > 0$ and $l = l(\epsilon)$ such that each interval $[\alpha, \alpha+l]$ contains an ω such that

$$\|x(t+t_k+\omega) - x(t+t_k)\|_{\{\xi,1\}} < \epsilon$$

for all $k \geq K$ and $t \geq 0$. As $k \to \infty$, we conclude that $\|x^*(t+\omega) - x^*(t)\|_{\{\xi,1\}} < \epsilon$ for all $t \geq 0$. This implies that $x^*(t)$ is an almost periodic function. The proof is completed.   □

Now, we can prove the main Theorem 7.45.

Proof By Lemma 7.50, we know that there exists an almost periodic solution for the system (7.74) in the sense of (7.59). By Lemma 7.46, we have $\|x(t) - x^*(t)\|_{\{\xi,1\}} = O(e^{-\delta t})$.

Finally, we prove that the almost periodic solution of the system (7.74) is unique. In fact, suppose that $x^*(t)$ and $v^*(t)$ are two almost periodic solutions of the system (7.74). Applying Lemma 7.46 again, we have $\|v^*(t) - x^*(t)\|_{\{\xi,1\}} = O(e^{-\delta t})$. From [61], one can conclude that $v^*(t) = x^*(t)$. Therefore, the almost periodic solution of the system (7.74) is unique. This completes the proof. □

Since any periodic function can be regarded as an almost periodic function, all the results apply to periodic case. Now, replacing assumption $\mathcal{B}_2$ with $\mathcal{B}_1$, we have the following result.

Corollary 7.51 *Suppose that the discontinuous activations satisfy assumptions $\mathcal{C}_{1,2}$, and that the hypotheses $\mathcal{B}_1$ hold. Suppose further that there exist positive constants ξ_i, $i = 1,\ldots,n$, and $\delta > 0$ such that $d_i(t) \geq \delta$ and*

$$\xi_i a_{ii}(t) + \sum_{j=1,j\neq i}^{n} \xi_j |a_{ji}(t)| + \sum_{j=1}^{n} \xi_j \int_0^\infty e^{\delta s} |d\bar{K}_{ji}(s)| < 0$$

for all $t \geq 0$ and $i = 1,\ldots,n$. Then, (1) for each initial data with assumption $\mathcal{A}_3$, the system (7.74) has a unique solution in the sense of (7.59) and (2) there exists a unique periodic solution $x^*(t)$ for system (7.74), which is globally exponentially stable.

Furthermore, a constant can be regarded as a periodic function with any period. Therefore, for the delayed system

$$\frac{dx_i(t)}{dt} = -d_i x_i(t) + \sum_{j=1}^{n} a_{ij} g_j(x_j(t))$$

$$+ \sum_{j=1}^{n} \int_0^\infty g_j(x_j(t-s)) d_s K_{ij}(s) + I_i, \quad i = 1,\ldots,n \qquad (7.79)$$

we have the following result.

Corollary 7.52 *Suppose that the discontinuous activations satisfy the assumptions $\mathcal{C}_{1,2}$, and suppose that there exist positive constants ξ_i, $i = 1,\ldots,n$, and $\delta > 0$ such that $d_i \geq \delta$ and*

$$\xi_i a_{ii} + \sum_{j=1,j\neq i}^{n} \xi_j |a_{ji}| + \sum_{j=1}^{n} \xi_j \int_0^\infty e^{\delta s} |d\bar{K}_{ji}(s)| \leq 0$$

for all $t \geq 0$ and $i = 1,\ldots,n$. Then, (1) for each initial data satisfying the stated assumptions, the system (7.79) has a unique solution in sense of (7.59) and (2) the system (7.79) has a unique equilibrium x^*, which is globally exponentially stable.

7.5 Review and Comparison of Literature

In the past decades, global stability analysis has been a focal topic in neural network theory and dynamical systems, with a large literature devoted to it. In this section, we give a brief review of selected papers and compare them with the results in this chapter.

The stability of equilibrium of delayed neural networks has been studied in many papers. For example, [9, 10, 14, 17, 18, 62, 74, 86] and many others. For more general functional differential equations, see the early works [49, 73] and others. The approach used in these papers consists of two steps: (1) prove the existence of the equilibrium and (2) prove its stability. In theorems in Sect. 7.2.2, we unify two types of delayed dynamical systems and investigate their dynamical behavior and global convergence. We consider the derivative of the state variable and prove that it converges to zero exponentially. This implies that the state trajectory converges to a certain equilibrium exponentially according to the Cauchy convergence principle.

Moreover, in most papers dealing with time-varying delays, the assumption of bounded delays is necessary, i.e., $\tau_{ij}(t) \leq \tau$ for all $i,j = 1,\ldots,n$ and $t \in \mathbb{R}$, or $\dot{\tau}_{ij}(t) \leq \mu$ for some $0 \leq \mu < 1$, which can guarantee exponential stability under some additional conditions. However, in this chapter, we have studied stability in the power rate, which is weaker than exponential rate, but under a milder condition for the unbounded delays, namely, $\tau_{ij}(t) \leq \mu t$ for some $0 \leq \mu < 1$.

As for the delayed Cohen–Grossberg neural network (7.19), there is also a large literature concerned with global stability. However, all the results obtained in these papers were based on the assumption that amplifier function $a_i(\cdot)$ is always *positive* (see [28, 29]; or even greater than some positive number $a_i(\cdot) \geq \underline{a}_i > 0$ (see [16, 71, 82]). In their original papers [35, 46, 47], the authors proposed this model as a kind of competitive-cooperation dynamical system for decision rules, pattern formation, and parallel memory storage. Hereby, each state of neuron x_i might be the population size, activity, or concentration, etc., of the i-th species in the system, which is nonnegative for all time. Theorem 7.20 gives a sufficient condition guaranteeing stability in the first orthant.

Periodicity and almost periodicity of delayed neural networks with time-varying coefficients have attracted much research attention [15, 24, 45, 72, 87, 88]. It should be pointed out that [87] studied the periodicity of delayed neural network via a L^p-norm-like Lyapunov functional and proved that among the sufficient conditions according to parameter $p \in [1, \infty]$, the condition with L^1-norm-like Lyapunov functional would be the best one, i.e., the mildest condition. Most of these papers concerned with periodic delayed neural networks use the Mawhin coincidence degree theory [44]. We use two different methods to prove existence, as mentioned in Sect. 7.3.3. In [24, 55], the authors presented some results on almost periodic trajectories and their attractivity of shunting inhibitory cellular neural networks (CNNs) with delays. In [24], authors proved existence and attractivity of almost periodic solutions for CNNs with distributed delays and variable coefficients.

In the last few years, several papers have appeared studying neural networks with discontinuous activations. Reference [40] discussed the absolute stability of

Hopfield neural networks with bounded and discontinuous activations. Reference [64] proved the global convergence for Cohen–Grossberg neural networks with unbounded and discontinuous activations. Also, [42] studied the dynamics of delayed neural networks and [78] discussed periodic solutions of periodic delayed neural networks with discontinuous activations and periodic parameters. In all these papers, the authors use the solution in the Filippov sense to handle differential equations with discontinuous right-hand side. The concept of the solution in the sense of Filippov is useful in engineering applications. Since a Filippov solution is a limit of the solutions of a sequence of ordinary differential equations with continuous right-hand side, we can model a system which is near a discontinuous system and expect that the Filippov trajectories of the discontinuous system will be close to the real trajectories. This approach is of significance in many applications, for instance, variable structure control, nonsmooth analysis [4, 77, 85]. In fact, the solution in the Filippov sense satisfies the corresponding differential inclusion induced by the convex extension of discontinuity.

The generalized viability of differential inclusions was investigated in the textbooks [4, 6]. Periodicity and almost periodicity for differential inclusions or Filippov systems have been studied in the recent decades. Methodologically, the existence of a periodic solution of a differential inclusion or differential system with discontinuous right-hand side (despite that some researchers did not study the Filippov solution) can be proved by fixed point theory, i.e., the periodic boundary condition can be regarded as a fixed point of a certain evolution operator [12, 38, 56, 70, 72, 89]. Several authors constructed a sequence of differential systems with continuous right-hand sides having periodic solutions and proved that the solution sequence converges to a periodic solution of the original differential inclusion [43, 48]. As for stability, the first approximation was used to deal with the local asymptotical stability for periodic differential inclusions [81], and Lyapunov method was extended to handle the global stability [7, 8]. Furthermore, similar methods were utilized to study the almost periodic solution of almost periodic differential inclusions, especially with delays. See [3] and [58] for references.

Appendix

Proof of Theorem 7.18

Proof Let

$$f_i(x) = d_i(x_i) - \sum_{j=1}^{n} (a_{ij} + b_{ij})g_j(x_j), \quad i = 1, \ldots, n,$$

$$f(x) = (f_1(x), \ldots, f_n(x))^\top,$$

$$F(x) = f(x^+) + x^-,$$

where x^+ and x^- are defined in Definition 7.6.

According to Lemma 7.7, we only need to prove that $F(x)$ is norm-coercive and locally univalent (one-to-one). First, we prove $F(x)$ is locally univalent. Let $x = (x_1, \ldots, x_n) \in \mathbb{R}^n$. Without loss of generality, by some rearrangement of the x_i, we can assume $x_i > 0$ if $i = 1, \ldots, p$, $x_i < 0$ if $i = p + 1, \ldots, m$, and $x_i = 0$ if $i = m + 1, \ldots, n$, for some integers $p \leq m \leq n$. Moreover, if $y \in \mathbb{R}^n$ is sufficiently close to $x \in \mathbb{R}^n$, without loss of generality, we can assume

$$
\begin{cases}
y_i > 0, \, i = 1, \ldots, p \\
y_i < 0, \, i = p + 1, \ldots, m \\
y_i > 0, \, i = m + 1, \ldots, m_1 \\
y_i < 0, \, i = m_1, \ldots, m_2 \\
y_i = 0, \, i = m_2 + 1, \ldots, n,
\end{cases}
$$

for some integers $m \leq m_1 \leq m_2 \leq n$. It can be seen that

$$(x_i^+ - y_i^+)(x_i^- - y_i^-) = 0, \quad i = 1, \ldots, n, \tag{7.80}$$

and

$$
\begin{aligned}
F(x) - F(y) &= d(x^+) - d(y^+) - (A + B)[g(x^+) - g(y^+)] + (x^- - y^-) \\
&= [\bar{D} - (A + B)K](x^+ - y^+) + (x^- - y^-),
\end{aligned}
$$

where $\bar{D} = diag\{\bar{d}_i, \ldots, \bar{d}_n\}$ and $K = diag\{K_1, \ldots, K_n\}$ with

$$
\bar{d}_i = \begin{cases} \dfrac{d_i(x_i^+) - d_i(y_i^+)}{x_i^+ - y_i^+}, & x_i^+ \neq y_i^+ \\ D_i, & \text{otherwise} \end{cases}, \qquad
K_i = \begin{cases} \dfrac{g_i(x_i^+) - g_i(y_i^+)}{x_i^+ - y_i^+}, & x_i^+ \neq y_i^+ \\ G_i, & \text{otherwise} \end{cases}.
$$

Then, $\bar{d}_i \geq D_i$ and $K_i \leq G_i$ because $d(\cdot) \in \mathcal{D}$ and $g(\cdot) \in H_2\{G_1, \ldots, G_n\}$. If $F(x) - F(y) = 0$, then we have

$$x^- - y^- = -[\bar{D} - (A + B)K](x^+ - y^+). \tag{7.81}$$

By (7.80), without loss of generality, we can assume

$$x^+ - y^+ = \begin{bmatrix} z_1 \\ 0 \end{bmatrix}, \quad x^- - y^- = \begin{bmatrix} 0 \\ z_2 \end{bmatrix},$$

where $z_1 \in \mathbb{R}^k$ and $z_2 \in \mathbb{R}^{n-k}$, for some integer k. Write

$$\bar{D} - (A + B)K = \begin{bmatrix} R_{11} & R_{12} \\ R_{21} & R_{22} \end{bmatrix},$$

where $R_{11} \in \mathbb{R}^{k,k}$, $R^{12} \in \mathbb{R}^{k,n-k}$, $R_{21} \in \mathbb{R}^{n-k,k}$, and $R_{22} \in \mathbb{R}^{n-k,n-k}$. The equation (7.81) can be rewritten as

$$\begin{bmatrix} 0 \\ z_2 \end{bmatrix} = - \begin{bmatrix} R_{11} & R_{12} \\ R_{21} & R_{22} \end{bmatrix} \begin{bmatrix} z_1 \\ 0 \end{bmatrix},$$

which implies $R_{11}z_1 = 0$. From Lemma 7.3, we can conclude that R_{11} is nonsingular, which implies $z_1 = 0$ and $x^+ = y^+$. Similarly, we can prove $x^- = y^-$. Therefore, $x = y$, which means that $F(x)$ is locally univalent.

Second, we will prove that $F(x)$ is norm-coercive. Suppose that there exists a sequence $\{x_m = (x_{m,1}, \ldots, x_{m,n})^\top\}_{m=1}^\infty$ such that $\lim_{m \to \infty} \|x_m\|_2 = \infty$. Then, there exists some index i such that $\lim_{m \to \infty} |d_i(x_{m,i}^+) + x_{m,i}^-| = \infty$, which implies that $\lim_{m \to \infty} \|g(x_m^+)\|_2 = \infty$.

Some simple algebraic manipulations lead to

$$g(x^+)^\top PF(x) = \sum_{i=1}^n g_i(x_i^+) P_i d_i(x_i^+) - g(x^+)^\top P(A + B)g(x^+) + \sum_{i=1}^n g_i(x_i^+) P_i x_i^-$$
$$\geq g(x^+)^\top \{P[DG^{-1} - (A + B)]\}^s g(x^+) \geq \alpha g(x^+)^\top g(x^+),$$

where $\alpha = \lambda_{\min}(\{P[DG^{-1} - (A + B)]\}^s) > 0$. Therefore,

$$\|F(x_m)\|_2 \geq \alpha \|P\|_2^{-1} \|g(x_m^+)\|_2 \to \infty,$$

which implies that $F(x)$ is norm-coercive. Combining with Lemma 7.7 proves the theorem.   □

Proof of Lemma 7.40

We will prove the existence of equilibrium of the system (7.53) under the assumption $\mathcal{C}_1$. We will prove existence of equilibrium for the system (7.53) by the Equilibrium Theorem [5]. First, we give some necessary definitions concerned with the equilibrium of a set-valued map.

Definition 7.53 For a convex subset K of $\mathbb{R}^n$, the tangent cone $T_K(x)$ to K at $x \in K$ is defined as

$$T_K(x) = \overline{\bigcup_{h>0} \frac{K - x}{h}}, \tag{7.82}$$

where $\overline{\bigcup}$ is the closure of the union set.

Proposition 7.54 *The necessary and sufficient condition for $v \in T_K(x)$ is that there exist $h_n \to 0^+$ and $v_n \to v$ as $n \to +\infty$, such that $x + h_n v_n \in K$ for all n. Moreover, if $x \in \text{int}(K)$, where $\text{int}(K)$ is the set of the interior points of K, then $T_k(x) = \mathbb{R}^n$.*

Definition 7.55 (Viability Domain) Let $F:X \to X$ be a non-trivial set-valued map. We say that a subset $K \subset \mathrm{Dom}(F)$ is a viability domain of F, if for all $x \in K$, we have $F(x) \cap T_K(x) \neq \emptyset$ where $\mathrm{Dom}(F)$ is the domain of F.

Definition 7.56 (Equilibrium) x^* is said to be an equilibrium of a set-valued map $F(x)$ if $0 \in F(x^*)$.

The following theorem is used below.

Lemma 7.57 (Equilibrium Theorem) (See p. 84 in [5]) *Assume that X is a Banach space and $F:X \to X$ is an upper semicontinuous set-valued map with closed convex image. If $K \subset X$ is a convex compact viability domain of $F(x)$, then K contains an equilibrium x^* of $F(x)$, i.e., $0 \in F(x^*)$.*

Now, we use the Equilibrium Theorem to prove the existence of the equilibrium of the system (7.53).

Lemma 7.58 *Suppose C_1 satisfied, and each $g_i(\cdot)$ is non-trivial, $P_i > 0$, for $i = 1, 2, \ldots, n$. Define*

$$\bar{V}(x) = \sum_{i=1}^{n} P_i \int_0^{x_i} g_i(\rho)\, d\rho. \tag{7.83}$$

For any $M > 0$, define $\Omega_M = \{x : \bar{V}(x) \leq M\}$, $\partial \Omega_M = \{x : \bar{V}(x) = M\}$, and

$$K_1 = \left\{ v = (v_1, v_2, \ldots, v_n)^\top \in \mathbb{R}^n : \sum_{i=1}^{n} v_i P_i \gamma_i \leq 0, \text{ for all } \gamma_i \in \overline{co}[g_i(x_i)] \right\}. \tag{7.84}$$

Then $K_1 \subset T_{\Omega_M}(x)$ whenever $x \in \partial \Omega_M$.

Proof For each $x \in \partial \Omega_M$, i.e., $\bar{V}(x) = M$, and $v \in \mathrm{int}(K_1)$ satisfying $\sum_{i=1}^{n} v_i P_i \gamma_i < 0$ for all $\gamma_i \in \overline{co}[g_i(x_i)]$. Let $y_n = x + h_n v$, where $0 < h_n \to 0$, as $n \to +\infty$. We will prove that $\bar{V}(y_n) \leq M$, namely, $y_n \in \Omega_M$.
    Denote

$$\gamma_i^e = \begin{cases} g_i(x_i^+), & \text{if } v_i > 0 \\ g_i(x_i^-), & \text{if } v_i < 0 \\ \text{any value,} & \text{if } v_i = 0. \end{cases} \tag{7.85}$$

Then we have $\sum_{i=1}^{n} v_i P_i \gamma_i \leq \sum_{i=1}^{n} v_i P_i \gamma_i^e$ for all $\gamma_i \in \overline{co}[g_i(x_i)]$. Thus, let $\epsilon = -\sum_{i=1}^{n} v_i P_i \gamma_i^e$, which is positive. We have

$$\bar{V}(y_n) - \bar{V}(x) = \sum_{i=1}^{n} P_i \int_{x_i}^{y_{n_i}} g_i(\rho)\, d\rho = \sum_{i=1}^{n} P_i \int_{x_i}^{x_i+h_n v_i} g_i(\rho)\, d\rho$$

$$= \left(\sum_{i=1}^{n} v_i P_i \gamma_i^e\right) h_n + o(h_n) = -\epsilon h_n + o(h_n). \tag{7.86}$$

If n is large enough, we obtain $\bar{V}(y_n) < \bar{V}(x) = M$, which implies $v \in T_{\Omega_M}(x)$, i.e., $\text{int}(K_1) \subset T_{\Omega_M}(x)$. Since $T_{\Omega_M}(x)$ is closed, $K_1 \subset T_{\Omega_M}(x)$. $\qquad\square$

Lemma 7.59 (Ky Fan Inequality [5]) *Let K be a compact convex subset in a Banach space X and $\varphi:X \times X \to \mathbb{R}$ be a function satisfying the following conditions:*

(1) For all $y \in K$, $x \mapsto \varphi(x,y)$ is lower semicontinuous;
(2) For all $x \in K$, $y \mapsto \varphi(x,y)$ is concave, i.e., for all $\lambda_i > 0$ satisfying $\sum_i^n \lambda_i = 1$ and $y_i \in K$,

$$\varphi\left(x, \sum_{i=1}^{n} \lambda_i y_i\right) \ge \sum_{i=1}^{n} \lambda_i \varphi(x, y_i); \tag{7.87}$$

(3) For all $y \in K$, $\varphi(y,y) \le 0$.

Then, there exists $\bar{x} \in K$ such that, for all $y \in K$, $\varphi(\bar{x}, y) \le 0$.

Theorem 7.60 *Assume $\mathcal{C}_1$ and let $-T$ be a Lyapunov diagonally stable (LDS) matrix. Then there exists an equilibrium x^* of system (7.53), i.e.,*

$$0 \in F(x^*), \tag{7.88}$$

where $F(x^*) = [-d(x^*) + T\,\overline{co}\,[g(x^*)] + J]$.

Proof Because $-T$ is LDS, there exists a diagonal matrix $P = diag\{P_1, P_2, \ldots, P_n\}$, with $P_i > 0$, $i = 1, 2, \ldots, n$, such that $(PT)^s < 0$. Let

$$\bar{V}(x) = \sum_{i=1}^{n} P_i \int_0^{x_i} g_i(\rho)\, d\rho. \tag{7.89}$$

Case 1: All $g_i(\cdot)$, $i = 1, 2, \ldots, n$, are non-trivial.

It is easy to see that Ω_M is a convex compact subset of $\mathbb{R}^n$. Let $\alpha = \min \lambda$ $(\{-PT\}^s) > 0$, $I = \sum_{i=1}^{n} [1/(2\alpha)] P_i^2 J_i^2$, $l = \min_i D_i$, and $M_0 = I/l$. In the following, we will prove that if $M > M_0$, then Ω_M is a viability domain of $F(x)$.

In fact, if $x \in \text{int}(\Omega_M)$, then $T_{\Omega_M}(x) = \mathbb{R}^n$ and it is easy to see that $F(x) \cap T_{\Omega_M}(x) = \emptyset$.

Now, we will prove that if $x \in \partial \Omega_M$, then $F(x) \cap T_{\Omega_M}(x) = \emptyset$. For this purpose, we define $\varphi(g_1, g_2):\overline{co}[g(x)] \times \overline{co}[g(x)] \mapsto \mathbb{R}$, as follows:

$$\varphi(g_1, g_2) = \sum_{i=1}^{n} g_{1,i} P_i \left[-d_i(x_i) + \sum_{j=1}^{n} t_{ij} g_{2,j} + J_i \right], \tag{7.90}$$

where $g_1 = (g_{1,1}, g_{1,2}, \dots, g_{1,n})^\top$ and $g_2 = (g_{2,1}, g_{2,2}, \dots, g_{2,n})^\top$. If we can find $g_2 \in \overline{co}[g(x)]$, such that $\varphi(g_1, g_2) \leq 0$ for all $g_1 \in \overline{co}[g(x)]$, then by Lemma 7.57, we have $F(x) \bigcap T_{\Omega_M}(x) \neq \emptyset$.

It can be seen that for each $g_1 \in \overline{co}[g(x)]$, $g_2 \mapsto \varphi(g_1, g_2)$ is continuous; for each $g_2 \in \overline{co}[g(x)]$, $g_1 \mapsto \varphi(g_1, g_2)$ is concave. Moreover, let $f = (f_1, f_2, \dots, f_n)^\top$, where $f_i \in \overline{co}[g_i(x)]$. Then it is easy to see that $f_i x_i \geq \int_0^{x_i} g_i(\rho) \, d\rho$, which implies

$$\begin{aligned}
\varphi(f, f) &= -\sum_{i=1}^{n} f_i P_i \frac{d_i(x_i)}{x_i} x_i + f^\top PTf + f^\top PJ \\
&\leq -lf^\top Px - \alpha f^\top f + f^\top PJ \leq -lf^\top Px - \frac{\alpha}{2} f^\top f + I \\
&\leq -lM + I \leq 0. \tag{7.91}
\end{aligned}$$

By Lemma 7.59, we can find $\bar{g} \in \overline{co}[g(x)]$ such that $\varphi(g, \bar{g}) \leq 0$ for all $g \in \overline{co}[g(x)]$. Therefore, for each $x \in \Omega_M$, we have $F(x) \bigcap T_{\Omega_M}(x) \neq \emptyset$. According to Lemma 7.57, in this case, Ω_M contains an equilibrium of $F(x)$.

Case 2: There exist some indices i such that $g_i(s) = 0$ for all $s \in \mathbb{R}$.

Without loss of generalization, we can assume that $g_n(s) = 0$ for all $s \in \mathbb{R}$ and $g_1, \dots, g_{n-1}$ are non-trivial. Considering $\tilde{x} = (x_1, x_2, \dots, x_{n-1})^\top$, by the discussion in Case 1, there exists an equilibrium $\tilde{x}^* = (x_1^*, x_2^*, \dots, x_{n-1}^*)^\top$, such that $0 \in -d_i(x_i^*) + \sum_{j=1}^{n-1} t_{ij} \overline{co}[g_j(x_j)] + J_i$ for $i = 1, \dots, n-1$. That is, there exist $\gamma_i \in \overline{co}[g_i(x_i^*)]$, for $i = 1, 2, \dots, n-1$, such that $0 = -d_i(x_i^*) + \sum_{j=1}^{n-1} t_{ij} \gamma_j + J_i$, $i = 1, 2, \dots, n-1$.

It can also be seen that there exists x_n^* such that $-d_n(x_n^*) + \sum_{j=1}^{n-1} t_{nj} \gamma_j + J_n = 0$. Therefore, $x^* = (\tilde{x}^*, x_n^*)^\top$ is an equilibrium of $F(x)$. The theorem is proved. $\qquad \square$

References

1. D. H. Ackley, G. E. Hinton, and T. J. Sejnowski. A learning algorithm for Boltzmann machines. *Cognit. Sci.*, 9: 147–169 1985.
2. I. Aleksander and H. Morton An Introduction to Neural Computing. Lodon, Chapman & Hall 1990.
3. J. Andres. Almost-periodic and bounded solutions of Carathéodory differential inclusions. *Differential Integral Equations*, 12(6): 887–912, 1999.
4. J. P. Aubin and A. Cellina. *Differential Inclusions*. Berlin, Springer-Verlag, 1984.
5. J. P. Aubin and H. Frankowska. *Set-valued Analysis*. Boston, Birhauser, 1990.
6. J. P. Aubin. *Viability Theory*. Boston, Birhauser, 1991.
7. A. Bacciotti, et.al. Discontinuous ordinary differential equations and stabilization. Tesi di Dottorato di Ricerca in Mathematica, Universita degli STUDI di Firenze, 2000.
8. A. Bacciotti. Generalized solutions of differential inclusions and stability. *Ital. J. Pure Appl. Math.*, 17: 183–192, 2005.

9. J. Belair. Stability in A Model of Delayed Neural Networks. *J. Dynam. Differential Equations*, 5: 607–623, 1993.
10. J. Belair, S. A. Campbell, and P. Driessche, Van Den. Stability and delay-induced oscillations in a neural network model. *SIAM J. Appl. Math.*, 56: 245–255, 1996.
11. A. Berman. Completely positive matrices. New Jersey, World Scientific Publishing, 2003.
12. R. Bader and W. Kryszewski. On the solution sets of differential inclusions and the periodic problem in Banach space. *Nonlinear Analysis*, 54: 707–754, 2003.
13. S. Boyd et al. *Linear Matrix Inequalities in System and Control Theory*. Philadelphia, SIAM, 1994.
14. J. Cao. On exponential stability and periodic solution of CNNs with delay. *Phys. Lett.* A, 267(5–6): 312–318, 2000.
15. J. Cao. New results concerning exponential stability and periodic solutions of delayed cellular neural networks. *Phys. Lett. A*, 307: 136–147, 2003.
16. J. Cao and J. Liang. Boundedness and stability for Cohen-Grossberg neural network with time-varying delays. *J. Math. Anal. Appl.*, 296: 665–685, 2004.
17. Y. J. Cao and Q. H. Wu. A note on stability of analog neural networks with time delays. *IEEE Trans. Neural Networks*, 7: 1533–1535, 1996.
18. T. Chen. Convergence of delayed dynamical systems. *Neural Proc. Lett.*, 10(3): 267–271, 1999.
19. T. Chen. Global exponential stability of delayed hopfield neural networks. *Neural Networks*, 14(8): 977–980, 2001.
20. T. Chen and S. Amari. Stability of asymmetric hopfield neural networks. *IEEE T. Neural Networks*, 12(1): 159–163, 2001.
21. T. Chen and Y. Bai. Stability of Cohen-Grossberg neural networks with nonnegative periodic solutions. *Proceedings of International Joint Conference Neural Networks (IJCNN 2007)*, 242–247, 2007.
22. T. Chen and H. Chen. Universal approximation to nonlinear operators by neural networks and its applications to dynamical systems. *IEEE Trans. Neural Networks*, 6(4): 911–917, 1995.
23. T. Chen and H. Chen. Approxiamtion capability functions of several variables, nonlinear functionals and operators with radial basis function neural networks. *IEEE Trans. Neural Networks*, 6(4): 904–910, 1995.
24. A. Chen and J. Cao. Existence and attractivity of almost periodic solution for cellular neural networks with distributed delays and variable coefficients. *Appl. Math. Comp.*, 134(1): 125–140, 2003.
25. P. P. Civalleri, L. M. Gilli, and L. Pabdolfi. On stability of cellular neural networks with delay. *IEEE Trans.Circuits Syst.*, 40: 157–164, 1993.
26. T. Chen and W. Lu. Stability analysis of dynamical neural networks. *Proceedings of International Conference Neural Networks and Signal Processing*, 14–17, 2003.
27. T. Chen, W. Lu, and G. R. Chen. Dynamical behaviors of a large class of general delayed neural networks. *Neural Comput.*, 17: 949–968, 2005.
28. T. Chen and L. Rong. Delay-independent stability analysis of Cohen-Grossberg neural networks. *Phys. Lett. A*, 317: 436–449, 2003.
29. T. Chen and L. Rong. Robust global exponential stability of Cohen-Grossberg neural networks with time delays. *IEEE Trans. Neural Networks*, 15(1): 203–206, 2004.
30. T. Chen and L. Wang. Power-rate stability of dynamical systems with unbounded time-varying delays. *IEEE T. CAS-II: Express Briefs*, 54(8): 705–709, (2007).
31. L. O. Chua, C. A. Desoer, and E. S. Kuh. *Linear and Nonlinear Circuits*. New York, Macgraw-Hill, 1987.
32. L. O. Chua and L. Yang. Cellular neural networks: Theory. *IEEE Trans.Circuits Syst.*, 35: 1257–1272, 1988.
33. L. O. Chua and L. Yang. Cellular neural networks: Application. *IEEE Trans.Circuits Syst.*, 35: 1273–1290, 1988.
34. F. H. Clarke. *Optimization and Nonsmooth Analysis*. Pliladelphia, SIAM 1983.

35. M. A. Cohen and S. Grossberg. Absolute stability and global pattern formation and parallel memory storage by competitive neural networks. *IEEE Trans. on Man, Syst. Cybern.*, 13: 815–826, 1983.

36. G. Cybenko. *Approximation by Superpositions of a Sigmoidal Function.* Urbana, University of ILLinois, 1988.

37. G. Cybenko. Approximation by superpositions of a sigmoidal function. *Math. Cont., Sig., Sys.*, 2: 303–314, 1989.

38. B. C. Dhage. Existence of extremal solutions for discontinuous functional integral equations. *Appl. Math. Lett.*, 19: 881–886, 2006.

39. A. F. Filippov. Classical solution of differential equations with multivalued right-hand side. *SIAM J. Control*, 5(4): 609–621, 1967.

40. M. Forti and P. Nistri. Global convergence of neural networks with discontinuous neuron activations. *IEEE Trans. Circuits Syst.I*, 50(11): 1421–1435, 2003.

41. M. Forti and A. Tesi. New condition for global stability of neural networks with application to linear and quadratic programming problems. *IEEE Trans. Circiuts Syst.-1*, 42: 354–366, 1995.

42. M. Forti, P. Nistri, and D. Papini. Global exponential stability and global convergence of delayed neural networks with infinite gain. *IEEE Trans. Neural Networks*, 16(6): 1449–1463, 2005.

43. M. E. Filippakis and N. S. Papageorgiou. Periodic solutions for differential inclusions in $\mathbb{R}^N$. *Arch. Math. (Brno)*, 42: 115–123, 2006.

44. R. E. Gaines and J. L. Mawhin. *Coincidence Degree and Nonlinear Differential Equations. Lecture Notes on Mathematics*, Berlin, Springer, 568: 10–35, (1977).

45. K. Gopalsamy and S. Sariyasa. Time delays and stimulus-dependent pattern formation in periodic environments in isolated neurons. *IEEE Trans. Neural Networks*, 13(2): 551–563, 2001.

46. S. Grossberg. Biological competition: Decision rules, pattern formation, and oscillations. *Proc. Natl. Acd. Sci. USA*, 77(4): 2338–2342, 1980.

47. S. Grossberg. Nonlinear neural networks: principles, Mechanisms, and architectures. *Neural Networks*, 1: 17–61, 1988.

48. G. Haddad. Monotone viable trajectories for functional differential inclusions. *J. Diff. Equ.*, 42: 1–24, 1981.

49. J. K. Hale. *Theory of Functional Differential Equations.* New York, Springer-Verlag 1977.

50. H. Harrer, J. A. Nossek, and R. Stelzl. An analog implementation of discrete-time neural networks. *IEEE Trans. Neural Networks*, 3: 466–476, 1992.

51. S. Haykin. *Neural Networks: A Comprehensive Foundation.* New York, Macmillan Publishing Company, 1994.

52. D. O. Hebb. *The Orgnization of Behavior: A Neuropsuchological Theory.* New York, Wiley, 1949.

53. R. Hecht-Nielsen. *Neurocomputing, Reading.* MA, Addison-Wesley, 1990.

54. J. J. Hopfield. Neurons with graded response have collective computational properties like thoseof of two-stage neurons. *Proc. Nat. Acad. Sci -Biol.*, 81: 3088–3092, 1984.

55. X. Huang and J. Cao. Almost periodic solution of shunting inhibitory cellular neural networks with time-varying delays. *Phys. Lett. A*, 314(3): 222–231, 2003.

56. S. Hu and N. S. Papageorgiou. On the existence of perioidic solutions for nonconvex-valued differential inclusions in $\mathbb{R}^N$. *Proc. Amer. Math. Soc.*, 123(10): 3043–3050 1995.

57. J. J. Hopfield, and D. W. Tank. Computing neural circuits: a model. *Science*, 233: 625–633 1986.

58. A. G. Ivanov. On the equivalence of differential inclusions and controlled almost periodic systems. (Russian) *Differ. Uravn.* 33(7): 876–884, 1997; translation in *Differ. Equ.*, 33(7): 879–887, 1997.

59. M. P. Kennedy and L. O. Chua. Neural networks for nonlinear programming. *IEEE Trans. Circuits Syst.-1*, 35: 554–562, 1988.

60. W. A. Kirk and B. Sims. *Handbook of Metric Fixed Point Theory.* Berlin, NY, Springer-Verlag, 2001.

61. B. M. Levitan and V. V. Zhikov. *Almost Periodic Functions and Differential Equations.* New York, Cambriadge University Press, 1982.
62. H. Lu. On stability of nonlinear continuous-time neural networks with delays. *Neural Networks*, 13(10): 1135–1144, 2000.
63. W. Lu and T. Chen. On periodic dynamical systems. *Chin. Ann. Math.*, 25B(4): 455–462, 2004.
64. W. Lu and T. P. Chen. Dynamical behaviors of Cohen-Grossberg neural networks with discontinuous activation functions. *Neural Networks*, 18: 231–242, 2005.
65. W. Lu and T. Chen. Global exponential stability of almost periodic trajectory of a large class of delayed dynamical systems. *Sci. China A: Math.*, 48(8): 1015–1026, 2005.
66. W. Lu and T. P. Chen. Dynamical Behaviors of delayed neural network systems with discontinuous activation functions. *Neural Comput.*, 18(3): 683–708, 2006.
67. W. Lu and T. Chen. $\mathbb{R}^n_+$ global stability of Cohen-Grossberg neural network system with nonnegative equilibrium. *Neural Networks*, 20: 714–722, 2007.
68. W. Lu and T. Chen. Almost periodic dynamics of a class of delayed neural networks with discontinuous activations. *Neural Comput.*, 20: 1065–1090, 2008.
69. W. Lu, L. Rong, and T. Chen. Global convergence of delayed neural networks systems. *Int. J. Neural Sys.*, 13(3): 193–204, 2003.
70. D. Li and P. E. Kloeden. On the dynamics of nonautonomous periodic general dynamical systems and differential inclusions. *J. Diff. Equ.*, 224: 1–38, 2006.
71. X. Liao, C. Li, and K. Wong. Criteria for exponential stability of Cohen-Grossberg neural networks. *Neural Networks*, 17: 1401–1414, 2004.
72. G Li and X. Xue. On the existence of periodic solutions for differential inclusions. *J. Math. Anal. Appl.*, 276: 168–183, 2002.
73. R. K. Miller. Asymptotic behavior of nonlinear delayed-differential equations. *J. Differ. Equ.*, 1: 293–305, 1995.
74. C. M. Marcus and R. M. Westervelt. Stability of analog neural networks with *delay. Phys. Rev. A.*, 39: 347–359, 1989.
75. J. M. Mendel and K. S. Fu. *Adaptive, Learning, and Pattern Recognition Systems: Theory and Applications.* New York, Academic Press, 1970.
76. N. Megiddo and M. Kojima. On the existence and uniqueness of solutions in nonlinear complementarity theory. *Math. Prog.*, 12: 110–130, 1977.
77. B. E. Paden and S. S. Sastry. Calculus for computing Filippov's differential inclusion with application to the variable structure control of robot manipulator. *IEEE Trans. Circuits Syst.*, 34: 73–82, 1987.
78. D. Papini and V. Taddei. Global exponential stability of the periodic solution of the delayed neural networks with discontinuous activations. *Phys. Lett. A.*, 343: 117–128, 2005.
79. C. S. Ramóny. Histologie du système nerveux de l'homme et des vertébrés. Paris: Maloine; Edition Francaise Revue: Tome I (1952); Tome II (1955); Madrid: Consejo Superior de Inverstigaciones Cientificas 2001.
80. T. Rockafellar and R.J.-B. Wets. *Variational Analysis.* Berlin Heidelberg, Springer-Verlag, 1998.
81. G. V. Smirnov. Weak asymptotic stability at first approximation for periodic differential inclusions. *NoDEA*, 2: 445–461, 1995.
82. L. Wang and X. Zou. Exponential stability of Cohen-Grossberg neural networks. *Neural Networks*, 15: 415–422, 2002.
83. K. Yoshida. *Functional Analysis. Grundlehren der Mathematicchen Wissenschaften.* New York: Springer-Verlag, 1978.
84. T. Yosizawa. *Stability Theory and The Existence of Periodic Solutions and Almost Periodic Solutions.* New York, Springer-Verlag, 1975.
85. V. I. Utkin. *Sliding Modes and Their Applications in Variable Structure Systems.* Moskow, MIR Publishers, 1978.

86. J. Zhang and X. Jin. Global stability analysis in delayed hopfield neural models. *Neural Networks*, 13(7): 745–753, 2000.

87. Y. Zheng and T. Chen. Global exponential stability of delayed periodic dynamical systems. *Phys. Lett. A*, 322(5–6): 344–355, 2004.

88. J. Zhou, Z. Liu, and G. Chen. Dynamics of delayed periodic neural networks. *Neural Networks*, 17(1): 87–101, 2004.

89. A. V. Zuev. On periodic solutions of ordinary differential equations with discontinuous right-hand side. *Math. Notes*, 79(4): 518–527, 2006.

Chapter 8
Stability and Hopf Bifurcation for a First-Order Delay Differential Equation with Distributed Delay

Fabien Crauste

8.1 Introduction

The present chapter is devoted to the stability analysis of linear differential equations with continuous distributed delay

$$\frac{dx}{dt}(t) = -Ax(t) - B \int_0^\infty F(\theta)x(t - \theta)\,d\theta, \qquad t > 0, \tag{8.1}$$

where A and B are real coefficients and F is an integrable function on $(0, +\infty)$, that can possibly have a compact support. Note that, except when it is mentioned, F will not be supposed to be a density function.

This class of equations is widely used in many research fields—it can be obtained through the linearization of different nonlinear problems (see, for example, Sect. 8.5)—such as automatic, economic, and, for our purpose, in biological modeling because it can be associated with problems in which it is important to take into account some history of the state variable (e.g., gestation period, cell cycle durations, or incubation time [23, 34]). When few data are available, this history is usually assumed to be discrete. Yet, in most cases very little is known about it, and how it is distributed; so, very abstract assumptions lead to equations in the form of (8.1).

We are interested in stability properties of (8.1), that is, under which conditions, on the parameters A and B or the function F, do all solutions of (8.1) converge toward zero? And, as a consequence, how can (8.1) be destabilized? Can oscillating or periodic solutions appear? All these questions arise from a need to understand how systems destabilize, or how can they stay stable for a long time. Partial answers are brought up in Sect. 8.4.

F. Crauste (✉)
Université de Lyon; Université Lyon 1, CNRS UMR5208 Institut Camille Jordan,
43 blvd du 11 novembre 1918, F-69622 Villeurbanne-Cedex, France
e-mail: crauste@math.univ-lyon1.fr

F.M. Atay (ed.), *Complex Time-Delay Systems,* Understanding Complex Systems,
DOI 10.1007/978-3-642-02329-3_8, © Springer-Verlag Berlin Heidelberg 2010

Many studies, as mentioned hereafter, tried to bring clear answers to these questions, yet only partial results have been proved up to now. Mainly, only sufficient conditions for stability—which are sometimes all one needs—have been obtained, or particular functions F have been used so as to simplify in some way the study of the stability of (8.1).

To make this latter argument clearer, let us recall that the stability of (8.1) is related to the sign of the real parts of its eigenvalues (see Sect. 8.2, Proposition 8.3). Indeed, it is sufficient to determine the sign of real parts of all eigenvalues to deduce the stability or instability of (8.1). With well-chosen functions F, the integral term in (8.1), or in its characteristic equation, can be explicitly calculated and stability can be determined. For instance, when F is the Dirac measure δ_τ defined by

$$\delta_\tau(\theta) = \begin{cases} 1, & \text{if } \theta = \tau, \\ 0, & \text{otherwise.} \end{cases}$$

Equation (8.1) reduces to the classical *linear discrete delay differential equation*

$$\frac{dx}{dt}(t) = -Ax(t) - Bx(t - \tau), \qquad t > 0. \tag{8.2}$$

This equation has been widely studied (see, for example, [18]) and its stability has been fully determined. This will be recalled in Sect. 8.3 where the importance of (8.2) is stressed. In particular, necessary and sufficient conditions for the existence of a Hopf bifurcation have been obtained.

When F is a sum of Dirac measures, (8.1) is a differential equation with several discrete distributed delays. Such equations may display very interesting behavior, and their analysis is much more difficult than for (8.2). It is not the aim of this work to deal with such equations, but the interested reader is referred to the works in [3, 12, 35, 36, 39] and the numerous references therein.

Other types of density functions F that have been often used, applied to different situations but mainly in biological modeling (e.g., to describe the distribution of cell cycle transit times [25]), are Gamma distributions. They consist in taking F as

$$F(\theta) = \frac{\sigma^{k+1}}{\Gamma(k+1)} \theta^k e^{-\sigma\theta}, \tag{8.3}$$

the parameters σ and k being related to experimental data. One can also consider that σ and k are defined by the following relations:

$$E = \frac{k+1}{\sigma} \qquad \text{and} \qquad V = \frac{k+1}{\sigma^2},$$

with E and V defining, respectively, the expectation and variance of F. In the stability analysis of (8.1), Gamma distributions have a technical advantage: they allow to 'transform' a transcendental characteristic equation into a polynomial equation, thus facilitating the analysis.

Studies for more general functions F have been done during the last 20 years and different achieved results may be mentioned. These results will be explained in more details in Sect. 8.3. Anderson [4, 5], between 1991 and 1992, considered a different form for (8.1), where F is a density function, and he obtained stability results related to the different moments (especially the expectation and the variance) of the distribution. These results stress the importance of the shape of the function F in the stability of (8.1). Kuang [24], in 1994, also obtained general stability results for systems of delay differential equations. More recently, sufficient conditions for the stability of delay differential equations with distributed delay have been obtained by Bernard et al. [8], when F is a probability density. The authors used some properties of the distribution to prove these results. However, it is noticeable to mention that in all the cited works the authors only focused on sufficient conditions for the stability, and no necessary condition has yet been proved.

In the next section we present some useful and important definitions for the stability of differential equations that will make the reading of this chapter easier. Then, in Sect. 8.3, we briefly summarize the state of the art in the stability of differential equations with distributed delay, starting with the case of the stability of (8.2) which, without falling in the framework of this chapter, is very useful in determining strategies to analyze the stability of (8.1). In Sect. 8.4, we present the main result of this chapter which consists in finding a critical value of the parameters that would destabilize (8.1). We show that this destabilization occurs through a Hopf bifurcation. We conclude with an application of the results of Sect. 8.4 to a model of hematopoietic stem cells dynamics, pointing out how the mathematical study allows to determine the existence of oscillating solutions in this model that can be related to chronic myelogenous leukemia, a severe blood disease.

8.2 Definitions and Hopf Bifurcation Theorem

Let's consider the delay differential equation

$$y'(t) = f(y_t), \quad t > 0, \tag{8.4}$$

where $f : \varphi \in \mathcal{C}((-\infty, 0], \mathbb{R}^n) \mapsto f(\varphi) \in \mathbb{R}^n$ has continuous first and second derivatives for all $\varphi \in \mathcal{C}((-\infty, 0], \mathbb{R}^n)$, the space of continuous functions mapping $(-\infty, 0]$ into $\mathbb{R}^n$, $f(0) = 0$, and y_t is a continuous function defined, for $\theta \leq 0$, by

$$y_t(\theta) = y(t + \theta).$$

Define $L : \mathcal{C}((-\infty, 0], \mathbb{R}^n) \to \mathbb{R}^n$ by

$$L\varphi = \frac{\mathrm{d}f}{\mathrm{d}\phi}(0)\varphi,$$

and consider the linear differential equation with delay

$$y'(t) = Ly_t, \quad t > 0. \tag{8.5}$$

Equation (8.5) is the linearized equation of (8.4) at $\varphi = 0$.

Example 8.1 If, for continuous functions $\varphi : (-\infty, 0] \mapsto \mathbb{R}$,

$$L\varphi = -A\varphi(0) - B \int_0^\infty F(\theta)\varphi(-\theta)\, d\theta,$$

then (8.5) is (8.1).

Definition 8.2 The trivial solution $y \equiv 0$ of (8.5) is said to be *asymptotically stable* if all solutions of (8.5) converge toward zero when t tends to infinity.

Instead of "The trivial solution of (8.5) is stable or unstable," one may find the expression "Eq. (8.5) is stable or unstable." One must note that asymptotic stability of (8.5) implies *local* asymptotic stability of (8.4), and instability of (8.5) implies instability of (8.4).

Different techniques may be used to determine the asymptotic stability of (8.5), although two approaches are usually privileged. The first technique, which is usually applied to nonlinear delay differential equations, aims to determine global asymptotic stability (the distinction vanishes for linear equations) and is based on the use of Lyapunov functions (see Hale [18]). Without giving too much details (we invite the interested reader to look through the book of Hale [18] for the application of Lyapunov functions to delay differential equations), the main difficulty lies in finding a good function, which is not an easy task. The second way of finding the stability properties of (8.5) is to study its eigenvalues.

Consider the equation

$$\det\left(\lambda - Le^{\lambda \cdot}\right) = 0, \quad \lambda \in \mathbb{C}. \tag{8.6}$$

This equation is called the *characteristic equation* of (8.5). When the linear function L is known, (8.6) is obtained by searching for solutions of (8.5) in the form $y(t) = Ce^{\lambda t}$, $C \in \mathbb{R}^n$. The solutions λ of (8.6) are called the *characteristic roots*, and they are the *eigenvalues* of (8.5).

Proposition 8.3 *The trivial solution of (8.5) is asymptotically stable if all characteristic roots of (8.6) have negative real parts, and unstable if (8.6) has a characteristic root with positive real part.*

Note that when $\lambda = 0$ is an eigenvalue of (8.5) and all other eigenvalues have negative real parts, then one cannot immediately conclude the stability or instability of (8.5). A more detailed analysis is then necessary. It is also important to note that the stability of (8.5) can only be lost if eigenvalues cross the imaginary axis from left to right. That is, if purely imaginary eigenvalues appear (see Cooke and Grossman [14] and the numerous generalizations of their result based on Rouché's Theorem [16, p. 248]).

Stated in the next theorem, the Hopf bifurcation theorem describes the instability of (8.4) through the appearance of periodic solutions, related to the existence of purely imaginary eigenvalues of (8.5). As mentioned in Hale [18], the Hopf bifurcation is one of the simplest way for (nonconstant) periodic solutions to arise in delay differential equations.

Let us suppose that the function f in (8.4) depends on a real parameter, say $\alpha \in \mathbb{R}$, and f has continuous first and second derivatives in α and φ for all $\alpha \in \mathbb{R}$ and $\varphi \in \mathcal{C}((-\infty, 0], \mathbb{R}^n)$. Then the linear function L in (8.5) also depends on α and (8.5) can be written as

$$y'(t) = L(\alpha)y_t, \quad t > 0. \tag{8.7}$$

If λ is an eigenvalue of (8.7), we denote by $\mathrm{Re}(\lambda)$ and $\mathrm{Im}(\lambda)$ the real and imaginary parts of λ, respectively.

Theorem 8.4 (Hopf Bifurcation Theorem for DDEs, Hale [18]). *Suppose that (8.7) has a simple nonzero purely imaginary eigenvalue λ_0 for $\alpha = 0$, and that all other eigenvalues are not integer multiples of λ_0. In addition, suppose that the branch of eigenvalues $\lambda(\alpha)$ which satisfies $\lambda(0) = \lambda_0$ is such that $\mathrm{Re}(\lambda'(0)) \neq 0$. Then, for α close to zero, (8.4) has nontrivial periodic solutions, with period close to $2\pi/\mathrm{Im}(\lambda_0)$.*

The assumption $\mathrm{Re}(\lambda'(0)) \neq 0$ in the above theorem is called the *transversality condition*, and α is called the *bifurcation parameter*. When all assumptions of Theorem 8.4 are fulfilled, one says that (8.4) undergoes a Hopf bifurcation for $\alpha = 0$.

Example 8.5 Through this example, we illustrate the results stated in Proposition 8.3 as well as in Theorem 8.4 on a differential equation with discrete delay. Consider the DDE

$$y'(t) = -\alpha \frac{y(t-1)}{1+y(t-1)}, \quad \alpha > 0, \, t > 0. \tag{8.8}$$

The linearization of (8.8) around $y \equiv 0$ gives the linear DDE

$$y'(t) = -\alpha y(t-1), \quad \alpha > 0. \tag{8.9}$$

This equation can be written as (8.7) with $L(\alpha)$ given by $L(\alpha)\varphi = -\alpha\varphi(-1)$. Looking for solutions of (8.9) in the form $y(t) = Ce^{\lambda t}$, we deduce the characteristic equation

$$\lambda + \alpha e^{-\lambda} = 0. \tag{8.10}$$

For the sake of simplicity, we write $\lambda = \mu + i\omega$, and we separate real and imaginary parts in the above equation to obtain

$$\mu + \alpha e^{-\mu}\cos(\omega) = 0 \quad \text{and} \quad \omega - \alpha e^{-\mu}\sin(\omega) = 0. \tag{8.11}$$

F. Crauste

One can notice that if (μ, ω) is a solution, then so is $(\mu, -\omega)$. Consequently, we will only consider $\omega > 0$.

Let us suppose that there exists a positive value of α, say α^*, for which $\mu = 0$. Then, from (8.11),

$$\alpha^* \cos(\omega) = 0 \qquad \text{and} \qquad \omega = \alpha^* \sin(\omega).$$

Since $\alpha^* > 0$, it follows that $\cos(\omega) = 0$, and so $\omega = \pi/2 + k\pi$. Using the second equation, we deduce that for $\alpha^* = \pi/2$, (8.9) has purely imaginary eigenvalues; these eigenvalues are $\pm \pi/2$.

Let us check first that (8.8) is locally asymptotically stable for $0 < \alpha < \pi/2$, and, second, that (8.8) undergoes a Hopf bifurcation when $\alpha = \pi/2$.

Suppose that $0 < \alpha < \pi/2$ and that $\lambda = \mu + i\omega$ is an eigenvalue of (8.9). By contradiction, assume $\mu > 0$. Then $e^{-\mu} < 1$ and we deduce, from (8.11), that $\omega = \alpha e^{-\mu} \sin(\omega) < \pi/2$. Consequently, with $\omega > 0$, we obtain that $\cos(\omega) > 0$ and, still from (8.11), that $\mu < 0$. There is a contradiction, so $\mu \le 0$. Since $\alpha < \pi/2$, μ cannot be zero, so $\mu < 0$ and all characteristic roots have negative real parts. From Proposition 8.3, we conclude that (8.9) is asymptotically stable when $0 < \alpha < \pi/2$, and consequently (8.8) is locally asymptotically stable for $0 < \alpha < \pi/2$.

Now let $\alpha = \pi/2$. We are going to check that all assumptions of Theorem 8.4 are satisfied, in particular that (8.9) has a periodic solution, so that (8.8) undergoes a Hopf bifurcation when $\alpha = \pi/2$.

First, we already checked that, when $\alpha = \pi/2$, (8.9) has a pair of purely imaginary eigenvalues, and that they are the only imaginary eigenvalues. Let us consider the branch of eigenvalues $\lambda(\alpha) = \mu(\alpha) + i\omega(\alpha)$, solutions of (8.10), such that $\lambda(\pi/2) = i\pi/2$. Differentiating (8.10) with respect to α, we obtain

$$\left(\alpha e^{-\lambda(\alpha)} - 1 \right) \lambda'(\alpha) = e^{-\lambda(\alpha)}.$$

It is straightforward to check that $\pm i\pi/2$ are simple eigenvalues. Indeed, if $\lambda'(\pi/2) = 0$ then $e^{-\lambda(\pi/2)} = 0$, which is impossible. Thus the first assumption in Theorem 8.4 is satisfied. Moreover, using the fact that $e^{-\lambda(\alpha)} = -\lambda(\alpha)/\alpha$, we deduce that

$$\lambda'(\alpha) = \frac{e^{-\lambda(\alpha)}}{\alpha e^{-\lambda(\alpha)} - 1} = \frac{\lambda(\alpha)}{\alpha(\lambda(\alpha) + 1)}.$$

Consequently,

$$\lambda'\left(\frac{\pi}{2}\right) = \frac{i\dfrac{\pi}{2}}{\dfrac{\pi}{2}(i\dfrac{\pi}{2} + 1)} = \frac{\dfrac{\pi}{2} + i}{\dfrac{\pi^2}{4} + 1},$$

and $\mathrm{Re}(\lambda'(\pi/2)) \neq 0$. The second assumption of Theorem 8.4 is then satisfied. It follows that (8.8) undergoes a Hopf bifurcation when $\alpha = \pi/2$, and periodic solutions with periods close to 4 exist.

8.3 State of the Art and Objectives

Differential equations with continuous distributed delay such as (8.1), or in a modified form, have not been the object of much attention, at least when one is interested in necessary and sufficient conditions for asymptotic stability. Mainly, particular cases have been investigated (when F is a Gamma distribution (8.3)), and sufficient stability conditions have been obtained.

Yet, to understand the lack of results concerning the asymptotic stability of (8.1), one has to have in mind the known results about the stability of the classical discrete delay differential equation (8.2)—which can logically be considered as the simplest delay differential equation—and the techniques used to obtain these results. In the next section, we recall stability results for (8.2). Then, we will focus on actual known results of stability for (8.1) and their limitations.

8.3.1 The Classical Linear Discrete Delay Differential Equation

Let us focus, for a while, on stability properties of (8.2), which is a particular case of (8.1), known as the *discrete delay differential equation*. Asymptotic properties of this equation, in terms of the coefficients A and B, and of the time delay τ have been established in [14, 18] using a well-known result by Hayes [20]. We state and prove these results, using an alternative proof, based on the one given in [14], which will be useful later in this chapter.

Stability of differential equations is related to the sign of the real parts of their eigenvalues (Proposition 8.3). The equation is *asymptotically stable* if all eigenvalues have negative real parts and *unstable* if eigenvalues with positive real parts exist. The stability can be lost only if purely imaginary eigenvalues appear. In the case of ordinary differential equations, the characteristic equation is a polynomial function; so, easy-to-find criteria can be used to locate the eigenvalues. In the case of delay differential equations, the characteristic equation is transcendental, making it difficult to locate the characteristic roots. In particular, the characteristic equation associated with (8.2), obtained by searching for solutions $x(t) = Ce^{\lambda t}$, $C \in \mathbb{R}$, is

$$\lambda + A + Be^{-\lambda \tau} = 0. \tag{8.12}$$

Let us set

$$\Delta(\lambda, \tau) = \lambda + A + Be^{-\lambda \tau}.$$

In the next theorem, we state and prove stability conditions for (8.2). We make a difference between delay-independent stability conditions (in *(a)* and *(b)*) and delay-dependent ones (in *(c)*). The latter are related to the existence of a Hopf bifurcation.

Theorem 8.6 *The stability properties of* (8.2) *are as follows:*
(a) *If* $B \leq 0$, (8.2) *is asymptotically stable for all* $\tau \geq 0$ *when* $A + B > 0$ *and unstable for all* $\tau \geq 0$ *when* $A + B < 0$;
(b) *If* $B > 0$, (8.2) *is asymptotically stable for all* $\tau \geq 0$ *when* $A > B$ *and unstable for all* $\tau \geq 0$ *when* $A + B < 0$;
(c) *If* $B > 0$ *and* $B > |A|$, *then* (8.2) *is asymptotically stable for* $\tau \in [0, \tau^*)$ *and unstable for* $\tau \geq \tau^*$, *where*

$$\tau^* = \frac{\arccos\left(-\dfrac{A}{B}\right)}{\sqrt{B^2 - A^2}}.$$

When $\tau = \tau^*$, *a Hopf bifurcation occurs.*

Proof Assertions in (a) and (b) follow immediately from Theorem 8.12, with $F = \delta_\tau$, the Dirac measure in τ. We refer to the proof of Theorem 8.12.

Let us focus on the case $B > |A|$ (that is, $A + B > 0$ and $A < B$). The stability in this case will depend on the delay. Indeed, we are going to prove the existence of a unique Hopf bifurcation that can destabilize the equation for some value of the delay.

First, notice that $\Delta(\lambda, 0) = \lambda + A + B$, so, when $\tau = 0$, $\bar{\lambda} = -(A + B)$ is the only eigenvalue of (8.2). Since $A + B > 0$, $\bar{\lambda} < 0$ and the equation is then asymptotically stable when $\tau = 0$.

Suppose $\tau > 0$, and search for purely imaginary eigenvalues $\lambda = i\omega$. Separating real and imaginary parts of $\Delta(i\omega, \tau)$, one finds that ω and τ must satisfy

$$\begin{cases} A + B\cos(\omega\tau) = 0, \\ \omega - B\sin(\omega\tau) = 0. \end{cases} \tag{8.13}$$

One can easily check that $\omega = 0$ cannot be a solution, since $A + B > 0$, and if (ω, τ) is a solution of (8.13), then so is $(-\omega, \tau)$. Hence, we only focus on positive values of ω.

Since $B > 0$, we rewrite (8.13) as

$$\cos(\omega\tau) = -\frac{A}{B} \quad \text{and} \quad \sin(\omega\tau) = \frac{\omega}{B}.$$

Then one easily checks that the only ω that can satisfy these conditions is given by

$$\omega^2 = B^2 - A^2.$$

Of course, this definition of ω is valid only if $B > |A|$, which is the case here.

We then deduce that

$$\tau\sqrt{B^2 - A^2} = \arccos\left(-\frac{A}{B}\right),$$

and the value of τ for which $i\omega$ is a root of Δ must be

$$\tau = \tau^* := \frac{\arccos\left(-\dfrac{A}{B}\right)}{\sqrt{B^2 - A^2}}.$$

The converse is straightforward: If $\tau = \tau^*$, then $\pm i\sqrt{B^2 - A^2}$ are purely imaginary eigenvalues of (8.2). And indeed they are the only ones. So there is only one change of stability and it occurs for $\tau = \tau^*$. Let us check that this change of stability occurs through a Hopf bifurcation.

Let us set $\omega^* = \sqrt{B^2 - A^2}$. In order to show that a Hopf bifurcation occurs when $\tau = \tau^*$, we must verify that $\pm i\omega^*$ are simple eigenvalues and

$$\frac{d}{d\tau}\mathrm{Re}(\lambda(\tau))\bigg|_{\tau=\tau^*} > 0.$$

Consider a branch of eigenvalues $\lambda(\tau) = \mu(\tau) + i\omega(\tau)$ such that

$$\mu(\tau^*) = 0 \quad \text{and} \quad \omega(\tau^*) = \omega^*.$$

Then, for all $\tau \geq 0$, $\Delta(\lambda(\tau), \tau) = 0$. Hence,

$$\frac{d\Delta}{d\tau}(\lambda(\tau), \tau) := \frac{d\lambda}{d\tau}(\tau)\frac{\partial\Delta}{\partial\lambda}(\lambda(\tau), \tau) + \frac{\partial\Delta}{\partial\tau}(\lambda(\tau), \tau) = 0. \tag{8.14}$$

First, notice that

$$\frac{\partial\Delta}{\partial\lambda}(\lambda(\tau), \tau) - 1 - B\tau e^{-\lambda(\tau)\tau}.$$

Therefore, using (8.13), we obtain

$$\frac{\partial\Delta}{\partial\lambda}(\lambda(\tau^*), \tau^*) = 1 - B\tau^*\cos(\omega^*\tau^*) + iB\tau^*\sin(\omega^*\tau^*) = 1 + A\tau^* + i\omega^*\tau^*.$$

Since ω^* and τ^* are strictly positive, it follows that $\partial\Delta(\lambda(\tau^*), \tau^*)/\partial\lambda \neq 0$, and $\pm i\omega^*$ are simple eigenvalues of (8.2).

Now, from (8.14), we deduce that

$$\frac{d\lambda}{d\tau}(\tau^*) = \frac{\dfrac{\partial \Delta}{\partial \tau}(i\omega^*, \tau^*)}{\dfrac{\partial \Delta}{\partial \lambda}(i\omega^*, \tau^*)} = \frac{iB\omega^* e^{-i\omega^* \tau^*}}{1 + A\tau^* + i\omega^* \tau^*}.$$

This yields

$$\frac{d}{d\tau}\mathrm{Re}(\lambda(\tau))\bigg|_{\tau=\tau^*} = \mathrm{Re}\left(\frac{d\lambda}{d\tau}(\tau^*)\right) = \frac{B\omega^* \sin(\omega^* \tau^*)}{(1 + A\tau^*)^2 + (\omega^* \tau^*)^2}.$$

Using the fact that (ω^*, τ^*) satisfies (8.13), we deduce

$$\frac{d}{d\tau}\mathrm{Re}(\lambda(\tau))\bigg|_{\tau=\tau^*} = \frac{(\omega^*)^2}{(1 + A\tau^*)^2 + (\omega^* \tau^*)^2} > 0.$$

Hence the branch of eigenvalues crosses the imaginary axis from left to right, and the transversality condition (see Theorem 8.4) is satisfied: a Hopf bifurcation occurs when $\tau = \tau^*$. This completes the proof of (c) and ends the proof of the theorem. □

Remark 8.7 In statement (c) of Theorem 8.6, and more particularly in the proof of (c), the critical value τ^* of the time delay appears to be unique, implying that a unique bifurcation can change the stability of (8.2). It is, however, important to notice that in practical cases the linear delay differential Equation (8.2) is obtained through the linearization of a nonlinear delay differential equation about one of its steady states, and coefficients A and B may depend, explicitly or not, on the time delay. Then the critical values of τ (here τ^*) for which a stability switch may occur may not be unique, but can be the solutions of a fixed point problem,

$$\tau = \frac{\arccos\left(-\dfrac{A(\tau)}{B(\tau)}\right)}{\sqrt{B(\tau)^2 - A(\tau)^2}}.$$

Consequently, stability switches may occur (see Beretta and Kuang [7]).

Remark 8.8 Hayes [20] proved the following result (whose proof can be found in [14] or [18]), which enables to locate the roots of (8.12), and thus allows to draw the stability diagram for (8.2). Yet, in contrast to Theorem 8.6, this result does not explain how the equation becomes unstable (e.g., whether through a Hopf bifurcation).

Theorem 8.9 *All roots of* (8.12) *have negative real parts if and only if* $A\tau > -1$, $A + B > 0$, *and* $B\tau < \zeta \sin(\zeta) - A\tau \cos(\zeta)$, *where* $\zeta \in [0, \pi]$ *is the root of* $\zeta = -A\tau \tan(\zeta)$ *if* $A \neq 0$ *and* $\zeta = \pi/2$ *if* $A = 0$.

When comparing results about distributed delay differential equations with the ones obtained in Theorem 8.6 for the discrete delay differential equations (see

Sect. 8.4), it will be important to notice that the main difficulty lies in the existence of purely imaginary eigenvalues. Most of other properties, such as the transversality condition, will be obtained without too much difficulty, in a similar manner to Theorem 8.6.

In the case of discrete delay, it is possible to obtain necessary and sufficient conditions for the stability (see Theorem 8.6) because, in the proof of the theorem, usual properties of cosine and sine functions (precisely, that $\cos^2 + \sin^2 = 1$) have been used to determine the exact values of the purely imaginary eigenvalues. This will be impossible for the case with distributed delay due to the presence of weighed integrals of cosine and sine functions. One can see that the so-called delay-independent stability results in Theorem 8.6 will hold for the distributed delay case, but as soon as stability results depend on the delay difficulties appear and we are obliged to develop new techniques to obtain stability or bifurcation conditions. A brief but wide summary of known results dealing with the stability of some distributed delay differential equations is presented in the following section.

8.3.2 Known Results About Stability of Distributed Delay Differential Equations

As mentioned in the Introduction, the stability analysis of DDEs with distributed delay has not received so much attention in the literature, except for some particular cases, when F, in (8.1), is a Gamma distribution for instance. In 1989, Boese [11] analyzed the stability of (8.1) when F is a Gamma distribution given by (8.3). The characteristic function of (8.1) is then a $(k + 1)$th degree polynomial function. The author determines sufficient conditions for the asymptotic stability of the trivial solution of (8.1), which are rather technical.

Kuang [24], in 1994, considered a system of two differential equations with continuous distributed delay, possibly infinite. He concentrates on purely imaginary eigenvalues and determines conditions for their nonexistence, obtaining sufficient conditions for the asymptotic stability of his system.

One particularly interesting result has been published in 2001 by Bernard et al. [8], where the authors considered (8.1) and determined sufficient conditions for its stability. Their main result is stated hereafter.

Theorem 8.10 (Bernard et al., 2001 [8]) *Suppose $B > |A|$, F is a density function, so that F is nonnegative and $\int_0^\infty F(\theta)\, d\theta = 1$, and let E be the expectation of F defined by $\int_0^\infty \theta F(\theta)\, d\theta = E$. The following assertions hold:*

(a) Eq. (8.1) is asymptotically stable if

$$E < \frac{\pi\left(1 + \dfrac{A}{B}\right)}{c\sqrt{B^2 - A^2}},$$

where $c := \sup\{\bar{c} \mid \cos(x) = 1 - \bar{c}x/\pi, x > 0\} \approx 2.2704$.

(b) *If F is symmetric, that is, $F(E + \theta) = F(E - \theta)$, then (8.1) is asymptotically stable if*

$$E < \frac{\arccos\left(-\dfrac{A}{B}\right)}{\sqrt{B^2 - A^2}}. \tag{8.15}$$

The more interesting result in Theorem 8.10 is cited in (b). One can note that (8.15) corresponds to the necessary and sufficient condition that gives the stability in Theorem 8.6, in the case when F is a Dirac measure δ_τ. It is noticeable to mention that Bernard et al. [8] suggest, as a conjecture, that the single Dirac measure δ_τ would be the most destabilizing distribution of delays for (8.1). This conjecture is still unproved, yet a recent result (Atay [6]) gave arguments in its favor.

In [6], Atay focuses on the stability of delay differential equations near a Hopf bifurcation and particularly on the respective influence of discrete and distributed delays on the stability of such equations. For linear delay differential equations, he shows that if the delay has a destabilizing effect, then the discrete delay is locally the most destabilizing delay distribution (one may note that when the delay has a stabilizing effect, then the discrete delay is locally the most stabilizing delay distribution). When the distribution is symmetric, as in Theorem 8.10 (b), he shows that the result is global: the discrete delay is the most destabilizing delay distribution.

When one deals with a nonsymmetric distribution F, the sufficient condition for the stability of (8.1) provided by Theorem 8.10 reveals bad results: the limit given for the stability is far from the exact stability boundary that can be obtained through numerical simulations on some easy-to-handle examples. Up to now, however, these results are probably the best obtained for the stability of a large class of differential equations with distributed delay.

More recently, Huang and Vandewalle [21] and Tang [38] analyzed the stability of equations similar to (8.1). The first authors were interested in the numerical stability of differential equations with distributed delay, but they proposed an interesting geometrical approach to determine conditions for the stability of (8.1) when F is defined by

$$F(\theta) = \begin{cases} 1, & \text{if } 0 < \theta < \tau, \\ 0, & \text{otherwise.} \end{cases}$$

Unfortunately, the way they proceed to obtain their stability conditions cannot be extended to general functions F.

In [38], Tang determines sufficient stability conditions for very general differential equations with distributed delay, but his results, that can be seen as a generalization of the works of Boese [11], are very technical and not easy to handle in particular nontrivial examples.

In [31], Ozbay et al. investigate the stability of linear systems of equations with distributed delays and apply their results to a model of hematopoietic stem cell dynamics. Considering an exponential distribution of delays, they obtain necessary and sufficient conditions for the stability using the small gain theorem and Nyquist stability criterion.

In the above-mentioned works, either particular functions F have been used to obtain stability conditions (usually Gamma distributions) or only sufficient stability conditions have been obtained. In the next section, we are going to present a way to obtain stability conditions and, more particularly, conditions for the loss of the stability. These conditions will not be, however, necessary and sufficient ones. But they will allow to describe, through a Hopf bifurcation, the appearance of periodic solutions for (8.1) and may lead to interesting results in practical cases. This will be detailed in Sect. 8.5.

Before turning to the main point of this chapter, let us mention a last contribution to the study of the stability for differential equations with distributed delay. In 1991 and 1992, Anderson [4, 5] focused on the stability of delay differential equations, called *regulator models*, in the form

$$\frac{dx}{dt}(t) = -h \int_0^\infty x(t-u)\,\mu(du), \tag{8.16}$$

where h is a parameter known as the *amplitude of the regulation* and $\mu(du)$ is a probability measure supported on $[0, \infty)$. By taking

$$h = A + B \int_0^\infty F(\theta)\,d\theta \quad \text{and} \quad \mu(du) = A\delta_0 + BF(u)\,du,$$

it is easy to check that (8.16) becomes (8.1). The theory developed by Anderson [4, 5] focuses on the properties of the probability measure $\mu(du)$, related to its expectation E_μ and its relative variance R_μ, defined by $R_\mu = V_\mu/E_\mu^2$, V_μ being the variance of $\mu(du)$.

Results established by Anderson in [4] and [5] are stated in the next theorem. Suppose that h_0 is the largest number such that for $0 \le h < h_0$ the trivial solution of (8.16) is asymptotically stable. This number is called the *threshold amplitude of the probability delay measure* $\mu(du)$. Note that h_0 may be infinite, and in this case (8.16) is always stable.

Theorem 8.11 *Suppose* $\mu(du)$ *is a probability measure of finite expectation* E_μ. *Denote by* R_μ *its relative variance.*
If R_μ *satisfies*

$$R_\mu < \frac{2}{\pi^2},$$

then h_0 *is finite, and there is a change of stability for (8.16).*
If $\mu(du) = k(u)\,du$ *where* k *is continuous, convex, and not everywhere piecewise linear on* $[0, +\infty)$, *then* $h_0 = +\infty$ *and the stability of (8.16) never changes.*

Although the results of Anderson are only valid for some class of probability measures (in particular, the second result in Theorem 8.11), they stress the importance of the shape of the delay distribution. This is an idea that can be found in [8], as stated in Theorem 8.10. Moreover, Anderson mentions that "the more concentrated the probability measure, the worse the stability property of the model" [5]. This goes in the sense of Bernard et al.'s conjecture [6, 8].

However, despite the interest of Theorem 8.11, in order to apply the theory of Anderson, one needs to know a lot about the probability measure properties, what is rarely the case, for example, in biological modeling. I would quote MacDonald [25] writing that, in the case of maturation times for some cells, too little is known to fix a form for their distribution. This was true in 1989, and it is unfortunately usually still true.

8.4 Stability Analysis and Hopf Bifurcation for a Delay Differential Equation with Distributed Delay

We now focus on the main objective of this chapter: the stability analysis of (8.1) and the existence of a Hopf bifurcation that would destabilize (8.1). Let us recall that (8.1) is

$$\frac{dx}{dt}(t) = -Ax(t) - B \int_0^\infty F(\theta)x(t-\theta)\,d\theta, \qquad t > 0,$$

where A and B are real coefficients, with $A \neq 0$ and $B \neq 0$, and F is a nonnegative integrable function on $(0, +\infty)$. We do not suppose that F is a density function. Without loss of generality one could assume F is a density function, yet to avoid introduction of more notations in the following section and to stay close to applications (where kernel functions are not necessarily density functions, see Sect. 8.5), this assumption will not be made.

Looking for solutions of (8.1) in the form $Ce^{\lambda t}$, $C \in \mathbb{R}$ and λ a complex number, we find the characteristic equation associated with (8.1),

$$\lambda + A + B \int_0^\infty F(\theta)e^{-\lambda\theta}\,d\theta = 0. \tag{8.17}$$

In the following, we denote by $\Delta(\lambda)$ the complex function

$$\Delta(\lambda) := \lambda + A + B \int_0^\infty F(\theta)e^{-\lambda\theta}\,d\theta. \tag{8.18}$$

If $\lambda = \mu + i\omega$ is an eigenvalue of (8.1), then separating real and imaginary parts in (8.17) leads to

$$\begin{cases} \mu + A + B \displaystyle\int_0^\infty F(\theta)e^{-\mu\theta}\cos(\omega\theta)\,d\theta = 0, \\ \omega - B \displaystyle\int_0^\infty F(\theta)e^{-\mu\theta}\sin(\omega\theta)\,d\theta = 0. \end{cases} \tag{8.19}$$

We can state and prove the following theorem on the stability of (8.1).

Theorem 8.12 *Equation* (8.1) *is unstable if*

$$A + B \int_0^\infty F(\theta)\, d\theta < 0, \qquad (8.20)$$

and asymptotically stable if

$$B \le 0 \quad \text{and} \quad A + B \int_0^\infty F(\theta)\, d\theta > 0 \qquad (8.21)$$

or

$$B > 0 \quad \text{and} \quad A > B \int_0^\infty F(\theta)\, d\theta. \qquad (8.22)$$

Proof Consider Δ, given by (8.18), as a function of real λ. Then Δ is differentiable and

$$\frac{d\Delta}{d\lambda}(\lambda) = 1 - B \int_0^\infty \theta F(\theta) e^{-\lambda\theta}\, d\theta. \qquad (8.23)$$

Moreover,

$$\lim_{\lambda \to +\infty} \Delta(\lambda) = +\infty.$$

First assume (8.20) holds. Then

$$\Delta(0) = A + B \int_0^\infty F(\theta)\, d\theta < 0.$$

Since Δ is a continuous function which tends to infinity when λ tends to infinity, there exists at least one $\lambda_0 \in \mathbb{R}$, $\lambda_0 > 0$, such that $\Delta(\lambda_0) = 0$. Consequently, (8.1) has at least one eigenvalue with positive real part and is unstable.

Suppose now that (8.21) holds. Since $B \le 0$, it follows from (8.23) that Δ is increasing. Moreover, since

$$\lim_{\lambda \to -\infty} \Delta(\lambda) = -\infty \quad \text{and} \quad \lim_{\lambda \to +\infty} \Delta(\lambda) = +\infty,$$

we deduce that there exists a unique $\lambda_0 \in \mathbb{R}$, such that $\Delta(\lambda_0) = 0$. In addition,

$$\Delta(0) = A + B \int_0^\infty F(\theta)\, d\theta > 0,$$

so $\lambda_0 < 0$.

Moreover, if $\lambda = \mu + i\omega$ is a root of Δ, $\lambda \ne \lambda_0$, then from (8.19) we obtain

$$\mu + A + B \int_0^\infty F(\theta) e^{-\mu\theta} \cos(\omega\theta)\, d\theta = 0.$$

Hence,

$$\lambda_0 - \mu = -B \int_0^\infty F(\theta) \left[e^{-\lambda_0 \theta} - e^{-\mu\theta} \cos(\omega\theta) \right] d\theta,$$

$$\geq -B \int_0^\infty F(\theta) \left[e^{-\lambda_0 \theta} - e^{-\mu\theta} \right] d\theta.$$

If one supposes that $\lambda_0 < \mu$, then one obtains a contradiction. Consequently, $\mu \leq \lambda_0$.

Since $\lambda_0 < 0$, we deduce that all roots of Δ have negative real parts. Hence, all eigenvalues of (8.1) have negative real parts and (8.1) is asymptotically stable.

Eventually, assume condition (8.22) is fulfilled. By contradiction, let us suppose that Δ has a root $\lambda = \mu + i\omega$ with $\mu > 0$. Then, from (8.19),

$$\mu = -A - B \int_0^\infty F(\theta) e^{-\mu\theta} \cos(\omega\theta) d\theta.$$

Since

$$-B \int_0^\infty F(\theta) e^{-\mu\theta} \cos(\omega\theta) d\theta \leq \left| -B \int_0^\infty F(\theta) e^{-\mu\theta} \cos(\omega\theta) d\theta \right|,$$

$$\leq B \int_0^\infty F(\theta) e^{-\mu\theta} |\cos(\omega\theta)| d\theta,$$

$$\leq B \int_0^\infty F(\theta) d\theta,$$

then, with (8.22), one finds that

$$\mu \leq -A + B \int_0^\infty F(\theta) d\theta < 0.$$

This gives a contradiction. Therefore, we deduce that $\mu \leq 0$. Let us show that $\mu \neq 0$.

Assume $\mu = 0$. Then, from (8.19), ω must satisfy

$$\begin{cases} A + B \int_0^\infty F(\theta) \cos(\omega\theta) d\theta = 0, \\ \omega - B \int_0^\infty F(\theta) \sin(\omega\theta) d\theta = 0. \end{cases} \tag{8.24}$$

Note that for all $\omega \in \mathbb{R}$

$$-\int_0^\infty F(\theta) d\theta \leq \int_0^\infty F(\theta) \cos(\omega\theta) d\theta \leq \int_0^\infty F(\theta) d\theta,$$

so a necessary condition for the existence of purely imaginary roots of Δ is

$$-\int_0^\infty F(\theta) d\theta \leq -\frac{A}{B} \leq \int_0^\infty F(\theta) d\theta.$$

Note that (8.22) implies that

$$-\frac{A}{B} < -\int_0^\infty F(\theta)\,d\theta.$$

Consequently, system (8.24) cannot have solutions, and $\mu \neq 0$. This yields that $\mu < 0$. Hence all eigenvalues of (8.1) have negative real parts and (8.1) is asymptotically stable. This ends the proof. □

Remark 8.13 If F is a density function, then $\int_0^\infty F(\theta)\,d\theta = 1$, and all conditions for the stability or instability of (8.1) in Theorem 8.12 are expressed in terms of A and B. Moreover, if F is the Dirac measure in τ denoted by δ_τ, with $\tau > 0$, then Theorem 8.12 reduces to Theorem 8.6. (a) and (b).

From the results established in Theorem 8.12, one can notice that the only parameter region for which the behavior of (8.1) is unknown is

$$B \int_0^\infty F(\theta)\,d\theta > |A|. \tag{8.25}$$

In this case, different behavior can be observed depending on the properties of the function F. Yet, no necessary and sufficient condition has been proved for the stability of (8.1). Most results only give sufficient conditions for the stability of (8.1) by finding conditions for the nonexistence of purely imaginary eigenvalues (see, for instance, Bernard et al. [8] or Theorem 8.10 in Sect. 8.3.2).

We investigate, in the following, the existence of purely imaginary eigenvalues of (8.1). We follow the idea developed in [1].

Assume (8.25) holds. Let $\lambda = i\omega$, with $\omega \in \mathbb{R}$, be a purely imaginary eigenvalue of (8.1). Then $i\omega$ is a root of Δ, and separating real and imaginary parts of $\Delta(i\omega)$ one obtains system (8.24), which we rewrite as

$$\begin{cases} A + BC(\omega) = 0, \\ \omega - BS(\omega) = 0, \end{cases} \tag{8.26}$$

where

$$C(\omega) = \int_0^\infty F(\theta)\cos(\omega\theta)\,d\theta \quad \text{and} \quad S(\omega) = \int_0^\infty F(\theta)\sin(\omega\theta)\,d\theta. \tag{8.27}$$

First, one can notice that $\omega = 0$ is not a solution of (8.26) under assumption (8.25), since

$$A + BC(0) = A + B\int_0^\infty F(\theta)d\theta > 0.$$

Moreover, if ω satisfies (8.26), then so does $-\omega$. Hence we only focus, in the following, on positive solutions of (8.26).

The main issue here is to prove the *existence* of purely imaginary eigenvalues. We will prove in Lemma 8.16 that a pair of purely imaginary eigenvalues is simple, and it is easy to determine conditions for the transversality condition to hold, in order to verify the Hopf Theorem (see Theorem 8.4).

To obtain the existence of purely imaginary eigenvalues, one must be able to solve system (8.26). To do that, and to determine a stability region, we have to choose a parameter that will be used as the *bifurcation parameter*. One can choose a quantity related to the function F (its expectation or its relative variance, as used by Anderson [4, 5]) or one of the parameters A or B. We study the loss of stability of (8.1) with respect to the parameter B. Hence, we look for (ω, B) solution of (8.26), with $\omega > 0$ and $B \int_0^\infty F(\theta) d\theta > |A|$.

In order to solve (8.26), we want to eliminate the parameter B to obtain an equation on ω. Thus we would find successively critical values of ω and B that would destabilize (8.1). The parameter B could be expressed from (8.26) as $\omega / S(\omega)$, provided that the division by $S(\omega)$ is allowed. We prove the next lemma.

Lemma 8.14 *Suppose that F is decreasing. Then, for $\omega > 0$,*

$$S(\omega) \geq \chi(\omega) > 0, \tag{8.28}$$

where

$$\chi(\omega) := \int_0^{\frac{2\pi}{\omega}} F(\theta) \sin(\omega\theta) d\theta, \quad \omega > 0. \tag{8.29}$$

Proof Let $N > 0$ be fixed. Define the truncated functions F_N by

$$F_N(\theta) = \begin{cases} F(\theta), & \text{if } \theta < N, \\ 0, & \text{if } \theta \geq N, \end{cases} \tag{8.30}$$

and the functions S_N and χ_N, for $\omega > 0$, by

$$S_N(\omega) = \int_0^N F_N(\theta) \sin(\omega\theta) d\theta \quad \text{and} \quad \chi_N(\omega) = \int_0^{\frac{2\pi}{\omega}} F_N(\theta) \sin(\omega\theta) d\theta. \tag{8.31}$$

Note that

$$\chi_N(\omega) = \begin{cases} S_N(\omega), & \text{if } \omega N \leq 2\pi, \\ \chi(\omega), & \text{if } \omega N > 2\pi. \end{cases} \tag{8.32}$$

Indeed, if $\omega N < 2\pi$, then from (8.30) one obtains

$$F_N(\theta) = 0 \quad \text{for } N < \theta < \frac{2\pi}{\omega},$$

so

$$\chi_N(\omega) = \int_0^N F_N(\theta) \sin(\omega\theta) d\theta = S_N(\omega).$$

If $\omega N = 2\pi$, then obviously $\chi_N(\omega) = S_N(\omega)$.

If $\omega N > 2\pi$, then, for all $\theta < 2\pi/\omega$, $\theta < N$ and $F_N(\theta) = F(\theta)$, so

$$\chi_N(\omega) = \int_0^{\frac{2\pi}{\omega}} F(\theta)\sin(\omega\theta)\, d\theta = \chi(\omega).$$

Let us check that $\chi_N(\omega)$ is positive for all $\omega > 0$. Let $\omega > 0$ and $N > 0$ be fixed. Then, using a simple change of variables,

$$\chi_N(\omega) = \int_0^{\frac{2\pi}{\omega}} F_N(\theta)\sin(\omega\theta)\, d\theta,$$

$$= \frac{1}{\omega}\int_0^{2\pi} F_N\left(\frac{\sigma}{\omega}\right)\sin(\sigma)d\sigma,$$

$$= \frac{1}{\omega}\int_0^{\pi} F_N\left(\frac{\sigma}{\omega}\right)\sin(\sigma)d\sigma + \frac{1}{\omega}\int_\pi^{2\pi} F_N\left(\frac{\sigma}{\omega}\right)\sin(\sigma)d\sigma,$$

$$= \frac{1}{\omega}\int_0^{\pi}\left[F_N\left(\frac{\sigma}{\omega}\right) - F_N\left(\frac{\sigma + \pi}{\omega}\right)\right]\sin(\sigma)d\sigma.$$

Since F is supposed to be decreasing, then so is F_N. Consequently, for $\sigma \in (0, \pi)$,

$$F_N\left(\frac{\sigma}{\omega}\right) - F_N\left(\frac{\sigma + \pi}{\omega}\right) > 0,$$

and

$$\chi_N(\omega) > 0.$$

It follows, from (8.32), that $S_N(\omega) > 0$ for $0 < \omega N < 2\pi$.
Let us show that $S_N(\omega) > \chi_N(\omega)$ for $\omega N > 2\pi$.
Assume $2\pi < \omega N \leq 3\pi$. From (8.31),

$$S_N(\omega) \quad \chi_N(\omega) = \int_{\frac{2\pi}{\omega}}^N F_N(\theta)\sin(\omega\theta)\, d\theta.$$

Using a simple change of variables ($\sigma = \omega\theta$), we obtain

$$S_N(\omega) - \chi_N(\omega) = \frac{1}{\omega}\int_{2\pi}^{\omega N} F_N\left(\frac{\sigma}{\omega}\right)\sin(\sigma)d\sigma.$$

Since $2\pi < \omega N \leq 3\pi$ and $2\pi < \sigma < \omega N$, then $\sin(\sigma) > 0$ and $S_N(\omega) - \chi_N(\omega) > 0$.

Assume now $3\pi < \omega N \leq 4\pi$. Then, similarly,

$$S_N(\omega) - \chi_N(\omega) = \int_{\frac{2\pi}{\omega}}^{N} F_N(\theta) \sin(\omega\theta) \, d\theta,$$

$$= \frac{1}{\omega} \int_{2\pi}^{\omega N} F_N\left(\frac{\sigma}{\omega}\right) \sin(\sigma) \, d\sigma,$$

$$= \frac{1}{\omega} \int_{2\pi}^{3\pi} F_N\left(\frac{\sigma}{\omega}\right) \sin(\sigma) \, d\sigma + \frac{1}{\omega} \int_{3\pi}^{\omega N} F_N\left(\frac{\sigma}{\omega}\right) \sin(\sigma) \, d\sigma.$$

Noting that, from (8.30), $F_N(\theta) = 0$ for $\theta \geq N$, then

$$\int_{3\pi}^{\omega N} F_N\left(\frac{\sigma}{\omega}\right) \sin(\sigma) \, d\sigma = \int_{3\pi}^{4\pi} F_N\left(\frac{\sigma}{\omega}\right) \sin(\sigma) \, d\sigma.$$

Thus,

$$S_N(\omega) - \chi_N(\omega) = \frac{1}{\omega} \int_{2\pi}^{3\pi} F_N\left(\frac{\sigma}{\omega}\right) \sin(\sigma) \, d\sigma + \frac{1}{\omega} \int_{3\pi}^{4\pi} F_N\left(\frac{\sigma}{\omega}\right) \sin(\sigma) \, d\sigma,$$

$$= \frac{1}{\omega} \int_{2\pi}^{3\pi} \left[F_N\left(\frac{\sigma}{\omega}\right) - F_N\left(\frac{\sigma + \pi}{\omega}\right) \right] \sin(\sigma) \, d\sigma.$$

Since F is supposed to be decreasing, we deduce that

$$S_N(\omega) - \chi_N(\omega) > 0.$$

By induction, we can prove that, for $\omega N \in (k\pi, (k+1)\pi], k \geq 2, S_N(\omega) > \chi_N(\omega)$. Therefore, $S_N(\omega) > \chi_N(\omega)$ for $\omega N > 2\pi$.

Now let $\omega > 0$ be fixed. There exists $N > 0$ large enough such that $\omega N > 2\pi$. Consequently, from the above result, $S_N(\omega) > \chi_N(\omega) > 0$.

Since $\omega N > 2\pi$, then (8.32) implies that $\chi_N(\omega) = \chi(\omega)$, where $\chi(\omega)$ is given by (8.29) and $\chi_N(\omega)$ is then independent of N. Thus,

$$S_N(\omega) > \chi(\omega) > 0.$$

Taking the limit when N tends to infinity in the above inequality, one obtains

$$S(\omega) \geq \chi(\omega) > 0.$$

This concludes the proof.   □

Under the assumption of Lemma 8.14, $S(\omega) > 0$ for $\omega > 0$. Thus, we can rewrite the second equation of (8.26) as

$$B = \frac{\omega}{S(\omega)}.$$

Then this expression of B can be used in the first equation of (8.26) to obtain an equation satisfied only by ω. This equation would be

$$\omega \frac{C(\omega)}{S(\omega)} = -A. \tag{8.33}$$

However, it is not easy to solve (8.33) and neither it is to determine properties of the function $\omega C(\omega)/S(\omega)$. That is why we introduce a new variable, $\overline{A}$, defined by

$$\overline{A} = A + B \int_0^\infty F(\theta)\, d\theta. \tag{8.34}$$

For the sake of simplicity, we denote, in the following, by $\overline{F}$ the integral of the function F, so that

$$\overline{A} = A + B\overline{F}.$$

Assumption (8.25) is then equivalent to $2B\overline{F} > \overline{A} > 0$.
    Writing $A = \overline{A} - B\overline{F}$ in (8.26), we obtain

$$B = \frac{\omega}{S(\omega)} \quad \text{and} \quad \omega \frac{\overline{F} - C(\omega)}{S(\omega)} = \overline{A}. \tag{8.35}$$

By solving (8.35), we will obtain a solution (ω^*, B^*) for which $\pm i\omega^*$ is a pair of purely imaginary eigenvalues of (8.1) when $B = B^*$.

Lemma 8.15 *Suppose $S(\omega) > 0$ for $\omega > 0$. Then there exists $B^* > 0$ satisfying (8.25) such that, when $B = B^*$, (8.1) has a pair of purely imaginary eigenvalues $\pm i\omega^*$, where $B^* = \overline{A}/(\overline{F} - C(\omega^*))$ and ω^* satisfies (8.35).*

Proof Define the function $\xi : (0, +\infty) \to \mathbb{R}$ by

$$\xi(\omega) = \omega \frac{\overline{F} - C(\omega)}{S(\omega)}, \quad \omega > 0,$$

where C and S are given by (8.27).
    Since

$$\lim_{\omega \to 0} \frac{S(\omega)}{\omega} = \lim_{\omega \to 0} \int_0^\infty \theta F(\theta) \frac{\sin(\omega\theta)}{\omega\theta}\, d\theta = \int_0^\infty \theta F(\theta)\, d\theta,$$

and $C(0) = \overline{F}$, we deduce

$$\lim_{\omega \to 0} \xi(\omega) = 0.$$

In addition, from Riemann–Lebesgue's Lemma,

$$\lim_{\omega \to +\infty} S(\omega) = \lim_{\omega \to +\infty} C(\omega) = 0,$$

so

$$\lim_{\omega \to +\infty} \xi(\omega) = +\infty.$$

Since ξ is a continuous function on $(0, +\infty)$, we obtain the existence of at least one $\omega > 0$, denoted by ω^*, such that

$$\xi(\omega^*) = \overline{A}, \qquad (8.36)$$

where we recall that $\overline{A} > 0$.

Note that $C(\omega^*) \neq \overline{F}$, otherwise $\xi(\omega^*) = 0 < \overline{A}$. Then, we set

$$B^* = \frac{\omega^*}{S(\omega^*)} = \frac{\overline{A}}{\overline{F} - C(\omega^*)} > 0.$$

Since $|C(\omega)| \leq \overline{F}$, then

$$B^* > \frac{\overline{A}}{2\overline{F}},$$

and $2B^*\overline{F} > \overline{A}$. Therefore B^* satisfies (8.25). Moreover, (ω^*, B^*) satisfies (8.35), which is equivalent to (8.26), so $\pm i\omega^*$ is a pair of purely imaginary eigenvalues of (8.1) when $B = B^*$. This ends the proof. $\qquad \square$

Lemma 8.15 gives a condition for the existence of purely imaginary eigenvalues of (8.1). In the next lemma, we show that purely imaginary eigenvalues of (8.1) are always simple and we determine a condition for the transversality condition to hold.

Lemma 8.16 *Suppose $\pm i\omega^*$, with $\omega^* > 0$, is a pair of purely imaginary eigenvalues of (8.1) that appears when $B = B^*$. Then $\pm i\omega^*$ is a simple pair of characteristic roots such that*

$$\mathrm{Re}\left(\frac{d\lambda}{dB}(B^*)\right) > 0 \quad \text{if and only if} \quad \frac{\omega^* S'(\omega^*)}{C(\omega^*)S(\omega^*)} > \frac{d}{d\omega}\left(\frac{\omega}{C(\omega)}\right)\bigg|_{\omega=\omega^*}. \qquad (8.37)$$

Proof Consider a branch $\lambda(B)$ of eigenvalues of (8.1), given by $\lambda(B) = \mu(B) + i\omega(B)$, such that

$$\mu(B^*) = 0 \quad \text{and} \quad \omega(B^*) = \omega^*, \ \omega^* > 0.$$

From now on, we explicitly write the dependence of the characteristic equation (8.17) on the bifurcation parameter B, so we write $\Delta(\lambda, B)$ instead of $\Delta(\lambda)$, as defined in (8.18). Since $\Delta(\lambda(B), B) = 0$, then differentiating $\Delta(\lambda(B), B)$ with respect to B gives

$$\frac{d\lambda}{dB}(B)\frac{\partial\Delta}{\partial\lambda}(\lambda(B), B) + \frac{\partial\Delta}{\partial B}(\lambda(B), B) = 0. \qquad (8.38)$$

One can check that

$$\frac{\partial \Delta}{\partial \lambda}(\lambda(B), B) = 1 - B \int_0^\infty \theta F(\theta)e^{-\lambda(B)\theta}\, d\theta, \tag{8.39}$$

and

$$\frac{\partial \Delta}{\partial B}(\lambda(B), B) = \int_0^\infty F(\theta)e^{-\lambda(B)\theta}\, d\theta.$$

Since $\lambda(B)$ is an eigenvalue of (8.1), it satisfies (8.17) and so

$$\int_0^\infty F(\theta)e^{-\lambda(B)\theta}\, d\theta = -\frac{\lambda(B) + A}{B}.$$

Hence,

$$\frac{\partial \Delta}{\partial B}(\lambda(B), B) = -\frac{\lambda(B) + A}{B}. \tag{8.40}$$

Suppose, by contradiction, that $\pm i\omega^*$ are not simple eigenvalues of (8.1). Then

$$\frac{\partial \Delta}{\partial \lambda}(\lambda(B^*), B^*) = 0.$$

From (8.38), we deduce that

$$\frac{\partial \Delta}{\partial B}(\lambda(B^*), B^*) = 0,$$

that is, from (8.40),

$$\frac{\lambda(B^*) + A}{B^*} = 0.$$

Separating real and imaginary parts and taking into account that $B^* > 0$, this yields

$$\omega^* = 0 \quad \text{and} \quad A = 0.$$

This is impossible since $\omega^* > 0$ and we assumed, at the beginning of this section, that $A \neq 0$, so a contradiction holds. We conclude that $\pm i\omega^*$ are simple eigenvalues of (8.1).

From (8.38), (8.39), and (8.40), we obtain

$$\left(\frac{d\lambda}{dB}(B)\right)^{-1} = -\frac{\dfrac{\partial \Delta}{\partial \lambda}(\lambda(B), B)}{\dfrac{\partial \Delta}{\partial B}(\lambda(B), B)} = B\frac{1 - B\displaystyle\int_0^\infty \theta F(\theta)e^{-\lambda(B)\theta}\, d\theta}{\lambda(B) + A}.$$

Therefore,

$$\left(\frac{\mathrm{d}\lambda}{\mathrm{d}B}(B^*)\right)^{-1} = B^* \frac{1 - B^*S'(\omega^*) + iB^*C'(\omega^*)}{A + i\omega^*},$$

where C and S are defined by (8.27). Hence,

$$\mathrm{Re}\left(\frac{\mathrm{d}\lambda}{\mathrm{d}B}(B^*)\right)^{-1} = \frac{B^*}{A^2 + (\omega^*)^2}\left[A(1 - B^*S'(\omega^*)) + \omega^*B^*C'(\omega^*)\right].$$

Remembering that ω^* and B^* satisfy (8.26), we have

$$B^* = \frac{\omega^*}{S(\omega^*)} \quad \text{and} \quad A = -B^*C(\omega^*).$$

Using these expressions, we obtain

$$\mathrm{Re}\left(\frac{\mathrm{d}\lambda}{\mathrm{d}B}(B^*)\right)^{-1} = \frac{(B^*)^2}{A^2 + (\omega^*)^2}\left[-C(\omega^*) + \frac{\omega^*C(\omega^*)S'(\omega^*)}{S(\omega^*)} + \omega^*C'(\omega^*)\right],$$

that we rewrite

$$\mathrm{Re}\left(\frac{\mathrm{d}\lambda}{\mathrm{d}B}(B^*)\right)^{-1} = \frac{(B^*)^2C^2(\omega^*)}{A^2 + (\omega^*)^2}\left[\frac{\omega^*S'(\omega^*)}{C(\omega^*)S(\omega^*)} - \frac{C(\omega^*) - \omega^*C'(\omega^*)}{C^2(\omega^*)}\right],$$

$$= \frac{A^2}{A^2 + (\omega^*)^2}\left[\frac{\omega^*S'(\omega^*)}{C(\omega^*)S(\omega^*)} - \frac{\mathrm{d}}{\mathrm{d}\omega}\left(\frac{\omega}{C(\omega)}\right)\Big|_{\omega=\omega^*}\right].$$

Since

$$\mathrm{sign}\left\{\frac{\mathrm{d}\mathrm{Re}(\lambda)}{\mathrm{d}B}(B^*)\right\} = \mathrm{sign}\left\{\mathrm{Re}\left(\frac{\mathrm{d}\lambda}{\mathrm{d}B}(B^*)\right)^{-1}\right\},$$

we conclude to (8.37). This concludes the proof.   □

From the results established in Lemmas 8.14, 8.15, and 8.16, we are able to prove the existence of a Hopf bifurcation that would destabilize (8.1) for a certain value of the parameter B. This is proved in the next theorem.

Theorem 8.17 *Assume*

$$B\int_0^\infty F(\theta)\,\mathrm{d}\theta > |A|,$$

and F is a decreasing function. Then there exists $B^* > 0$ satisfying (8.25) such that (8.1) is asymptotically stable for

$$\frac{|A|}{\int_0^\infty F(\theta)\,\mathrm{d}\theta} < B < B^*, \tag{8.41}$$

and (8.1) *becomes unstable for* $B \geq B^*$, *with a Hopf bifurcation occurring when* $B = B^*$, *provided that*

$$\frac{\omega^* S'(\omega^*)}{C(\omega^*) S(\omega^*)} \neq \frac{\mathrm{d}}{\mathrm{d}\omega}\left(\frac{\omega}{C(\omega)}\right)\Bigg|_{\omega=\omega^*}, \tag{8.42}$$

where $\pm i\omega^*$ *are purely imaginary eigenvalues of* (8.1) *when* $B = B^*$.

Proof Since F is supposed to be decreasing, $S(\omega) > 0$ for all $\omega > 0$ from Lemma 8.14. Thus, Lemma 8.15 gives the existence of (ω^*, B^*), with $\omega^* > 0$ and B^* satisfying (8.25), such that a pair of simple (see Lemma 8.16) purely imaginary eigenvalues $\pm i\omega^*$ of (8.1) exists when $B = B^*$. As a consequence of the proof of Lemma 8.15, (8.1) does not have purely imaginary eigenvalues when $B < B^*$.

From Rouché's Theorem [16, p. 248], it follows that (8.1) is asymptotically stable when B satisfies (8.41).

Assume (8.42) holds. Then, from Lemma 8.16, either

$$\mathrm{Re}\left(\frac{\mathrm{d}\lambda}{\mathrm{d}B}(B^*)\right) > 0 \quad \text{or} \quad \mathrm{Re}\left(\frac{\mathrm{d}\lambda}{\mathrm{d}B}(B^*)\right) < 0.$$

Suppose, by contradiction, that

$$\mathrm{Re}\left(\frac{\mathrm{d}\lambda}{\mathrm{d}B}\right) < 0$$

for $B < B^*$, B close to B^*. Then there exists an eigenvalue $\lambda(B)$ of (8.1) such that $\mathrm{Re}(\lambda(B)) > 0$ and $B < B^*$. This contradicts the stability of (8.1) for $B < B^*$. Thus,

$$\mathrm{Re}\left(\frac{\mathrm{d}\lambda}{\mathrm{d}B}(B^*)\right) > 0.$$

This implies the existence of a Hopf bifurcation when $B = B^*$ (see Theorem 8.4).

$\square$

Theorem 8.17 gives a condition for the existence of a Hopf bifurcation, which leads to the appearance of periodic solutions. This condition that the function F is decreasing can be relaxed to the condition that F is such that $S(\omega) > 0$ for $\omega > 0$, where S is defined by (8.27). However, up to now, no better condition has been found.

If F is a density function, it can be decreasing if it is piecewise constant, or if F is an exponential law, for example. If F is a Gamma distribution with parameters k and σ, given by (8.3), then it is a decreasing function if and only if $k = 0$, that is, if F is a *weak kernel*, which is an exponential distribution. However, given the fact that Gamma distributions are the more used density functions in the literature, it is not so worrying that Theorem 8.17 only works for the weak kernel, because if F has a Gamma distribution the characteristic equation can be reduced to a polynomial function and so other methods that are more interesting can be used to determine the stability of (8.1).

Eventually, the interest of Theorem 8.17 relies on the fact that it can be applied to functions F that are not *necessarily* density functions, but rather the product of a density function and a more general term (see the next section).

Another important point that deserves to be stressed is that Theorem 8.17 does not give information on an eventual bifurcation that could occur for larger values of the parameter B and lead to a stability switch. This can be observed for equations with delay-dependent coefficients (see [7]), and it can be assumed that, maybe under particular assumptions, the same behavior could be observed in differential equations with distributed delay.

In the next section, we consider a problem from population dynamics, the evolution of a stem cell population, modeled by a nonlinear delay differential equation with distributed delay, and we show that the existence of a Hopf bifurcation, given by Theorem 8.17, leads to interesting results related to some blood diseases, in particular to leukemias.

8.5 Application: Periodic Oscillations in a Stem Cell Population

Let's consider a population of hematopoietic stem cells (HSC) denoted by $S(t)$. These cells are at the root of the blood production process. They are located in the bone marrow, where they mature and differentiate through successive divisions to produce more and more differentiated cells, which will eventually give birth to blood cells. HSC can be divided into actively proliferating cells (which are in cell cycle where they synthesize DNA and divide at mitosis) and quiescent cells. The cell population $S(t)$ denotes the quiescent HSC population. It satisfies the nonlinear delay differential equation (see Adimy et al. [1, 2])

$$S'(t) = -[\alpha + \beta(S(t))]\, S(t) + 2 \int_0^\infty e^{-\gamma a} f(a)\beta(S(t-a))S(t-a)\, da. \quad (8.43)$$

In this model, HSC are assumed to differentiate with a constant rate $\alpha > 0$. Moreover, they can be introduced in the cell cycle whenever during their life with a rate β, depending on the HSC population itself. Typically, and from now on, β is chosen to be a Hill function,

$$\beta(S) = \beta_0 \frac{\theta^n}{\theta^n + S^n}, \quad \beta_0, \theta, n > 0. \quad (8.44)$$

It is a decreasing and bounded function of the HSC population, which tends to zero at infinity. The coefficient β_0 represents the maximum introduction rate, θ is the HSC population for which the introduction rate reaches half of its maximum, and n is the sensitivity of the introduction rate.

A proportion of cells that have been introduced in the cell cycle returns in the HSC population after division. The division occurs a certain time a after the introduction of cells in cycle. The time of division is distributed according to a density function f, which can usually be considered to have a compact support $[0, \tau]$, where $\tau > 0$ is the maximum age of division. Moreover, while in cycle, cells can die by apoptosis (a programmed cell death) with a rate $\gamma > 0$. The term $e^{-\gamma a}$ must then be understood as a survival rate of cells that have spent a time a in the cell cycle. Hence, the integral term in the right-hand side of (8.43) describes the amount of cells corresponding to cells introduced a time a earlier in cycle, that have survived, and divide according to the density f. Finally, the coefficient 2 in the last term of the right-hand side of (8.43) takes into account the division of each cell in two daughter cells. The reader interested in the mathematical modeling of hematopoiesis and stem cells dynamics is invited to study the works of Mackey [26, 27], Mackey and Rudnicki [29, 30], Mackey et al. [29], Bernard et al [9, 10], Pujo et al. [32, 33], Adimy et al [1, 2], Crauste [15], and the references therein.

Equation (8.43) is a nonlinear differential equation with distributed delay. From Hale and Verduyn Lunel [19], for every nonnegative and continuous initial condition φ defined on $(-\infty, 0]$ (or $[-\tau, 0]$ if f is supported on $[0, \tau]$), (8.43) has a unique nonnegative and continuous solution $S = S^\varphi$ defined for $t > 0$. In addition, if $\alpha > 0$, all solutions of (8.43) are bounded, and, if $\alpha = 0$ (8.43) may admit unbounded solutions. We refer to [1] for a detailed proof of these boundedness results.

We are going to apply the results obtained in Sect. 8.4, Theorems 8.12 and 8.17, to the linearized equation of (8.43) in order to determine the stability of the unique positive steady state of (8.43). We recall that a steady state of (8.43) is said to be *locally asymptotically stable*, or LAS, if the linearized equation of (8.43) about this steady state is asymptotically stable. The steady state is unstable if the linearized equation is unstable.

A steady state of (8.43) is a solution $\overline{S}$ satisfying $\overline{S}'(t) = 0$ for all $t > 0$. That is

$$\left[\alpha + \beta(\overline{S})\right]\overline{S} = 2\left(\int_0^\infty e^{-\gamma a}f(a)\,da\right)\beta(\overline{S})\overline{S}. \tag{8.45}$$

One easily checks that $\overline{S} = 0$ is always a steady state of (8.43). It describes the cell population extinction and, therefore, its study is not really biologically relevant.

Searching for nontrivial steady states of (8.43), one can see that (8.45) has a unique positive solution, denoted by S^*, satisfying

$$\left(2\int_0^\infty e^{-\gamma a}f(a)\,da - 1\right)\beta(S^*) = \alpha, \tag{8.46}$$

provided that

$$\left(2\int_0^\infty e^{-\gamma a}f(a)\,da - 1\right)\beta_0 > \alpha > 0. \tag{8.47}$$

From (8.46), using (8.44), we obtain

$$S^* = \theta \left(\frac{\kappa \beta_0 - \alpha}{\alpha} \right)^{\frac{1}{n}}, \tag{8.48}$$

where we have set, for the sake of simplicity,

$$\kappa = 2 \int_0^\infty e^{-\gamma a} f(a) \, da - 1.$$

Under assumption (8.47), $\kappa > 0$.

To analyze the behavior of the nontrivial steady state S^*, let us linearize (8.43) about S^* and study the eigenvalues of (8.43). We set

$$\beta^* := \frac{d}{dS} (S\beta(S))_{S=S^*} = \frac{\alpha}{\kappa^2 \beta_0} [\kappa \beta_0 - (\kappa \beta_0 - \alpha)n]. \tag{8.49}$$

Let $x(t) = S(t) - S^*$. The linearized equation of (8.43) about S^* is then given by

$$x'(t) = -(\alpha + \beta^*)x(t) + 2\beta^* \int_0^\infty e^{-\gamma a} f(a) x(t-a) \, da. \tag{8.50}$$

Setting

$$A = \alpha + \beta^*, \quad B = -\beta^*, \quad F(\theta) = 2e^{-\gamma \theta} f(\theta),$$

Equation (8.50) can be written in the form of (8.1). Moreover, the variable $\overline{A}$, defined by (8.34), is given here by

$$\overline{A} = \alpha - \kappa \beta^* = \alpha \frac{\kappa \beta_0 - \alpha}{\kappa \beta_0} n,$$

which is positive, thanks to (8.47). Note that $B > 0$ if and only if $\beta^* < 0$. Hence, we can state and prove the next proposition, using Theorems 8.12 and 8.17.

Proposition 8.18 *The nontrivial steady state S^* of (8.43) is locally asymptotically stable when*

$$0 < n \le \left[1 + \frac{\kappa}{\kappa + 2} \right] \frac{\kappa \beta_0}{\kappa \beta_0 - \alpha}. \tag{8.51}$$

In addition, assuming the function $a \mapsto e^{-\gamma a} f(a)$ is decreasing, there exists n^* satisfying

$$\left[1 + \frac{\kappa}{\kappa + 2} \right] \frac{\kappa \beta_0}{\kappa \beta_0 - \alpha} < n^*, \tag{8.52}$$

such that S^* is locally asymptotically stable for $0 < n < n^*$ and unstable for $n \geq n^*$, with a Hopf bifurcation occurring when $n = n^*$ provided that (8.42) holds.

Proof As recalled above, S^* is locally asymptotically stable if (8.50) is asymptotically stable. Then we use Theorem 8.12.

We already noticed that $\bar{A} = A + B \int_0^\infty F(\theta)\, d\theta > 0$, so S^* is LAS if $B \leq 0$ (see Theorem 8.12), that is, if $\beta^* \geq 0$. Using (8.44), this corresponds to

$$0 < n \leq \frac{\kappa\beta_0}{\kappa\beta_0 - \alpha}.$$

In addition, Theorem 8.12 says that S^* is LAS if $B > 0$ and $A > B \int_0^\infty F(\theta)\, d\theta$. After easy computations, it comes that these conditions are equivalent to

$$\frac{\kappa\beta_0}{\kappa\beta_0 - \alpha} < n \leq \left[1 + \frac{\kappa}{\kappa + 2}\right] \frac{\kappa\beta_0}{\kappa\beta_0 - \alpha}.$$

This proves (8.51) and the first point of this proposition.

Since finding a critical value of B is equivalent to finding a critical value of n, the second point of Proposition 8.18 is a trivial application of Theorem 8.17. □

The following corollaries are straightforward.

Corollary 8.19 *Suppose f has a uniform distribution on the interval $[0, \tau]$, $\tau > 0$. Then there exists n^* satisfying (8.52) such that S^* is locally asymptotically stable for $0 < n < n^*$ and unstable for $n \geq n^*$.*

Corollary 8.20 *Suppose f has an exponential distribution, as in (8.3) with $k = 0$ and $\sigma \in \mathbb{R}$. Then, if $\gamma + \sigma \geq 0$, there exists n^* satisfying (8.52) such that S^* is locally asymptotically stable for $0 < n < n^*$ and unstable for $n \geq n^*$.*

In order to numerically compute the solutions of (8.43), let us fix the values of the parameters β_0, θ, α, and γ, as given by Mackey [26],

$$\beta_0 = 1.77\ \text{d}^{-1}, \quad \theta = 1.68 \times 10^8\ \text{cells/kg}, \quad \alpha = 0.05\ \text{d}^{-1}, \quad \gamma = 0.2\ \text{d}^{-1}. \quad (8.53)$$

Using the MATLAB solver dde23 [37], we can compute the solutions of (8.43). We use the values of the parameters given by (8.53) and we consider two different types of density functions f, a first case when f has a uniform law on the interval $[0, \tau]$, with $\tau = 7$ days, and a second case when f is a *weak kernel*, that is, an exponential distribution (as in (8.3) with $k = 0$). The value of σ, in this latter case, is chosen as $\sigma = 2/7$ so that the expectation of f equals 3.5 days, similar to the uniform law case. Both distributions are depicted in Fig. 8.1.

When f has a uniform law on the interval $[0, 7]$, the Hopf bifurcation determined in Corollary 8.19 occurs when $n^* = 2.53$, whereas when f has an exponential distribution with $\sigma = 2/7$, the Hopf bifurcation occurs for $n^* = 3.44$ (see Fig. 8.2).

A bifurcation diagram showing values of the steady state S^*, explicitly given by (8.48), for both types of distributions is presented in Fig. 8.3. One can observe that

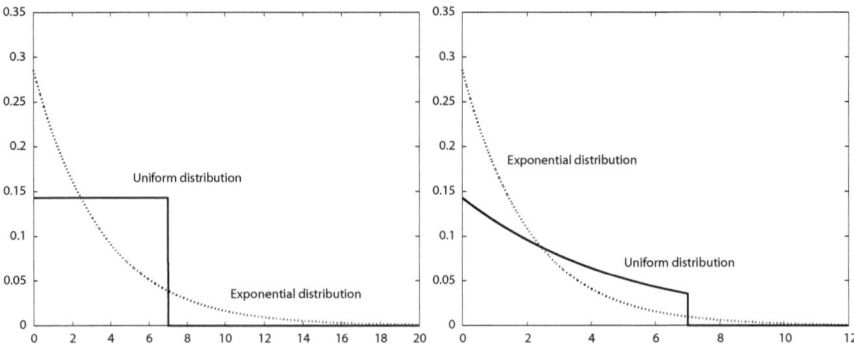

Fig. 8.1 *Left*: Density functions used in the simulations displayed in Figs. 8.2, 8.3, 8.4, and 8.5, a uniform distribution on the interval $[0, 7]$ and an exponential distribution, as defined in (8.3), with $\sigma = 2/7$ and $k = 0$. *Right*: One can observe shapes of the density functions multiplied by the survival rate $e^{-\gamma a}$, with $\gamma = 0.2$ days^{-1}

values of the steady state are different depending on the choice of the density function, especially for small values of n (that is, when the steady state is asymptotically stable). For large values of n, this phenomenon disappears.

When n increases away from the critical value n^*, oscillating solutions can be observed, with increasing periods. This is displayed in Figs. 8.4 and 8.5. However, it is noticeable that amplitudes and periods of the oscillations are different depending on the choice of the density function f. With a uniform distribution of cell cycle durations, long-period oscillations, in the order of 40–70 days, can be observed for values of n between 3 and 5 (see Fig. 8.4). In the case of exponential distribution, periods and amplitudes of the oscillations are shorter, in the order of 26–40 days, for larger values of the parameter n (see Fig. 8.5).

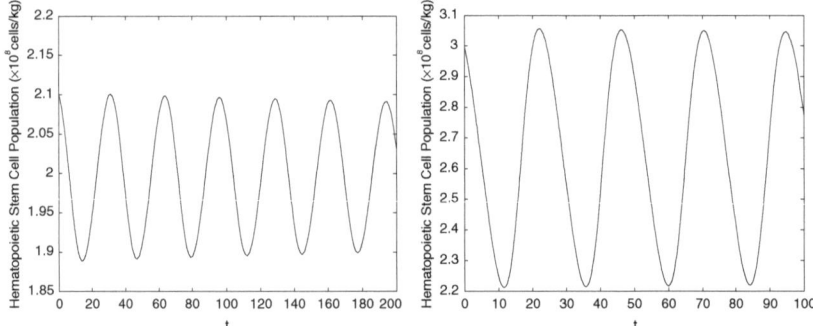

Fig. 8.2 Solutions $S(t)$ of (8.43) are drawn for parameters given by (8.53). For a critical value n^* of the parameter n, the steady state S^* undergoes a Hopf bifurcation and periodic solutions appear, with different periods and amplitudes depending on the density function. When division times are distributed uniformly on the interval $[0, 7]$ (*left*), the Hopf bifurcation occurs for $n^* = 2.53$. Oscillating solutions have periods about 33 days. When the density function is an exponential distribution (*right*), the Hopf bifurcation occurs for $n^* = 3.44$ and periods of the oscillations are in the order of 24 days

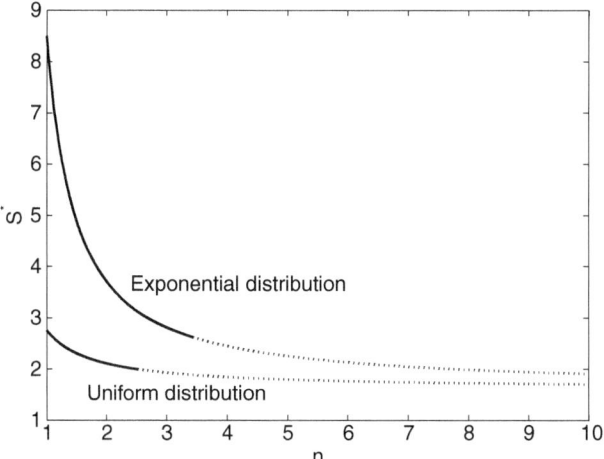

Fig. 8.3 Bifurcation diagram for the steady state S^* given by (8.48). All parameters are given by (8.53). The upper curve corresponds to an exponential distribution (8.3) with $\sigma = 2/7$. The steady state is asymptotically stable for $n < 3.44$ and unstable for $n \geq 3.44$. The lower curve corresponds to a uniform distribution of cell cycle times on $[0, 7]$. The steady state is asymptotically stable for $n < 2.53$ and unstable for $n \geq 2.53$

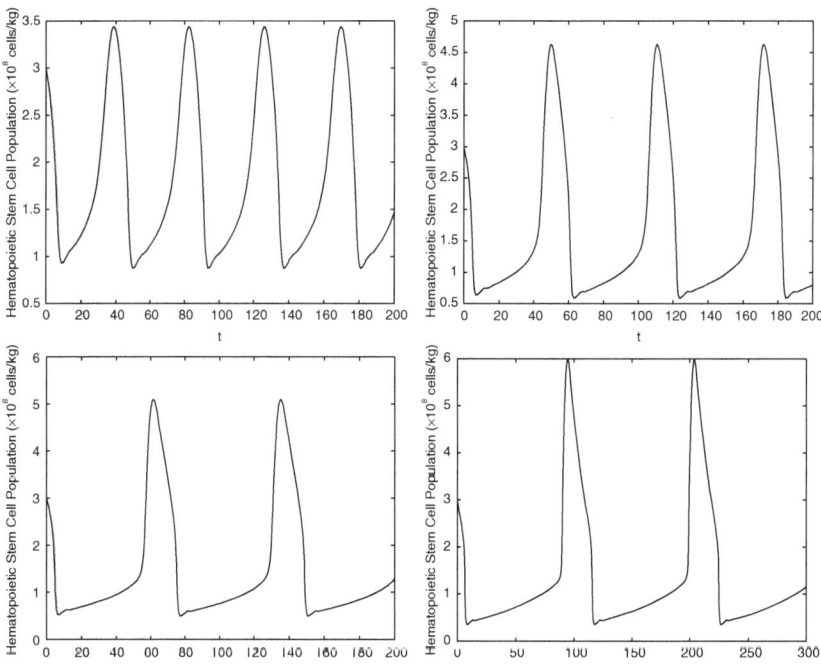

Fig. 8.4 Solutions of (8.43) are drawn for n larger than the critical value n^* given by Corollary 8.19, when the density function has a uniform distribution on the interval $[0, 7]$. From *left* to *right* and *top* to *bottom* n successively equals 3, 4, 5, and 10. Periods of the oscillations range from 40 to 110 days, with periods about 60–70 days when $n = 4$ or 5. Amplitudes increase as n increases, with very low values reached by the population when n is large enough

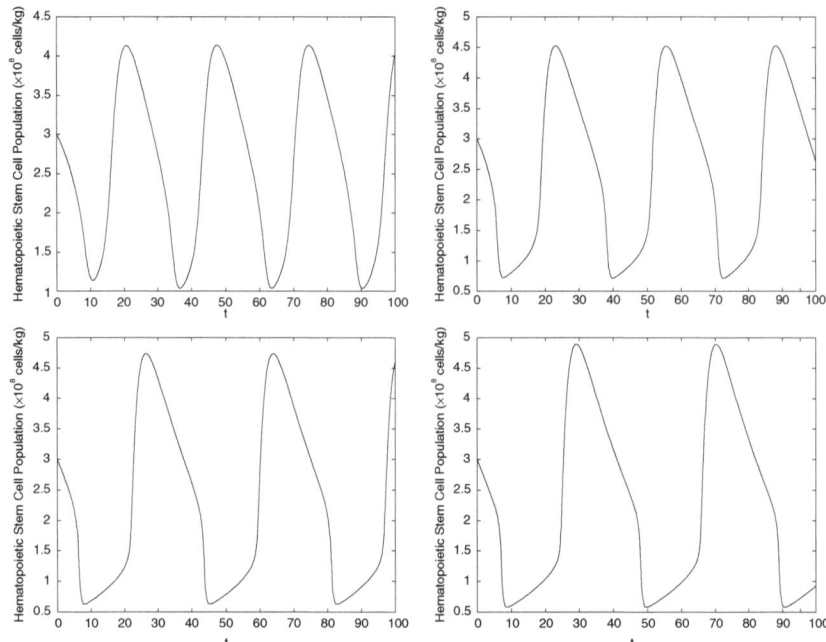

Fig. 8.5 Solutions of (8.43) are drawn for n larger than the critical value n^* given by Corollary 8.20, when the density function has an exponential distribution with parameter $\sigma = 2/7$. From *left* to *right* and *top* to *bottom* n successively equals 4, 6, 8, and 10. Periods of the oscillations range from 26 to 40 days. Amplitudes increase as n increases, with very low values reached by the population when n is large enough, yet the increase is less drastic than in the case of the uniform distribution

Periods of the oscillations for the uniform distribution increase rapidly with n to reach very long periods, whereas the exponential distribution slows down the increase of the periods, which quickly reach some limit value. The same effect is observed for amplitudes of the oscillations. Thus the shape of the density function— more particularly in this case the shape of the density function multiplied by the exponential survival rate—plays an important role not only in the appearance of oscillating solutions but also in the length of the periods and the range of the amplitudes (see Fig. 8.1).

Oscillating solutions in blood cell populations have been observed in some patients with blood diseases [22]. For example, experimental studies report cases of oscillations of all blood cell counts with average periods of 70–80 days in some patients with chronic myelogenous leukemia (see [17]), a widespread form of blood cancer. Cyclical neutropenia [22], another blood disease, characterized by a fall of blood cell counts every 3 weeks in human, may also exhibit stressed oscillations, with periods in the range of 20–30 days. Oscillations obtained in Figs. 8.4 and 8.5 by simulating the differential equation with distributed delay (8.43) for different density functions may contribute to the study of blood diseases characterized by oscillations of blood cell counts (such diseases are known as *periodic hematological diseases*). This can be found, for example, in [9] or [1, 2], where the results obtained by the authors stress the localization of these diseases in the pluripotent HSC compartment. This explains why oscillations are observed in all cell types (red blood cells, white cells, and platelets) though these diseases are associated with only one blood cell type (white cells for leukemia, for example).

References

1. M. Adimy, F. Crauste, and S. Ruan. A mathematical study of the hematopoiesis process with applications to chronic myelogenous leukemia. *SIAM J. Appl. Math.*, 65(4): 1328–1352 2005.
2. M. Adimy, F. Crauste, and S. Ruan. Stability and hopf bifurcation in a mathematical model of pluripotent stem cell dynamics. Nonlinear Analysis: Real World Applications, 6: 651–670 2005.
3. M. Adimy, F. Crauste, and S. Ruan. Periodic oscillations in leukopoiesis models with two delays. *J. Theor. Biol.*, 242: 288–299 2006.
4. R. F. V. Anderson. Geometric and probabilistic stability criteria for delay systems. *Math. Biosci.*, 105: 81–96 1991.
5. R. F. V. Anderson. Intrinsic parameters and stability of differential-delay equations. *J. Math. Anal. Appl.*, 163: 184–199 1992.
6. F. M. Atay. Delayed feedback control near hopf bifurcation. *Disc. Cont. Dyn. Syst. Ser. S* 1 (2): 197–205 2008.
7. E. Beretta, and Y. Kuang. Geometric stability switch criteria in delay differential systems with delay dependent parameters. *SIAM J. Math. Anal.* 33(5): 1144–1165 2002.
8. S. Bernard, J. Bélair, and M. C. Mackey. Sufficient conditions for stability of linear differential equations with distributed delay. *Disc. Cont. Dyn. Syst. Ser.*, B 1: 233–256 2001.
9. S. Bernard, J. Bélair, and M. C. Mackey. Oscillations in cyclical neutropenia: new evidence for origins based on mathematical modeling. *J. Theor. Biol.*, 223: 283–298 2003.
10. S. Bernard, J. Bélair, and M. C. Mackey. Bifurcations in a white blood cell production model. *Compt. R. Biol.* 327: 201–210 2004.
11. F. G. Boese. The stability chart for the linearized cushing equation with a discrete delay and gamma-distributed delays. *J. Math. Anal. Appl.*, 140: 510–536 1989.
12. F. G. Boese. Stability criteria for second-order dynamical systems involving several time delays. *SIAM J. Math. Anal.* 26: 1306–1330 1995.
13. F. J. Burns and I. F. Tannock. On the existence of a G_0 phase in the cell cycle. Cell. *Tissue Kinet*, 19: 321–334 1970.
14. K. L. Cooke and Z. Grossman. Discrete delay, distributed delay and stability switches. *J. Math. Anal. Appl.* **86**, 592–627 (1982)
15. F. Crauste. Global asymptotic stability and hopf bifurcation for a blood cell production model. *Math. Biosci. Eng.*, 3(2): 325–346 2006.
16. J. Dieudonné. *Foundations of Modern Analysis*. Academic Press, New-York, 1960.
17. P. Fortin and M. C. Mackey. Periodic chronic myelogenous leukemia: spectral analysis of blood cell counts and etiological implications. *Brit. J. Haematol.* 104: 336–345 1999.
18. J. Hale. *Theory of Functional Differential Equations*. Springer, New York, 1977.
19. J. K. Hale, and S. M. Verduyn Lunel. *Introduction to Functional Differential Equations*. Applied Mathematical Sciences 99, Springer-Verlag, New York, 1993.
20. N. D. Hayes. Roots of the transcendental equation associated with a certain differential difference equation. *J. London Math. Society*. 25: 226–232 1950.
21. C. Huang and S. Vandewalle. An analysis of delay-dependent stability for ordinary and partial differential equations with fixed and distributed delays. *SIAM J. Sci. Comput.*, 25 (5): 1608–1632 2004.

22. C. Haurie, D. C. Dale, and M. C. Mackey. Cyclical neutropenia and other periodic hemato-logical diseases: A review of mechanisms and mathematical models. *Blood*, 92: 2629–2640 1998.
23. Y. Kuang. *Delay Differential Equations with Applications in Population Dynamics*. Mathe-matics in Science and Engineering 191, Academic Press, 1993.
24. Y. Kuang. Nonoccurrence of stability switching in systems of differential equations with dis-tributed delays. *Quart. Appl. Math.*, LII(3): 569–578 1994.
25. N. MacDonalds. *Biological Delay Systems: Linear Stability Theory*. Cambridge Studies Math-ematical Biol. 8, Cambridge University Press, Cambridge, 1989.
26. M. C. Mackey. A unified hypothesis on the origin of aplastic anaemia and periodic hematopoiesis. *Blood*, 51: 946–956 1978.
27. M. C. Mackey. Dynamic hematological disorders of stem cell origin. In: J. G. Vassileva-Popova, and E. V. Jensen, (eds.) *Biophysical and Biochemical Information Transfer in Recog-nition*. Plenum Press, New York, 1979.
28. M. C. Mackey, C. Haurie, and J. Bélair. Cell replication and control. In: A. Beuter, L. Glass, M. C. Mackey, and M. S. Titcombe (eds.) *Nonlinear Dynamics in Physiology and Medicine*. Springer, New York 2003.
29. M. C. Mackey and R. Rudnicki. Global stability in a delayed partial differential equation describing cellular replication. *J. Math. Biol.*, 33: 89–109 1994.
30. M. C. Mackey and R. Rudnicki. A new criterion for the global stability of simultaneous cell replication and maturation processes. *J. Math. Biol.*, 38: 195–219 1999.
31. H. Ozbay, C. Bonnet, and J. Clairambault. Stability Analysis of Systems with Distributed Delays and Application to Hematopoietic Cell Maturation Dynamics. *Proceedings of the 47th IEEE Conference on Decision and Control*. Cancun, Mexico, December 2008 (to appear).
32. L. Pujo-Menjouet, S. Bernard, and M. C. Mackey. Long period oscillations in a G_0 model of hematopoietic stem cells. *SIAM J. Appl. Dynam. Sys.*, 4 (2): 312–332 2005.
33. L. Pujo-Menjouet, and M. C. Mackey. Contribution to the study of periodic chronic myeloge-nous leukemia. *C. R. Biologies*, 327: 235–244 2004.
34. S. Ruan. Delay differential equations in single species dynamics. In: O. Arino, M. Hbid, and E. Aitdads (eds.) *Delay Differential Equations with Applications*. Springer, Berlin 2006.
35. S. Ruan and J. Wei. Periodic solutions of planar systems with two delays. *Proc. Royal Soc. Edinburgh Ser.* 129A: 1017–1032 1999.
36. Ruan S., Wei J. On the Zeros of Transcendental Functions with Applications to Stability of Delay Differential Equations with Two Delays. *Dyn. Contin. Discrete Impuls. Syst. Ser. A Math. Anal.*, 10: 863–874 2003.
37. L. F. Shampine and S. Thompson. Solving DDEs in MATLAB. *Appl. Numer. Math.*, 37: 441–458 2001. http://www.radford.edu/~thompson/webddes/.
38. X. H. Tang. Asymptotic behavior of a differential equation with distributed delays. *J. Math. Anal. Appl.*, 301: 313–335 2005.
39. J. Wei and S. Ruan. Stability and bifurcation in a neural network model with two delays. *Physica D*, 130: 255–272 1999.

Chapter 9
Deterministic Time-Delayed Traffic Flow Models: A Survey

Rifat Sipahi and Silviu-Iulian Niculescu

Abstract Research in understanding traffic flow is conducted since 1930s in mathematics, physics, and engineering fields. The main interest is to reveal the characteristics of traffic dynamics and consequently propose ways to reduce undesirable impacts of traffic flow to social and economical life. This can be achieved only if rigorous and reliable *mathematical models* are constructed. The first part of this work covers the classification of such models as well as empirical and software tools used to study and predict traffic flow. The second part is devoted to a *critical* parameter in the traffic dynamics: *time delay*, which is recognized in this particular area as early as 1958. Delay originates from the time needed by human drivers to become conscious, make decision, and perform control actions in traffic. Such a definition states that human beings actively control the time evolution of traffic by their time-delayed behaviors (human as a controller/plant), and thus traffic dynamics becomes *inherently* time delayed. This dynamical structure, in a global sense, can also be seen as an interconnection of dynamics that transfer information/energy/momentum among each other, but under the presence of communication/transportation delays. For the specific problem considered, we first discuss the source of time delay, its physical interpretations, and mathematical nature, and next present a survey on mathematical models that explicitly account for delays. We conclude with interesting research topics at the intersection of control theory and time-delay systems. In this context, an example traffic flow scenario is covered to both demonstrate this intersection and show the consequences of delay presence in traffic flow dynamics, especially from the stability point-of-view.

R. Sipahi (✉)
Department of Mechanical and Industrial Engineering, 321 Snell Engineering Center,
Northeastern University, Boston, MA 02115, USA
e-mail: rifat@coe.neu.edu

F.M. Atay (ed.), *Complex Time-Delay Systems,* Understanding Complex Systems,
DOI 10.1007/978-3-642-02329-3_9, © Springer-Verlag Berlin Heidelberg 2010

9.1 Introduction to Traffic Flow Problem

Traffic flow problem has been a focus for researchers since 1930s. The main reason is that a mismanaged traffic flow results in undesirable effects in social and economical life. Starting from 1990s, the highways of rapidly developing cities especially in the USA, Germany, the UK, France, and Japan start to be a source of problems due to the irregular flow of traffic. This irregular traffic, mainly as *congestion*, in many cities causes high emissions and undesirable amounts of dust and smog. Moreover, for instance in Germany, the estimated financial costs of the time consumed due to traffic congestion and the impacts on the environment are about $100 billion each year (BMW study, 1997). Traffic congestion brings many disadvantages where human life is threatened, environment is irrecoverably damaged, and economical losses become substantial so that they cannot be ignored. These undesirable effects can be eliminated, when an efficiently functioning traffic flow is achieved in the vehicular network. For this, the ultimate goal is to understand first the *underlying mechanisms of traffic dynamics*, i.e., to create realistic *mathematical models*. Following this objective, in early 1990s, many researchers primarily from physics have initiated the development of mathematical models [55, 56, 76], which was followed by many publications in mathematics and engineering.

There are many parameters that play an important role in traffic models; for example, the physical conditions of highways, the mechanical properties of the vehicles, the psychological states of the drivers, traffic laws, on- and off-ramps, multiple lanes, and traffic lights. For a guided tour, see, for instance, [22] and the references therein. The *first part* of this work (Sect. 9.2) focuses on a brief classification of these traffic models in order to guide the reader on the problem. In general, this classification can be performed in two groups, *stochastic* and *deterministic* models. Here, for the sake of brevity, we only include deterministic models, which can be *continuous* or *discrete* in nature. Furthermore, we cover in Sect. 9.3 empirical studies and software developments that were conducted to predict the dynamical behavior of traffic flow. We stress that similar surveys in the above sense also exist in the literature, but we only aim the completeness of the work and wish to prepare the readers for the second part of the text.

Among the parameters that play major role in the behavior of traffic, there exists a critical one which is recognized in the traffic studies as early as 1958s [7]: it is the *time delay*. It mainly originates due to the time needed by human operators in sensing velocity and position variations of the vehicles in the traffic. Consequently, traffic dynamics and ultimately its mathematical models *inherently* carry time delays. In this sense, see, for instance, [5, 50, 71] for some models, related discussions, and interpretations.

Time delays may drastically deteriorate the interpreted characterizations of the delay-free dynamics. In traffic dynamics, for instance, a prediction of *homogeneously flowing* traffic of a mathematical model may become *congested* or may predict *accidents* when time delay is taken into account. For this reason, when time delay is present in the dynamics, a stability analysis with respect to time delay parameter becomes necessary in order to understand (a) if the flow dynamics may operate stably for a given time delay or (b) in order to find the maximum allowable time delay for maintaining the stability of the traffic dynamics. However, before going into the stability analysis, one first needs to answer the following question: *How does time delay appear in mathematical models?* In Sect. 9.4, we first answer this question by covering the *origin of time delay* and *time-delay modeling*. In general, time delay may be of stochastic and/or deterministic nature with constant and time-varying components which originate due to various factors. We discuss these factors and give an overview of how these factors may generate the components of time delay. Next, we present our survey on time-delayed mathematical models of traffic dynamics in a chronological order. As to our knowledge, a survey along this perspective has not been presented in the literature.

Based on our survey of time-delayed traffic models, we see that the stability analysis of these models and their controller design did not fully benefit from the *systems and control engineering perspective*, except in a few studies, [49, 50, 52, 53, 59] which are the first attempts along this direction to our best knowledge. This is an interesting observation especially because these two research areas have been developed in almost the same time frames. In Sect. 9.5, we outline some specific assumptions and approaches taken in the existing literature for studying traffic flow dynamics.

In Sect. 9.6, we merge *systems and control approach* to *time-delayed traffic problems*. This integration may produce new and beneficial results, as we demonstrate. We conclude in Sect. 9.7 with observations on the existing time-delayed models along with open problems for future work in this research field.

9.2 Classification of Traffic Models

A wide range of traffic models we find in the literature can be classified into two main sub-groups depending on the nature of their mathematical approaches. These two sub-groups are deterministic models and stochastic ones. In this text, for the sake of brevity we only focus on *deterministic models* leaving the discussions on stochastic models [30, 75] to another study.

Among many publications on deterministic mathematical models, we cite [9, 56, 77]. The major distinction between the models is on the *type of traffic flow parameters* that are used in their framework. While one stream of models considers *flow density* and *flow rate (flux)* as dependent variables, the other stream considers *dynamics of individual vehicles in traffic* as dependent variables.

An interesting observation is that many modeling techniques depart from heat transfer, fluid dynamics, thermodynamics, and granular media which, under certain *physical interpretations* and *assumptions*, serve very usefully in capturing a variety of phenomena of traffic flow dynamics. Many mathematical models are based on this philosophy, i.e., physical interpretations lead to models that evolve mathematically from *simplicity* to *complexity*.

In the following, we briefly explain the existing models by following a classification path usually preferred in the literature.

9.2.1 Macroscopic Models

With the analogy of hydrodynamic theory of fluids, macroscopic models are gathered from the fundamental equations of fluid dynamics and thus they are named as 'fluid dynamical theories' [8, 22]. In this approach, traffic is treated as a compressible fluid, and mathematical modeling becomes possible by writing the continuity and momentum equations of the flow. These models are usually in the form of partial differential equations whose two dependent variables are *traffic density* and *traffic flow rate* (flux). Two review papers [8, 22] on macroscopic traffic models are suggested for further details.

9.2.2 Microscopic Models

This type of models takes into account the dynamics of every vehicle in a traffic flow scenario. Thus, the interaction between the vehicles becomes the focus of the models. As a consequence, the traffic flow is a set of particle dynamics that interact with each other. According to this definition, some of the model parameters can be listed as *velocity of vehicles, headway between vehicles, response characteristics of human operators, safe following distance (desired headway), and the effects of single-/multi-lanes*.

In the following, a further classification of microscopic models is given. First, the *car following models* are presented. Next, we introduce *cellular automaton* (CA or particle hopping) models that are based on discrete time and discrete space. We include CA models here for completeness, although they are usually in stochastic nature. The third sub-category establishes an interesting link between *gas-kinetic theories* and the behavior of traffic flow.

9.2.2.1 Car Following Models

These types of models consider different driving strategies of the humans. The fundamental idea is that the drivers follow their neighboring vehicles to maintain safe driving conditions. In brief, drivers receive a certain *stimulus*, which can be in the form of headway error between two consecutive cars, and according to this 'stimulus', a certain 'response' is produced, which eventually leads to accelerations or decelerations of the vehicle. This type of modeling is also called 'follow-the-leader' model. While the model is able to represent individual vehicles and drivers in the form of continuous-time differential equations, the model may fall short in explaining multi-lane traffic dynamics, since drivers do not wish to change lanes because they are satisfied so long they maintain a safe velocity and headway in their lanes.

9.2.2.2 Cellular Automaton (CA) Models/Particle Hopping Models

CA models vary fundamentally from the follow-the-leader models in the sense that they are defined by discrete equations. These models are both spatially and temporally discrete; the space where the traffic flows is discretized into fixed-size cells and the model is simulated by discrete increments of time. For the behavior of the vehicles, certain 'if' conditions are set in order to represent a real traffic flow where these rules update the individual vehicles at every time step during the simulations. The advantage of CA models is that they lead to rapid computations of large-scale simulations, and it is convenient to incorporate stochastic effects into these models. A thorough discussion on various CA models can be found in the surveys [8, 22, 58].

9.2.2.3 Kinetic Theories

In this sub-class, the modeling of traffic is based on the assumption that the vehicles act as interacting particles in a gas flow, see [22]. For the analysis of traffic dynamics, the pertaining gas-kinetic equations should be adjusted so that the model is physically realistic. For example, in gas-kinetic theory, gas particles share their momentum by colliding with each other, however, this has to be completely avoided when modeling traffic dynamics via this approach.

9.2.3 Mesoscopic Models

These types of models, in essence, are obtained by departing from microscopic gas-kinetic models to arrive at macroscopic fluid dynamic traffic models. There are several studies along this line [8, 22], where Navier–Stokes-like continuity equations are derived by manipulating the Boltzmann-like equations. One example is given here to guide the reader. A thermodynamic approach by Nagatani is proposed in [42] by deriving macroscopic-governing dynamics departing from a car following model. The study analyzes the phase transitions of traffic dynamics based on the derivation of the macroscopic equations, and it shows that these results are in well agreement with the simulations. Finally, jamming phenomenon is proven via this thermodynamic approach, which is based on phase transitions.

9.3 Empirical and Simulation Studies

9.3.1 Experimental Studies

In parallel to the analytical studies, experimental work is also developed and reported in the literature. The main objective in the experimental studies is either to show their agreement with analytical results or to develop empirical–analytical models using measured experimental data. In 1961, Edie [14] investigated the experimental results taken from Lincoln Tunnel and realized that *density-flow* graphs are formed by two separate branches, one for free flow and the other for congested flow. Edie also reports that these two branches are separated by a discontinuity, indicating a *capacity drop* in the flowing traffic. Another capacity drop study is done by Koshi et al. in 1983 [32] by analyzing the data taken from Tokyo Expressway. Koshi et al. claim that *density-flow* graphs resemble a mirror image of a reversed λ symbol. In 1998, Kerner [26] using data collected from German Highways shows that flow rates out of a traffic jam are much smaller than the maximum flow rates possible in a free flow.

In 1995, Bando published a comparison study [3] between non-delayed optimal velocity model (OVM; see (9.4) below for delayed OVM) and experimental results that are taken from Japan Highway Public Corporation; compare also Figs. 9.1 and 9.2. It is claimed that OVM predicts satisfactorily the free flow, congested flow, and the discontinuity occurring between these two flows on *density-flow* graphs.

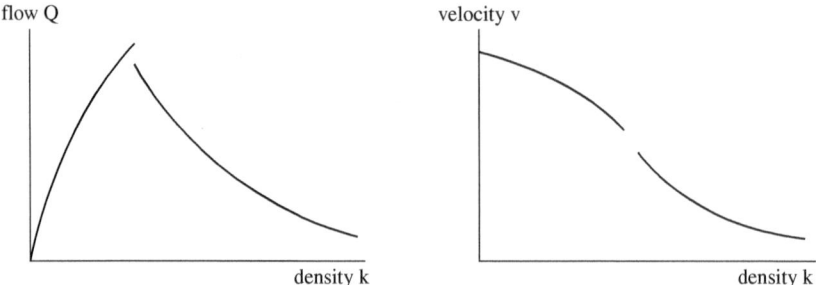

Fig. 9.1 (**a**) Flow vs. density and (**b**) velocity vs. density plots. Figure is borrowed from [3] with the permission of the publisher

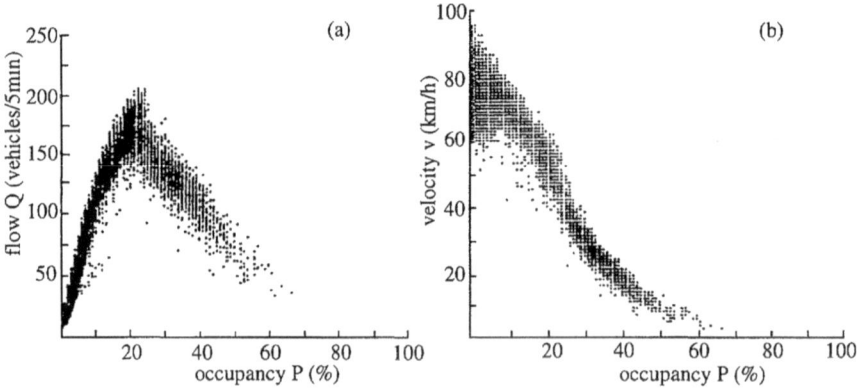

Fig. 9.2 Empirical (**a**) flow vs. occupancy and (**b**) velocity vs. occupancy plots obtained from Japan Highway Public Corporation. Figure is borrowed from [3] with the permission of the publisher

In [71] a memory effect is coupled to the governing traffic dynamics. This effect represents a measure of adaptation of drivers to their environment during a certain period of time. The authors implement this memory effect to intelligent driver model (IDM, see (9.10)) and simulate the traffic. It is shown that the simulation results are in good agreement with the empirical data obtained from German Freeway A5-South.

An in-depth study comparing experimental results taken from German Highways A5-South, A5-North, A9-South, and A8-East with numerical simulations is given by Treiber et al. [72]. For simulations, authors deploy the IDM along with empirical boundary conditions and physical obstacles (inhomogeneities). As the authors claim, the simulations show acceptable agreement with empirical results, and moreover they demonstrate that an *on-ramp* merging to traffic causes similar effects seen in the capacity drop of traffic. We note that the references [3, 26, 71, 72] depart fundamentally from microscopic models of OVM or IDM, and they reach to conclusions pertaining to macroscopic features of the traffic flow. A similar approach is also the basis of Kerner's work in [28].

Before we close this section, some of the work on the experimental results in macroscopic modeling is cited. For an analysis of experimental results obtained on German Highways and their consequences on the macroscopic properties, see [24]. Furthermore, the work in [25] investigates the experimental results to identify and characterize the phase transitions (various behavioral characters of traffic, such as congestion, free flow) occurring in traffic. Another work of interest is found in [27], which analyzes the bottleneck (on/off ramp) effects in traffic dynamics from *macroscopic features*. The author in this study investigates the *synchronized flow*, distinguishes *weak* and *strong congestion* properties of traffic dynamics, and establishes the link between *wide moving jams* and synchronized flow phenomenon.

We direct the readers to Helbing's thorough survey ([22] p. 1074) for further discussions and references in this research direction.

9.3.2 Software Development

Computer simulations are primarily needed for understanding the behavior of traffic dynamics through mathematical models. The simulation software, in essence, is designed to simulate a large-scale traffic dynamics which can arise from a particular mathematical model or a combination of various models.

Two open sources based on microscopic models can be mentioned, MITSIM and SUMO. The Microscopic Traffic Simulator (MITSIM) is developed at Massachusetts Institute of Technology (MIT) and it deploys the *non-integer car following model* [22] (see also (9.3)). Simulation of Urban Mobility (SUMO) utilizes the Krauss model, [33, 34], which is the extended version of the Gipps model, [17].

Another package we wish to mention is SIMONE, which is based on a discretized macroscopic model of Payne [54] who derived this model from a microscopic car following model [43]. After further modifications, Payne's model is also adapted in a freeway simulation package called FREFLO.

In 1999, Helbing and Treiber published an article [21] where they suggested the MASTER simulation package, which deploys *nonlocal, gas-kinetic-based traffic flow model*. For optimization of traffic lights and to perform simulations of city traffic, TRANSYT, SCOOT, CRONOS packages are mentioned. Further references about these packages can be found in [22].

For commercially available packages, we cite OLSIM and TRANSIMS (based on cellular automaton models), VISSIM (based on microscopic models), AIMSUN (advanced interactive microscopic simulator for urban and non-urban networks), and PARAMICS (parallel microscopic traffic simulator).

A list of micro-simulators is also available at the following internet link
http://www.its.leeds.ac.uk/projects/smartest/links.html.
On Martin Treiber's website, 3D simulations of traffic behavior are also presented:
http://www.mtreiber.de/movie3d/index.html,
http://vwisb7.vkw.tu-dresden.de/ treiber/MicroApplet/index.html.
Multi-lane traffic can also be simulated via a Java Applet designed and written by Martin Treiber. This simulator merges two algorithms: intelligent driver model (IDM) (see also (9.10)), [70] and lane-changing algorithm MOBIL to demonstrate various behaviors of traffic flow with roundabouts, lane closings, traffic lights, lane changes, and on-ramp effects.

For further interpretations of human behavior in traffic and in different social environments, we direct the readers to Dirk Helbing's web site at *www.helbing.org*.

9.4 Time-Delay Effects in Traffic Flow Models

In this section, we first present how time delays originate and interfere the traffic dynamics, and what their quantitative measures are. Next, we go over the historical development of time-delayed traffic models.

9.4.1 What is the Origin of Time Delay?

We classify time delay in traffic dynamics according to its origin. This effort carries importance since it sheds light on how time delays should appear in mathematical models. The major reason for time delay influencing the traffic flow dynamics is due to the reaction time of drivers in response to certain stimuli they receive while driving (see 'stimulus' discussion in the previous section), [1, 5, 10, 18].

Since we consider only deterministic traffic flow models, we do not discuss the stochastic component and/or character of the delay. Without any loss of generality, the time delay τ represents, in general, a time-varying physical entity $\tau = \tau(t)$. Although it is quantitatively different for every driver, for an individual driver it consists of two components: (a) time-invariant part, constant τ_{inv}, (b) time-varying part, $\tau_v(t)$, which leads to $\tau(t) = \tau_{inv} + \tau_v(t)$. Inspired by the work of Bando et al. [4, 5] and Green [18], the time delay in traffic dynamics can be classified into three main components:

- *Physiological lag, $\bar{\tau}(t)$*: This part of the time delay originates directly from human operators driving the vehicles, hence the attribute *physiological*. When a driver receives a stimulus, he performs an action in his vehicle in the way of acceleration or deceleration. However, for any human driver receiving a stimulus, processing it, and creating a decision are not instantaneous, but they require a finite period of time $\bar{\tau}$ called the *physiological lag*. Notice that this delay may also be decomposed into two parts: time-invariant $\bar{\tau}_{inv}$ and time-varying part $\bar{\tau}_v(t)$. Hence, $\bar{\tau}(t) = \bar{\tau}_{inv} + \bar{\tau}_v(t)$. $\bar{\tau}(t)$ is represented with *pure time delay* in the mathematical models. Its quantitative level is measurable and is very well established within a certain range, as we discuss in the next section.
- *Mechanical time lag, $\tilde{\tau}$*: This delay component is independent of human operators, and it is completely determined by the mechanical capabilities of the vehicles in the traffic. Mechanical delay is defined as the period of time between the action of the driver on the gas/brake pedal and the time the vehicle starts to accelerate/decelerate. Clearly this delay exists in all the vehicles. It can be seen that this time delay is *following* the physiological delay in an acceleration/deceleration action and it also appears as a pure time delay in mathematical models. We state that $\tilde{\tau}$ can be assumed to be time invariant, in general, for a particular vehicle.
- *Delay time of vehicle motion, $\bar{T}$*: This entity is not physically a pure time delay. It is defined as the period of time that a vehicle changes its velocity to the velocity of the preceding vehicle. In linear systems, this is the period of time $\bar{T}$ that corresponds to a certain phase difference between a sinusoidal excitation input to the vehicle $M \sin(\omega t)$ and its sinusoidal response $\bar{M} \sin(\omega t - \phi)$. In other words, $\bar{T} = \phi/\omega$. As very well known, $\bar{M}$ and ϕ are determined directly from the mechanical properties, which in turn determine $\bar{T}$ as a function of excitation frequency ω. For these reasons, we state that $\bar{T}$ does not appear as a pure time delay in mathematical models.

In some of his works, Bando calls the combination of physiological and mechanical time lags $\bar{\tau}(t) + \tilde{\tau}$ as the *delay time of response* [4, 5]. We do not get into the details about the link between τ and $\bar{T}$; we just comment that there exists a direct link on the bandwidth of a dynamics, which is linked to $\bar{T}$, and the quantitative level of $\tau(t)$.

One can further categorize *physiological lag* by following the work in [18] and the references therein:

1. *Sensing*. It is the time it takes to *detect an object in the roadway* (e.g., 'There is a shape in the road'). Under equal conditions, reaction time decreases with greater signal intensity (brightness, contrast, size, loudness, etc.), foveal viewing, and better visibility conditions. Reaction times are also *faster* for auditory signals than for visual ones.

2. *Perception.* It is the time needed to *recognize the meaning of the sensation* (e.g., 'The shape is a person'). Reaction time increases with low signal probability, uncertainty (signal location, time, or form), and surprise. Reaction time, when there are multiple possible signals and responses, is generally much slower than simple reaction time, which occurs when there is only one possible signal or response pair, or both.
3. *Response selection and programming.* It is the time necessary to decide *which, if any, response to make and to mentally program the movement* (e.g., 'I should steer left instead of braking'). Response selection slows when there are multiple possible responses. Conversely, *practice* decreases the required time.
4. *Movement time.* It is the time it takes the *responder's muscles to perform the programmed movement.* For example, it is the time required to lift the foot from the accelerator and then to touch the brake pedal. In general, the more complex the movement, the longer the movement time. Increased arousal level and practice decreases movement time.

Investigating the above categorization, it is clear to conclude that physiological delays are formed in a very complicated way. There exist many factors which determine the reaction delays of the drivers, and every factor has a different effect to alarm the driver for a required action (acceleration, deceleration, or steering). The reaction time changes according to visual or auditory signals, brightness and visibility of the environment, size of the objects moving in the path of vehicles, multiple reaction choices of drivers ('which choice to select?'), experience of drivers, age, and gender.

9.4.2 What is the Measure of Time Delay?

To give a measure of time delay, we cite [1, 5, 10, 18]. According to [10], the drivers act (to accelerate or decelerate) with a delay 0.75–1.0 s (which is in the same levels as in [5]) after they receive an instantaneous stimulus.

In [18], the braking response time (time delay) is defined at the level of 1.2–1.35 s with a standard deviation of 0.6 s depending on the drivers. Moreover, the author gives a measure of the reaction delays of the drivers by a classification of certain events occurring in traffic. For example, if the drivers are fully aware of the time and location they should act to accelerate/decelerate, the reaction delays are in the range of 0.70–0.75 s. The delays become larger if the drivers should act to an expected but common signals (such as when the brake lights of the preceding vehicle turns on). In this case, the reaction delays are in the order of 1.25 s, while sudden and uncommon signals (such as an object suddenly moving in front of the vehicle's path) require 1.5 s of period of time for the drivers to act for acceleration/deceleration.

In Green's work, a table listing the reaction times of the drivers under various conditions can be found (pp. 197–199 in [18]). Inspecting the table, the reaction times can be seen to be in the range of 0.6–1.9 s depending on the factors mentioned in the previous section. As discussed in the cited work, American Association of

Table 9.1 Some examples of delay measurement: Delayed reaction of human drivers in response to various stimuli

Measure of time delay [s]	Reference
0.496	[19]
0.73	[47]
0.7	[12]
1.16	[74]
1.13	[2]
0.75–1.0	[5, 10]
1.2–1.35	[18]
1.1	[29, 73]
0.70–0.75	[18]

State Highway and Transportation Officials suggest a 2.5 s braking response time of drivers under normal road conditions and visibility (p. 208). See also Table 9.1 for a list of measures of time delays obtained from road experiments or simulators. These measures vary from one reference to another mainly because they represent different reaction delays of drivers in response to various stimuli.

In [1], a comparison between longitudinal speed control (remaining in lane) and steering control (changing lanes) of a human operator is given. It is mentioned that steering control offers an order of magnitude higher bandwidth when compared to longitudinal speed control. Furthermore, a human operator is under influence of a 0.2–0.3 s time delay in steering control while the delay rises to 2 s in longitudinal control. Although steering control seems to be advantageous, it may not always be a safe preference in crowded traffic and at higher cruising speeds. On page 4 of this reference, a 1.5 s time delay is mentioned to be required by the human operators for responding to velocity changes. An interesting comparison is also given: a sensing time of 1.9 s is needed if a leading vehicle accelerates at 0.76 m/s^2, while sensing time increases to 2.5 s for sensing 0.5 m/s^2 acceleration of a lead vehicle. As a general remark, it is stated that response time of a human operator decreases by 0.8 s for each increase in acceleration by 0.3 m/s^2.

Remark 1 (Non-identical time delays) The claims above are interesting as they point out that human operators create their driving decisions depending on many physical factors that may be sensed by different quantitative measures of time delays. To our best knowledge, this fact is disregarded when analysis of mathematical models are carried out in the literature. This may be mainly due to the complicated nature of modeling the delays.

Remark 2 (Measure of time delay versus stability) Looking at the time-delayed traffic problem from *systems and control perspective*, it is important to note that time delays occurring from drivers' responses are substantial and the presence of delays may drastically affect the dynamic progression of traffic flow.

9.4.3 Development of Time-Delayed Traffic Models

In the following, a survey of time-delayed traffic models is presented in chronological order.

9.4.3.1 Chandler's Model (1958)

The first model with time delay has been proposed in [7] to the best of our knowledge. This model is a very simple linear delay differential equation with a single time delay:

$$\ddot{x}_n(t) = \kappa[\dot{x}_{n-1}(t - \tau) - \dot{x}_n(t - \tau)], \qquad (9.1)$$

where $x_n(t)$, $\dot{x}_n(t)$, and $\ddot{x}_n(t)$ are the displacement, velocity, and acceleration of the nth vehicle, respectively. $\kappa > 0$ is a constant representing the sensitivity of the driver against velocity-error readings and τ is the time delay, which physically corresponds to the time needed by the drivers to sense and act (in the way of accelerating or decelerating his vehicle) to the velocity differences.

It is interesting to note that this model is still used for traffic modeling in the literature. A recent paper [6] has shown that this model very well predicts the experimental measurements from human driving, and we see that a 2004 report from University of Southern California [78] deploys this model to simulate the behavior of trucks among the cars in a traffic flow scenario.

Remark 3 (Memory of human drivers) The model in (9.1) is the basis for further analysis. It is clear that right-hand side of this equation, the control action, is restricted to cases where stimuli received depend only on some information from a *point of time* in the past. With the fact that human drivers continuously observe the traffic flow, some particular functions *distributed* over the history may represent the received stimuli. Physically, such a mathematical model will correspond to taking into account the *memory* of human drivers [59].

9.4.3.2 Gazis' Models (1959, 1961)

The work in [15] offers a new model in a nonlinear form as follows:

$$\ddot{x}_n(t) = \frac{\kappa}{x_{n-1}(t - \tau) - x_n(t - \tau)}[\dot{x}_{n-1}(t - \tau) - \dot{x}_n(t - \tau)], \qquad (9.2)$$

where the denominator on the right-hand side takes into account the sensitivity of the driver as a function of headway sensed τ s earlier. This approach is physically sound: the *smaller* the headway, the *more sensitive* the driver becomes for his/her reactions.

Gazis et al. improved the above model by defining a new sensitivity factor κ in (9.1) in a complicated nonlinear form [16]. This formula is suggested in order to offer free parameters that need to be carefully selected by fitting data to the experimental measurements. It reads

$$\kappa_n = \frac{\lambda[\dot{x}_n(t)]^m}{[x_{n-1}(t-\tau) - x_n(t-\tau)]^\ell},$$ (9.3)

where λ, m, and ℓ are the constant fitting parameters. The traffic model with (9.1) and (9.3) together is attributed as *non-integer car following model* [22], and it is deployed in the simulation package MITSIM. The model in (9.2) alone is a simpler version of the non-integer car following model in which the special case $\lambda = \kappa$, $m = 0$, and $\ell = 1$ hold. See [38] for $m_1 \approx 0.8$, $m_2 \approx 2.8$, and [35, 36] for $m_1 \approx 0.953$, $m_2 \approx 3.05$.

9.4.3.3 Optimal Velocity Function (OVF)-Based Models

Optimal velocity model (OVM) originated from Bando's work in 1995 which was then followed in [5] by incorporating the time-delay effect in this OVM by adding a single time delay τ

$$\ddot{x}_n(t+\tau) = \kappa[V(x_{n-1}(t) - x_n(t)) - \dot{x}_n(t)],$$ (9.4)

where $V(x_{n-1}(t) - x_n(t))$ is called the optimal velocity function (OVF), which is, in general, a nonlinear hyperbolic function defining the desired velocity of the drivers in terms of headway $\Delta x(t) = x_{n-1}(t) - x_n(t)$.

The choice of OVF is ad hoc. In [69], it is taken as

$$V(\Delta x(t)) = \vartheta_{max}\Theta(\Delta x(t) - d),$$ (9.5)

where $d > 0$ is a constant and Θ is the Heaviside step function. The above OVF requires that the vehicle stops when headway is less than d, otherwise the vehicle accelerates until the maximum allowed speed ϑ_{max} is reached.

In [53], OVF is taken as

$$V(h) = \begin{cases} 0, & \text{if } 0 \le h \le 1, \\ v^0 \frac{(h-1)^3}{1+(h-1)^3}, & \text{if } \quad h > 1, \end{cases}$$ (9.6)

where v^0 is the desired speed and h is the normalized headway.

There also exist optimal velocity functions that are obtained by fitting of experimental measurements [5],

$$V(\Delta x(t)) - 16.8[\tanh{(0.086(\Delta x(t)) - 25)} + 0.913].$$ (9.7)

Bando [5] includes time-domain simulations of nonlinear dynamics (9.4) as well as a linear stability analysis around an equilibrium point. The author arrives at the conclusion that small delays (in the range of 0–0.2 s) do not affect the homogeneous flow of the dynamics with or without delays, however, larger delays (in the range of 0.2–0.4 s) cause congestion in the traffic flow. Notice that these quantitative levels of delay are about five times less than the claimed values for human beings, see Table 9.1.

9.4.3.4 Traffic Control over Optimal Velocity Models (2000)

With the work in [31], a continuous traffic dynamics model departing from (9.4) is developed. The linear stability analysis is pursued on this model first. Next, time delay is taken into account and a feedback controller that is based on the delayed headway information is added

$$\dot{x}_n(t+\tau) = \kappa[V(x_{n-1}(t) - x_n(t)) - \dot{x}_n(t)] + u_n(t),$$
$$u_n(t) = k(x_{n-1}(t) - x_n(t) + x_{n-1}(t-\tau) - x_{n-1}(t-\tau)), \tag{9.8}$$

where the controller is constructed via the error between the instantaneous and τ s earlier *headway*. Expanding on this dynamics, the linear stability analysis is pursued and the selection of the controller gain k versus the single delay τ is performed. For this, the stability and frequency response analyses are performed and simulation results are presented for specific selections of (k, τ) pairs.

9.4.3.5 Models Based on Davis' Work (2002, 2003)

The work of [9] takes a chain of 100 vehicles and tests Bando's delayed OVM, (9.4). Interestingly, even with delay levels of 0.3 s, only the first 14 vehicles avoid a collision under some small disturbances on the linearized model. This is an indication that the time-delayed OVM may not be completely realistic since accidents do not happen in reality even if the drivers are under larger time-delay influences (see the previous section for the quantitative measure of time delay in traffic dynamics).

Extending on the previous work, the study in [10] offers two modified OVMs in order to create reasonable traffic models that better match the reality. We present one of these modifications here. In this model, instead of $V(x_{n-1}(t) - x_n(t))$ in (9.4), the following nonlinear function is used:

$$V_{OV} = V(\Delta x_n(t-\tau) + \tau \Delta \dot{x}_n(t-\tau)). \tag{9.9}$$

Even though this modification can predict stable traffic dynamics with a time delay up to 1 s, it results in very high accelerations and decelerations that are located at the end of the platoon of vehicles. The author of the work removes this deficiency with another modification with which more realistic velocity profiles are produced. Moreover, the modifications can detect the *formation of traffic jams*; however, they still fall short in detecting the *sequence of jam and free-flow phenomena* that surfaced in [40], as [10] states. We should note that most of the results obtained in these work are based on numerical simulations. In 2004, Davis improved the models in order to cover multi-lane interactions of multi-lane traffic with on/off ramps [11].

9.4.3.6 Intelligent Driver Model (IDM, 2000, 2004) and Human Driver Model (HDM, 2006)

In [70] the intelligent driver model (IDM) is proposed in response to the need of a mathematical model that is robust, accident-free, numerically efficient, and easy to calibrate. It is shown that IDM also reproduces empirically obtained data and yields realistic acceleration/deceleration behavior. The model is given as

$$\ddot{x}_n(t) = \kappa_n \left[1 - \left(\frac{\dot{x}_n(t)}{\dot{x}_o^{(n)}} \right)^{\delta} - \left(\frac{s^*(\dot{x}_n(t), \Delta \dot{x}_n(t))}{s_n(t)} \right) \right], \qquad (9.10)$$

where δ is the acceleration exponent which is often taken as $\delta = 4$, $\dot{x}_o^{(n)}$ is the desired velocity, $\Delta \dot{x}_n(t) = d(\Delta x_n(t))/dt$ is the time derivative of headway, $s_n(t)$ is the actual headway, and s^* is the desired headway that is dynamically calculated by

$$s^*(\dot{x}_n(t), \Delta \dot{x}_n(t)) = s_0^{(n)} + s_1^{(n)} \sqrt{\frac{\dot{x}_n(t)}{\dot{x}_o^{(n)}}} + T^{(n)} \dot{x}_n(t) + \frac{\dot{x}_n(t) \Delta \dot{x}_n(t)}{2\sqrt{a^{(n)} b^{(n)}}}, \qquad (9.11)$$

where $s_0^{(n)}$ and $s_1^{(n)}$ are jam distances, $T^{(n)}$ is the safe time headway, $a^{(n)}$ is the maximum acceleration capability of the vehicle, and $b^{(n)}$ is the desired (comfortable) deceleration of the vehicle.

The IDM is later adapted for human drivers in the work of Treiber [73], giving rise to human driver model (HDM). This model incorporates the finite reaction time of drivers to the *right-hand* side of the IDM in (9.10). The reaction time of drivers, τ, is considered by taking the *right-hand* side of (9.10) at $t - \tau$ rather than at t. In [73], this reaction time is considered to be between 0 and 2 s.

9.4.3.7 Multiple Vehicle Following Models

The models presented so far consider that a driver only follows the preceding vehicle. One can extend this idea in the way that a driver observes not only the vehicle in front but also other vehicles further ahead. In [57], the model is proposed in a way that all drivers follow the two vehicles ahead of them with the influence of a single delay τ

$$\ddot{x}_n(t) = \kappa_1 [\dot{x}_{n-1}(t - \tau) - \dot{x}_n(t - \tau)] + \kappa_2 [\dot{x}_{n-2}(t - \tau) - \dot{x}_n(t - \tau)]. \qquad (9.12)$$

Departing from (9.12), one can express a more generalized delay differential equation set depending on how many vehicles the drivers follow in front of them. This basic idea can be used to extend the mathematical models discussed here, for instance, see [37, 73].

9.5 Assumptions and Analysis on Mathematical Models

The limitations of time-delayed microscopic models can be listed as follows. (a) Large-scale traffic simulation using microscopic models may be cumbersome, nevertheless, a traffic scenario with 1,000 vehicles [73] is within the computational capabilities. (b) Some microscopic models may fall short in explaining some phenomena of traffic flow, such as high-density flows. (c) A direct link may not always be possible to establish between the features of macroscopic and microscopic models; for example, *What does "viscosity" represent in a microscopic model?* (d) Fitting empirical data to a time-delayed model becomes more complicated since delay is an additional parameter that needs to be considered. (e) The time-varying nature of the time delay is not considered and data fitting in this case can be tremendously difficult. (f) Measurements of time delay and its varying component may be difficult or impractical, especially when time delay is modeled with stochastic components.

The commonly preferred simplifications appearing in mathematical modeling and ultimately in the stability analysis include *linearization* around an equilibrium and homogeneity in the sense that all vehicles, drivers, and driver delays are taken to be identical: $\kappa_n = \kappa$ and $\tau_n = \tau$. Although linearization is a simplification for the analysis, it offers insight about the behavior of the dynamics at an equilibrium point. In many studies in the literature, one observes that the linearized dynamics is studied for its stability via analytical tools [4, 5, 31] and/or simulations [9–11], whereas nonlinear stability analysis appears in [71–73, 78] by way of simulations and in [49, 50, 52] via analytical tools that analyze the bifurcation features of the nonlinear dynamics. Ref. [49] is one of the very few and earliest analytical works on the stability analysis of nonlinear time-delayed traffic flow dynamics. This study is based on the investigation of Hopf bifurcations of time-delayed systems with translational symmetry. These works and the existing limitations give rise to the motivational discussions in Sect. 9.6.

Identical vehicle dynamics is another simplification that is preferred in most of the models studied. Nevertheless, approaching the problem from systems and control perspective can eliminate this restriction [62, 65–67]; see also the next section.

Another *critical* simplification is made by taking only a *single delay τ*, in the differential equations, which assumes all drivers have *identical* reaction delays. This offers a very simplified stability analysis that has been addressed by a variety of systematic methods; see [44, 48, 61, 68] and the references therein. The main challenge lies in considering multiple (heterogeneous) delays due to the presence of different drivers and each driver's different sensing capabilities.

9.6 Interesting Research Topics

In this section, we present how systems and control approach [20, 39, 41, 44, 60, 61, 68] may shed light on the stability analysis of various linear/nonlinear time-delayed mathematical models representing traffic flow scenarios. See also [44, 68] on the links between the representation of traffic dynamics and many other real-life applications treated in the literature.

9.6.1 Linear Analysis with a Single Delay

9.6.1.1 Generalization to N Vehicles Around a Ring

In [52], a new optimal velocity function is suggested for (9.4) and a single delay is considered in the headway only:

$$\ddot{x}_n(t) = \kappa[V(x_{n-1}(t - \tau) - x_n(t - \tau)) - \dot{x}_n(t)], \qquad (9.13)$$

with $n = 1, \ldots, N$, where N is the number of vehicles and $x_0 = x_N$ formulates the traffic flow around a ring. To linearize (9.13), the equilibrium is considered, which is in the form of $x_n^{eq}(t) = V(L/N)t + x_n^0$, where $x_{n-1}^0 - x_n^0 = L/N$, L is the length of the ring. Dynamics of the perturbations $y_n(t)$ around this equilibrium is

$$\ddot{y}(t) = -\kappa \dot{y}_n(t) + b_1(y_{n-1}(t - \tau) - y_n(t - \tau)), \qquad (9.14)$$

where $b_1 = \kappa V'(L/N)$ and $\kappa = 1/T$. The characteristic equation corresponding to (9.14) is

$$f(s, \tau, N) = (s^2 + \kappa s + b_1 e^{-\tau s})^N - (b_1 e^{-\tau s})^N = 0, \qquad (9.15)$$

where b_1 and κ are constants. In [50, 53], (9.15) is studied for any selection of N. Specifically, in [50, 52], bifurcations are investigated by numerical continuation (using DDE-BIFTOOL [13]) and stopping and collision curves are obtained. In [49, 53], bifurcations are studied analytically via Hopf bifurcations and bifurcation branches for oscillatory solutions are detected. Also collision and stopping characteristics are identified on the bifurcation diagrams.

When $N > 2$, stability analysis on (9.15) becomes more challenging, but this does not constitute a major problem mathematically, since there exist well-established methods for handling the stability problem of this type of single-delayed characteristic equations; see [44, 48, 61, 68] and the references therein.

9.6.1.2 The case $N = 2$

Simplified forms of (9.15) are also considered in the literature, as it becomes easier to derive some analytical results and interpretations on traffic flow dynamics. Cases with $N = 2$ and $N = 3$ can be found in [49] and [52], respectively. The model in this section is borrowed from [49] (see Fig. 9.3), where a *single delay* τ is considered for the reaction time of the drivers. The dynamical model includes a time delay only in the headway readings, with which a nonlinear braking function is used to penalize the headway. The linearized dynamics around the equilibrium point is given by

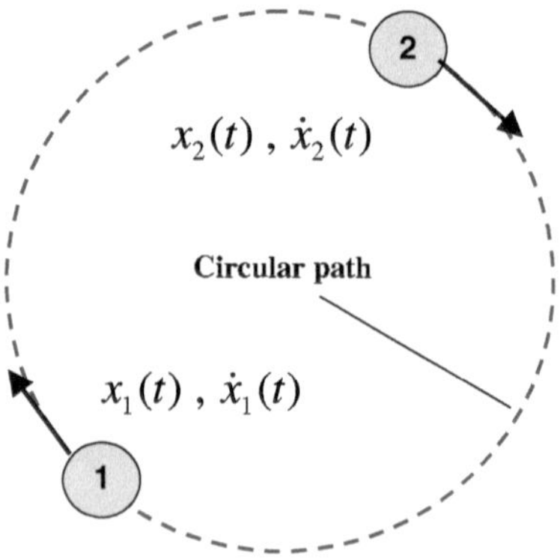

Fig. 9.3 A conceptual traffic
scenario: Two vehicles
traveling around a circular
path

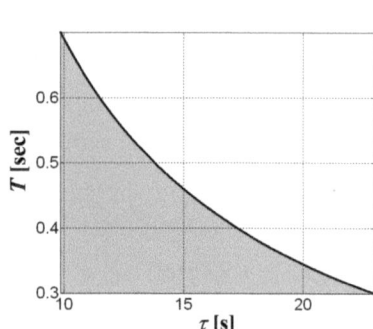

$x_2(t)$, $\dot{x}_2(t)$

Circular path

$x_1(t)$, $\dot{x}_1(t)$

Fig. 9.4 Stability regions
(*shaded*) of the traffic flow
dynamics (9.16) in the
parameter space of τ and T

$$\ddot{y}_1(t) = -\tfrac{1}{T}\dot{y}_1(t) + b_1(y_2(t-\tau) - y_1(t-\tau)),$$
$$\ddot{y}_2(t) = -\tfrac{1}{T}\dot{y}_2(t) + b_1(y_1(t-\tau) - y_2(t-\tau)),$$
(9.16)

where τ is the time delay in headway readings, $b_1 > 0$ is the local deceleration
sensitivity with respect to the headway, and $T > 0$ defines the *time constant*. Using
the methods presented in [44, 48, 68], one can obtain the so-called stability map of
the equilibrium dynamics in the parameter space of τ and T, as shown in Fig. 9.4.

9.6.2 Multiple Delays

Note that the mathematical models with a single delay τ can be further improved
by suggesting independent time delays for the reaction of the drivers in sensing the
headway and velocity. This can be demonstrated by taking the same mathematical
model in (9.16) with the consideration of another time delay which is independent
of the first one

$$\ddot{y}_1(t) = -\tfrac{1}{T}\dot{y}_1(t-\tau_2) + b_1(y_2(t-\tau_1) - y_1(t-\tau_1)),$$
$$\ddot{y}_2(t) = -\tfrac{1}{T}\dot{y}_2(t-\tau_2) + b_1(y_1(t-\tau_1) - y_2(t-\tau_1)).$$
(9.17)

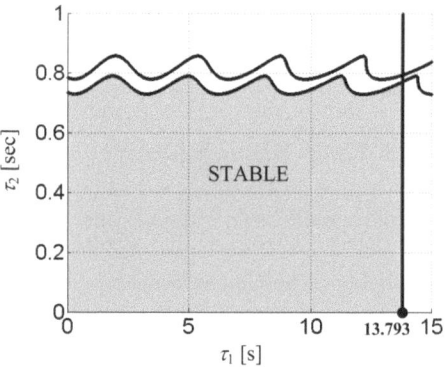

Fig. 9.5 Stability region (*shaded*) of the traffic flow dynamics in (9.17) for $T = 0.5$ and $b_1 = 0.145$

The stability question can be posed as follows: For which values of the time delays (τ_1 vs. τ_2) does the above dynamics preserve its stability? For asymptotic stability, the roots of the characteristic equation

$$f(s, \tau_1, \tau_2) = (s^2 + s/T\,e^{-\tau_2 s} + b_1 e^{-\tau_1 s})^2 - b_1^2$$
$$= (s^2 + s/T\,e^{-\tau_2 s} + b_1 e^{-\tau_1 s} - b_1)(s^2 + s/T\,e^{-\tau_2 s} + b_1 e^{-\tau_1 s} + b_1) = 0,$$
$$(9.18)$$

corresponding to the dynamics in (9.17) should all lie on the left half of the complex plane for a given pair of (τ_1, τ_2). Once this question is answered, it is also possible to find the maximum allowable τ_2 (delay margin), above which instability initiates. If this margin along τ_2 is less than the realistic values corresponding to human operators, then the dynamics may not be robustly stable, or the model may not be fully capturing the physics of the problem.

The stability analysis of the characteristic Equation (9.18) can be performed by the techniques proposed in [20, 60, 61]. In [60, 61], the method Cluster Treatment of Characteristic Roots (CTCR) is based on an idea of one to infinity mapping, and in [20], an interesting *geometric technique* forms the basis of the stability analysis approach. We take $b_1 = 0.145$ which is an acceptable entity as can be seen from Fig. 9.3 of [49]. When choosing T, one can use another reference [5], where $1/T = a = 2$. In order not to disrupt the flow of discussions, we omit the stability analysis and depict in Fig. 9.5 the stability region of the dynamics in the delay parameter space.

General remark 1 (Stability analysis results) We wish to stress once again that the available tools in systems and control perspective may shed further light on studying and analyzing the dynamics of traffic flow. The example case study is a good proof-of-concept of this effort. Notice that one of the main interests is stability robustness of the dynamics in (9.17) along the parameter τ_2. It is found that the maximum allowable τ_2 delay is less than 0.8 s, which is a comparable measure to time delays seen in the reactions of human drivers. Furthermore, there exists an order of magnitude larger stability robustness along the τ_1-axis. In the worst case when the reaction of human operators is time delayed for 2.5 s, this would not endanger the stability robustness along the τ_1-axis.

9.6.3 Time-Varying Delays

It is more realistic to consider time delay as time varying, $\tau(t) = \tau_0 + \delta(t)$ where $0 \le \delta(t) \le \epsilon$ and τ_0 is constant. In [23], it is shown empirically that variations of time delay of humans in time is very slow. In other words, $0 \le |\dot{\delta}(t)| \le \rho$, where ρ is sufficiently small. With this information, one can use the approach developed in [46] to obtain some robust stability criteria of the traffic flow with time-varying delays; see also [63].

9.6.4 Improved Traffic Stability with Multiple Vehicle Following

Intelligent driver model (IDM) in (9.10) incorporates four components (other than reaction times of the drivers) into the right-hand side of the model; (a) finite reaction time of drivers, (b) imperfect estimation capabilities, (c) temporal anticipation, (d) spatial anticipation for several vehicles. Item (a) is the focus here due to the scope of this survey. Reaction time of drivers, τ is considered by taking the right-hand side of (9.10) at $t - \tau$ rather than at t. In [29, 73], reaction time is considered between 0 and 2 s. Items (b) and (c) take into account drivers' imperfections (anticipation with some error) as well as drivers' corrective actions by anticipating the flow of the traffic. The last item (d) assumes that the drivers follow more than one vehicle ahead (similar to the idea in (9.12)) in order to compensate for their slow decision making due to their delayed reactions.

In [29, 73], the authors perform the stability analysis for a platoon of 100 vehicles and show that the stability region of the entire traffic dynamics is considerably enhanced when human drivers follow more cars ahead of them rather than following only one car. This is interpreted by saying that *the destabilizing effects of time-delayed reactions of human drivers are compensated by multiple vehicle following strategy*. Also, the authors determine the phase diagrams (stable, oscillatory, and unstable platoon) of the traffic flow as a function of the number of vehicles followed ahead vs. reaction delay of the drivers.

9.6.5 Multiple Vehicle Following Under Multiple Delays

A recent study [64] starts with a different assumption: A driver senses the motion of a nearer vehicle in front faster than sensing a vehicle further ahead. In other words, time-delayed information obtained from different vehicles ahead of a vehicle are *non-identical*, i.e., *independent* of each other. This assumption originates from the fact that drivers update their information more frequently from a nearer vehicle than from a further one. The main objective in this work is to understand how different control laws under the influence of single and heterogeneous time delays may reveal various stability features of the dynamics, and whether or not the stability would be improved when multiple car following strategy is chosen. On some fundamental models, it is shown in this work that, compared to only velocity feedback in the decision making of drivers, combination of velocity and position feedback is not suggesting an improvement over the stability of the traffic flow.

9.6.6 Nonlinear Time-Delayed Traffic Dynamics

This recent work [53] follows from [49]. To our best knowledge, this work presents one of the few analytical studies on the stability analysis of nonlinear car following dynamics in which time-delayed reactions of drivers is considered. Studies similar to this one form one of the reasons in motivating this survey. The study in [49, 53] analytically considers weak nonlinearities close to bifurcation point using Hopf normal form calculations which characterize different phases of traffic flow, such as bistability, uniform flow, and the stop-and-go traveling waves. In [50, 52], full nonlinearities are taken into account via a numerical continuation technique called DDE-BIFTOOL [13]. The main conclusion in this work is that regions in the parameter space may be found where the equilibrium and traffic jams may coexist and that (most importantly) delay makes this to be a robust feature of the system.

9.6.7 Optimal Velocity Model with Time Delay and Stochastic Process

An optimal velocity model with time delays is considered in [51], where the characteristics of the drivers change according to some stochastic process. Cases with and without such stochastic components are simulated in order to obtain features of flow motions into and out of traffic jams.

9.6.8 Effects of Drivers' Memory

The work in [59] focuses on understanding how human drivers utilizing their memory affect the stability of the traffic flow dynamics. A single-lane microscopic car following model is studied for this purpose in which human drivers perform control actions based on information distributed over an interval of time in history. This memory effect is characterized by standard distribution functions which define the memory size, particular memory distribution considered, and the memory horizon. Stability analysis is presented, via analytical tools, in the domain defining various distribution functions. Physical interpretations follow the analysis. See [65, 66] for large number of vehicles and the effects on the formation of stability regions.

9.7 Conclusion and Discussion

The literature offers a broad range of studies on the behavior of traffic flow dynamics. Many in-depth surveys also exist on the topic; however, none of these surveys approach the research topic with a *focus of time-delay effects*. A survey with this perspective is presented in this chapter, covering the widely studied time-delayed mathematical models of traffic flow dynamics.

It is seen that existing mathematical models did not fully benefit from systems and control engineering perspective. Being aware of the tools developed in the systems and control engineering, a new section aimed to link these developments for resolving some open problems in traffic flow studies is also presented. It is believed that this section will shed light on this application field from a different perspective.

Some observations on the studies of time-delayed mathematical models can be listed as follows:

- Results from the systems and control engineering field are not effectively integrated into the analysis of traffic dynamics.
- Most of the time-delayed models are in the form of coupled nonlinear differential equations which are simulated in time domain in order to understand their behavior.
- Linear analysis of traffic behavior is often pursued. The stability of linear dynamics is studied, mostly without delays, and stability criteria are given with respect to system parameters.
- Linearized differential equations are also used to assure string stability of a platoon of cars. Often, numerical frequency domain analysis are pursued for the purpose, especially when delay is present in the transfer function.
- The studies investigated concentrate on models with only a single delay. Some exceptions to this are coming from systems and control theory.
- Bifurcation analysis is not followed by many of the studies, except by a few publications.

We believe that there exists a variety of opportunities for further research on time-delayed traffic dynamics. We hope that this survey will serve as a good reference for many researchers from mathematics, physics, and engineering disciplines.

Acknowledgments A large portion of this work has been done by the authors at HeuDiaSyC (UMR CNRS 6599), Université de Technologie de Compiègne, Centre de Recherche de Royallieu, BP 20529, 60205, Compiègne, France. Rifat Sipahi's work was financially supported by Chateaubriand Bourse 2005-2006 of the French Government. Authors gratefully acknowledge the financial support of CNRS (National Scientific Research Center of France) for Rifat Sipahi's (a) visit to Technische Univeristät Dresden and Max Plank Institute where he had fruitful discussions with Prof. Dirk Helbing, Prof. Martin Treiber, Prof. Serge Hoogendoorn, and Prof. Fatihcan Atay on modeling traffic flow dynamics and behavior of human drivers and (b) attendance at IFAC Time-Delay Systems Workshop in L'Aquila, Italy, where authors had interactive communications on the topic with Prof. Gabor Stepan, Prof. Martin Treiber, and Dr. Gabor Orosz.

References

1. R. W. Allen, T. D. Marcotte, T. J. Rosenthal, and B. L. Aponso. Driver assessment with measures of continuous control behavior. Presented at *Proceedings of the Third International Driving Symposium on Human Factors in Driver Assessment*, Training and Vehicle Design, 2005.
2. H. Alm, and L. Nilsson. Changes in driver behaviour as a function of handsfree mobile phones—A simulator study. *Accid. Anal. Prev.*, 26: 441–451, 1994.
3. M. Bando, K. Hasebe, K. Nakanishi, A. Nakayama, A. Shibata, and Y. Sugiyama. Phenomeno-logical study of dynamical model of traffic flow. *J. de Physique I. France*, 5: 1389–1399, 1995.
4. M. Bando, K. Hasebe, K. Nakanishi, and A. Nakayama. *Delay of Vehicle Motion in Traffic Dynamics*. Internal Report Aichi University, 1996. http://arxiv.org/PS_cache/patt-sol/pdf/9608/9608002.pdf
5. M. Bando, K. Hasebe, K. Nakanishi, and A. Nakayama. Analysis of optimal velocity model with explicit delay. *Phys. Rev. E*, 58: 5429–5435, 1998.
6. A. Bose and P. A. Ioannou. Analysis of traffic flow with mixed manual and semiautomated vehicles. *IEEE Int. Trans. Sys.*, 4(4): 173–188, 2003.
7. R. E. Chandler, R. Herman, and E. W. Montroll. Traffic dynamics: Analysis of stability in car following. *Operat. Res.*, 7(1): 165–184, 1958.
8. D. Chowdhury, L. Santen, and A. Schadschneider. Statistical physics of vehicular traffic and some related systems. *Phys. Rep.*, 329: 199–329, 2000.
9. L. C. Davis. Comment on "Analysis of optimal velocity model with explicit delay". *Phys. Rev. E*, 66: Paper no. 038101, 2002.
10. L. C. Davis. Modifications of the optimal velocity traffic model to include delay due to driver reaction time. *Physica A*, 319: 557–567, 2003.
11. L. C. Davis. Multilane simulations of traffic phases. *Phys. Rev. E*, 69: Paper no. 016108, 2004.
12. E. Dureman and C. Boden. Fatigue in simulated car driving. *Ergonomics.* 15: 299–308, 1972.
13. K. Engelborghs. *DDE-Biftool: A Matlab Package for Bifurcation Analysis of Delay Differential Equations*. Department of Computer Science, Katholieke Universiteit Leuven, Leuven, Belgium, 2000.
14. L.C. Edie. Following and steady-state theory for non-congested traffic. *Oper. Res.*, 9: 66–76, 1961.
15. D. C. Gazis, R. Herman, and R. B. Potts. Car following theory of steady state traffic flow. *Oper. Res.*, 7: 499, 1959.
16. D. C. Gazis, R. Herman, and R. W. Rothery. Non-Linear follow the leader models of traffic flow. *Oper. Res.*, 9: 545–567, 1961.
17. P. G. Gipps. A behavioural car-following model for computer simulation. *Transportation Res*, Part B (Methodological), 15B: 105–111, 1981.
18. M. Green. How long does it take to stop? methodological analysis of driver perception-brake times. *Transport. Hum. Factors*, 2: 195–216, 2000.
19. B. Greenshields. Reaction time in automobile driving. *J. Appl. Psychol.*, 20: 353–357, 1936.
20. K. Q. Gu, S.-I. Niculescu, and J. Chen. On stability crossing curves for general systems with two delays. *J. Math. Anal. Appl.* 311: 231–253, 2005.
21. D. Helbing and M. Treiber. Numerical simulation of macroscopic traffic equations. *Comput. Sci. Eng.*, 1: 89, 1999.
22. D. Helbing. Traffic and related self-driven many-particle systems. *Rev. Mod. Phys.*, 73: 1067–1141, 2001.
23. S.P. Hoogendoorn and Ossen S. Parameter estimation and analysis of car-following models. In: H. S. Mahmassani (ed.) *Flow, Dynamics and Human Interaction* (Transportation and Traffic Theort, 16) Elsevier, The Netherlands, pp. 245–266 2005.
24. B. S. Kerner and H. Rehborn. Experimental properties of complexity in traffic flow. *Phys. Rev. E*, 53: R4275–R4278 1996.

25. B. S. Kerner and H. Rehborn. Experimental properties of phase transitions in traffic flow. *Phys. Rev. Lett.*, 79: 4030–4033, 1997.
26. B. S. Kerner. Experimental features of self-organization in traffic flow. *Phys. Rev. Lett.*, 81: 3797–3800, 1998.
27. B. S. Kerner. Empirical macroscopic features of spatial-temporal traffic patterns at highway bottlenecks. *Phys. Rev. E*, 65: Paper no. 046138, 2002.
28. B. S. Kerner and S. L. Klenov. Microscopic theory of spatial-temporal congested traffic patterns at highway bottlenecks. *Phys. Rev. E*, 68: 2003.
29. A. Kesting, M. Treiber. Influence of reaction times and anticipation on the stability of vehicular traffic flow. Invited session on *Traffic Dynamics Under the Presence of Time Delays*, 6th IFAC Workshop on Time-Delay Systems, L'Aquila, Italy, July 2006.
30. W. Knospe, L. Santen, A. Schadschneider, and M. Schreckenberg. Human behavior as origin of traffic phases. *Phys. Rev. E*, 65: 2001.
31. K. Konishi, H. Kokame, and K. Hirata. Decentralized delayed-feedback control of an optimal velocity traffic model. *Eur. Phys. J. B*, 15: 715–722, 2000.
32. M. Koshi, M. Iwasaki, I. Ohkura. Some findings and an overview on vehicular flow characteristic. *Proceedings 8th International Symposium on Transportation and Traffic Flow Theory* (V. Hurdle, E. Hauer, G. Stuart editors), pp. 403–426, 1983.
33. S. Krauss, P. Wagner, C. Gawron. Metastable states in a microscopic model of traffic flow. *Phys. Rev. E*, 55 (5): 5597–5602, 1997.
34. S. Krauss. Microscopic modeling of traffic flow: Investigation of collision free vehicle dynamics Ph.D. thesis, 1998.
35. R. D. Kuhne, and M. B. Rodiger. Macroscopic simulation model for freeway traffic with jams andstop-start waves. In: B. L. Nelson, W. D. Kelton, and G. M. Clark (eds.) *Proceedings of the 1991 Winter Simulation Conference* Society for Computer Simulation International, Phoenix, AZ, p. 762–770, 1991.
36. Kroen, A. and Kuhne, R. D. Knowledge-based Optimization of Line Control Systems for Freeways. International Conference on Artificial Intelligence Applications in Transportation Engineering, June 20–24, 1992 (San Buenaventura, CA), p. 173-192, 1992.
37. H. Lenz, C.K. Wagner, and R. Sollacher. Multi-anticipative car-following model. *Eur. Phys. J. B*, 7: 331–335, 1999.
38. A. D. May, Jr., and H. E. M. Keller. Non-integer car-following models *Highw. Res. Rec.*, 199: 19–32, 1967.
39. W. Michiels. *Stability and Stabilization of Time-Delay Systems*. Katholieke Universiteit Leuven, Belgium, May 2002.
40. N. Mitarai and H. Nakanishi. Spatiotemporal structure of traffic flow in a system with an open boundary. *Phys. Rev. Lett.*, 85: pp. 1766–1769, 2000.
41. R. M. Murray. Control in an information rich world Report of the Panel on Future Directions in Control, Dynamics and Systems, 30 June 2002. http://www.cds.caltech.edu/ murray/ cdspanel/report/cdspanel-15aug02.pdf
42. T. Nagatani. Thermodynamic theory for the jamming transition in traffic flow. *Phys. Rev. E*, 58: 1998.
43. G. F. Newell. Nonlinear effects in the dynamics of car following. *Oper. Res.* 9: 209–229, 1961.
44. S.-I. Niculescu. *Delay Effects on Stability*. Springer-Verlag, 2001.
45. S.-I. Niculescu and W. Michiels. Stabilizing a chain of integrators using multiple delays. *IEEE Transa. Autom. Cont.*, 49(5): 802–807, 2004.
46. S.-I. Niculescu and K. Gu. Robust stability of some oscillatory systems including time-varying delay with applications in congestion control. *ISA Trans.*, 42(4): 595–603, 2003.
47. O. Norman. Braking distance of vehicles from high speed. *Public Roads*. 27: 159–169, 1952.
48. N. Olgac and R. Sipahi. An exact method for the stability analysis of time delayed LTI systems. *IEEE Trans. Autom. Cont.*, 47: 793–797, 2002.
49. G. Orosz and G. Stepan. Hopf bifurcation calculations in delayed systems with translational symmetry. *J. Nonlinear Sci.*, 14(6): 505–528, 2004.

50. G. Orosz, B. Krauskopf, and R. E. Wilson. Bifurcations and multiple traffic jams in a car-following model with reaction-time delay. *Physica D*, 211(3–4): 277–293, 2005.
51. G. Orosz, B. Krauskopf, and R. E. Wilson. Traffic jam dynamics in a car-following model with reaction-time delay and stochasticity of drivers, Invited session on *Traffic Dynamics Under the Presence of Time Delays*, in 6*th* IFAC Workshop on Time-Delay Systems, L'Aquila, Italy, July 2006.
52. G. Orosz, R. E. Wilson, and B. Krauskopf. Global bifurcation investigation of an optimal velocity traffic model with driver reaction time. *Phys. Rev. E*, 70(2): 026207, 2004.
53. G. Orosz and G. Stepan. Subcritical Hopf bifurcations in a car-following model with reaction-time delay. *Proceedings of the Royal Society of London A*, 462(2073): 2643–2670, 2006.
54. H. J. Payne. *Research Directions in Computer Control of Urban Traffic Systems*. In W. S. Levine, E. Lieberman, and J. J. Fearnsides (eds.) (p. 251) American Society of Civil Engineers, New York 1979.
55. I. Prigogine. In R. Herman (ed.) Theory of Traffic Flow. Amsterdam, Elsevier, 1961.
56. I. Prigogine and R. Herman. *Kinetic Theory of Vehicular Traffic*. New York: Elsevier, 1971.
57. R. W. Rothery. Transportation Research Board (TRB) special report 165. In N. H. Gartner, C. J. Messner, and A. J. Rathi, (Eds.), *Traffic Flow Theory*, (2nd Ed.) Transportation Research Board of the National Academies. 1998.
58. V. I. Shvetsov. Mathematical modeling of traffic flows. Automation and Remote Control, 64: 1651–1689, 2003.
59. R. Sipahi, F.M. Atay, and S.-I. Niculescu. Stability of traffic flow with distributed delays modeling the memory effects of the drivers. *SIAM Appl. Math.*, 68(3): 738–759, 2007.
60. R. Sipahi and N. Olgac, Complete stability robustness of third-order LTI multiple time-delay systems. *Automatica*, 41: 1413–1422, 2005.
61. R. Sipahi. *Cluster Treatment of Characteristic Roots, CTCR, A Unique Methodology for the Complete Stability Robustness Analysis of Linear Time Invariant Multiple Time Delay Systems Against Delay Uncertainties*, Ph.D. Thesis, University of Connecticut, Mechanical Engineering Department, August 2005.
62. R. Sipahi and S.-I. Niculescu. Some remarks on the characterization of delay interactions in deterministic car following models. MTNS 2006, Kyoto, Japan.
63. R. Sipahi, and S.-I., Niculescu. *Slow Time-Varying Delay Effects – Robust Stability Characterization of Deterministic Car Following Models*. IEEE International Conference on Control Applications, Munich, Germany, October 2006.
64. R. Sipahi and S.-I. Niculescu. Stability study of a deterministic car following model under multiple delay interactions. Invited session on *Traffic Dynamics Under the Presence of Time Delays*, 6th IFAC Workshop on Time-Delay Systems, L'Aquila, Italy, July 2006.
65. R. Sipahi, S.-I., Niculescu, and F. Atay. *Effects of Short-Term Memory of Drivers on Stability Interpretations of Traffic Flow Dynamics*. American Control Conference, New York, 2007.
66. R. Sipahi, and S.-I. Niculescu. *Chain Stability in Traffic Flow with Driver Reaction Delays*. American Control Conference (ACC), Seattle, June 2008.
67. R. Sipahi, S.-I. Niculescu, and Delice, I.I. *Asymptotic Stability of Constant Time Headway Driving Strategy with Multiple Driver Reaction Delays*. American Control Conference, St Louis, June 2009.
68. G. Stepan. *Retarded Dynamical Systems: Stability and Characteristic Function*. New York: Longman Scientific & Technical,co-publisher John Wiley & Sons Inc., US., 1989.
69. Y. Sugiyama, and H. Yamada. Simple and exactly solvable model for queue dynamics. *Phys. Rev. E*, 55: 7749 7752, 1997.
70. M. Treiber, A. Hennecke, and D. Helbing. Congested traffic states in empirical observations and microscopic simulations. *Phys. Rev. E*, 62: 1805–1824, 2000.
71. M. Treiber and D. Helbing. Memory effects in microscopic traffic models and wide scattering in flow-density data. *Phys. Rev. E*, 68: 2003.

72. M. Treiber, A. Hennecke, and D. Helbing. *Congested traffic states in empirical observations and microscopic simulations.* University of Stuttgart, Stuttgart *http://www.theo2.physik.uni-stuttgart.de/treiber/,* August 7 2004.

73. M. Treiber, A. Kesting and D. Helbing. Delays, inaccuracies and anticipation in microscopic traffic models. *Physica A,* 360: 71–88, 2006.

74. T.J. Triggs. Driver brake reaction times: Unobtrusive measurement on public roads. *Public Health Rev.,* 15: 275–290, 1987.

75. B.-H. Wang, Y. R. Kwong, and P. M. Hui. Statistical mechanical approach to cellular automaton models of highway traffic flow. *Physica A,* 254: 122–134, 1998.

76. G. B. Whitham. *Linear and Nonlinear Waves.* New York: Wiley, 1974.

77. G. B. Whitham. *Lectures on Wave Propagation.* Berlin: Springer, 1979.

78. J. Zhang and P. Ioannou. *Control of Heavy-Duty Trucks: Environmental and Fuel Economy Considerations.* University of Southern California, Los Angeles TO4203, 2004.

Index

A

Almost periodic function, 218
Amplitude death, 6, 12, 46, 52
    partial, 46, 54
Asymptotic properties, 100

B

Bandpass filtering, 100, 124
Bifurcations, 312
Braking response time, 306

C

Car following model, 300
Cellular automaton model, 301
Chandler's model, 308
Characteristic equation, 13, 89, 95, 108, 137,
    141, 266
Chimera state, 26, 33
    clustered, 34
Clarke regular, 234
Clarke's generalized gradient, 234
Cluster treatment of characteristic roots
    (CTCR), 315
Complex Ginzburg-Landau equation (CGLE),
    20, 23, 33
Components of time delay, 304
Control amplitude, 117, 122
Control phase, 117, 122
Coupling
    diffusive, 55
    direct, 54
Critical fluctuations, 173, 174
Cyclical neutropenia, 294

D

Davis's model, 310
Death islands, 15, 18
Delay
    characteristic equation, 64

control law, 63
crossing characterization, 70
crossing direction, 71, 81
quasipolynomial, 65
stabilizability, 67, 73
stabilization, 66, 72, 73
Delay-induced bifurcation, 136
Delay-induced orbit, 113
Delay robustness, 315
Delay time of response, 305
Derivative control, 87
Deterministic traffic model, 298, 299
Diagonal coupling, 89
Dispersion relation, 22, 23
Distribution functions, 317

E

Eigenperiod, 92
Eigenvalue spectrum, 90, 100
ETDAS, 95
Exchange of stability, 110, 112
Excitable system, 131
Experimental traffic studies, 301
Extended time-delay autosynchronization, 95

F

Feedback gain, 48, 88
Feedback matrix, 108
Filippov solution, 236
Finite propagation speed, 152
FitzHugh-Nagumo model, 131, 133
Floquet exponent, 112
Floquet multiplier, 107, 110, 112
Fold bifurcation, 116, 121
Follow-the-leader strategy, 300

G

Gamma distribution, 264
Gazis's model, 308